Fundamentals of Linear Algebra for Signal Processing

James Reilly

Fundamentals of Linear Algebra for Signal Processing

James Reilly
Electrical and Computer Engineering
McMaster University
Hamilton, ON, Canada

ISBN 978-3-031-68917-8 ISBN 978-3-031-68915-4 (eBook)
https://doi.org/10.1007/978-3-031-68915-4

This Springer imprint is published by the registered company Springer Nature Switzerland AG
The registered company address is: Gewerbestrasse 11, 6330 Cham, Switzerland

Preface

This book is intended for graduate students whose research involves the processing of signals in some form, and hence requires some background in linear algebra. It is also suited for practitioners who are entering a new field related to signal processing and require some familiarity with this topic. The book will provide the reader with sufficient knowledge of the basic linear algebra toolkit that is required in many disciplines of modern engineering and science relating to signal processing, including machine learning, signal processing, control theory, process control, applied statistics, robotics, etc. Above all, this is a *teaching* text, where the emphasis is placed on understanding and interpretation of the material. The theoretical material is augmented with an array of examples and simulations that help the reader interpret and further understand the underlying concepts.

Pre-requisites We assume the reader has an equivalent background to a freshman course in linear algebra, so that they are familiar with the basic concepts of matrices and vectors, as follows:

- Familiarity with arithmetic operations on matrices and vectors, as well as transposes, inverses and other elementary knowledge.
- Solving linear systems of equations.
- Bases and change of bases.
- A fundamental knowledge of calculus.
- Some knowledge of probability and statistics, such as knowledge of the expectation operator, probability distributions, multi-variate probability distributions, and conditional probabilities is important. This material is part of the usual undergraduate curriculum in most disciplines in engineering, science, and mathematics and so does not require any supplementary material in this text that reviews these topics.
- A basic knowledge of linear system theory, including the Fourier and z-transforms, sampling phenomena in time and frequency, and the general topic of digital filters, is helpful in understanding several of the examples and problems presented in the text. These are fundamental topics in the undergraduate electrical engineering curriculum, but are not covered in most other engineering, science,

and math curricula. Therefore in order to make this text as accessible as possible to this broader audience, an Appendix on linear system theory is included that gives an overview of the elements necessary to understand all aspects of the examples and problems in this text.

- Familiarity with a high-level programming language such as Matlab®, Python, or R is assumed.

The material in the Appendix is not only useful in understanding the examples and problems in this text as previously indicated, but it also provides the intrinsic knowledge that is necessary for any researcher who deals with signals in some form. Typical signal processing algorithms that make use of linear algebra depend on the preliminary step where the signal is first converted into a digital format, where it can be stored and processed within a computer. However, the conversion of the signal into this digital format is not a trivial procedure and to be done properly requires knowledge on the part of the user. The Appendix provides supplementary material that deals with this conversion procedure. Thus the book as a whole covers a wide range of topics associated with signal processing, including the background necessary for acquisition of the signal within the computer, and then providing the algebraic background necessary for implementation of more sophisticated, modern signal processing methods.

There is a vast array of disciplines in engineering, science, and mathematics whose mission is related to signal processing. A few examples of such disciplines include physics, chemistry, neuroscience, and psychology (for processing EEG and fMRI signals), medical imaging/physics (for analysing signals received from various imaging modalities), and all the various engineering disciplines. A prominent example of a signal processing application, which requires a considerable degree of linear algebra and which encompasses virtually all the disciplines mentioned, is machine learning in all its forms. Other examples are high-speed communication devices such as cell phones, process control, seismology, vibration analysis, image processing, analysis of the EEG, and so on.

Thus we see that the field of signal processing embodies a very wide scope of disciplines and is very pertinent in this modern era of technology. This text provides all the necessary background in linear algebra for any researcher or practitioner within this wide range of disciplines to actively work and contribute within this field. The text has been written in a more informal style, with the objective of making the material as accessible and understandable to as wide an audience as possible.

The book may be divided loosely into two halves. The first half (Chaps. 1–6) puts in place all the requisite algebraic background for the second half. The reader should be aware that ALL the material in the first half is the required knowledge for the second half. The second half (Chaps. 7–10) applies the material from the first half to develop the least squares (LS) method and related topics. The LS method is a practical and widely used estimation procedure with a large variety of applications. A solid understanding of LS methods facilitates the learning of more advanced topics such as estimation theory and related fields. Various versions of the LS method are presented—these include ordinary least squares, least squares in the rank-deficient or almost rank-deficient cases, regularisation and Toeplitz systems. These topics serve as a foundation for the student if they wish to engage in more advanced study in topics such as estimation theory and related fields.

The first chapter sets up the foundation for the remaining portion of the book. We look at some elements of the foundations linear algebra such as linear independence, subspaces, rank, nullspace, range, etc., and how these concepts are interrelated. A review of matrix multiplication from a more advanced perspective is introduced and then vector norms are also addressed.

In Chap. 2, the most basic matrix decomposition, the so-called eigendecomposition, is presented. The focus of the presentation is to de-mystify the underlying eigen-analysis theory and give an interpretive insight into what this decomposition accomplishes. The ideas of autocorrelation and the covariance matrix of a stationary signal are discussed and interpreted. We then combine these two areas into the important topic of principal component analysis (PCA), *aka* the Karhunen-Loeve transform, and present a series of examples where PCA has proven useful in signal processing. In this way, the reader is made familiar with the many applications of the eigendecomposition and PCA.

In Chap. 3, we develop the *singular value decomposition* (SVD), which is closely related to the eigendecomposition of a matrix. We develop the relationships between these two decompositions and explore various properties of the SVD. Orthogonal projectors are also discussed. The chapter is then closed off with a discussion on the approximation of a matrix with another of lower rank. This is a key concept with respect to latent variable methods discussed in Chap. 8.

Chapter 4 deals with the quadratic form, positive definiteness, and their relation to the eigendecomposition. The relationship between the multi-variate Gaussian probability function and the concept of the joint confidence regions is presented. We also discuss methods for the calculation of a single eigenvector/value pair.

In Chap. 5, a brief introduction to numerical issues encountered when dealing with floating point number systems is presented. Then Gaussian elimination is discussed in some length. The Gaussian elimination process is described through a bigger-block matrix approach that leads to other useful decompositions such as the Cholesky decomposition of a square symmetric positive definite matrix. The treatment includes an analysis of the effect of error induced by finite precision floating point number systems. The condition number of a matrix, which is a critical part in determining a lower bound on the relative error in the solution of a system of linear equations, is also developed.

The QR decomposition is developed in Chap. 6. This decomposition yields an orthonormal basis for the range space of a matrix and is the key component of the so-called QR algorithm for computing the complete eigendecomposition. We discuss two forms of the Gram–Schmidt procedure as well as the Householder method for computing the QR decomposition. We also include a discussion on the rank-deficient case.

Chapters 7–10 deal with solving least-squares (LS) problems. The standard least squares problem, its solution, and its properties are developed in Chap. 7. Several examples where LS methods have been successfully applied in the field of signal processing are presented. We discuss performance of the LS estimator in the presence of white and coloured noise relative to the Cramer–Rao lower bound. We show that whitening the noise is required in order to achieve optimum performance of the LS estimator. In this respect we present several techniques that can be used to achieve this purpose.

In Chap. 8 we discuss methods for mitigating the adverse effects of poor conditioning when solving the least squares problem. These include the pseudo-inverse, which is a well-known approach for dealing with inversion of a rank-deficient matrix. We then discuss latent variable methods in some detail and establish the connection between these methods and the pseudo-inverse.

Chapter 9 deals with the least squares problem from the alternative perspective of regularisation, which is an additional means of introducing prior knowledge into the LS solution to mitigate the effects of poor conditioning. Regularisation is used extensively in training deep learning networks and in dealing with large systems. Several regularisation methods are presented and interpreted.

Chapter 10 deals with Toeplitz systems, which arise when solving stationary autoregressive (AR) systems with LS methods. The chapter discusses some clever ideas relating to AR analysis that result in faster execution, a more efficient parameterisation of the AR process, and fast algorithms for computing inverses and determinants.

Selected problems in Chaps. 1, 2, and 7 require specific data to solve them. This data is in the form of downloadable MATLAB files available from this website, provided by the publisher: https://link.springer.com/book/10.1007/978-3-031-68915-4. Furthermore, a GitHub repository is available at https://github.com/sn-code-inside/Linear-Algebra-for-Signal-Processing, which can be used to retrieve the code and store additional or modified MATLAB data should the need for such data arise in the future.

Course Delivery The author has used this book as a text for a course he teaches to what are mostly junior-level Ph.D. or Master's students. The course is approximately 12 weeks in duration, 3 hours per week. This amount of time permits coverage of only the material in Chap. 1–5, Sects. 6.1 and 6.2, Chap. 7, Sects. 8.1–8.3, and Chap. 9. It would be possible and even desirable to split the material into two courses— the first of which would cover a slightly reduced and therefore slower curriculum compared to the present course offering, and the second course would fill in the remaining sections and chapters. In the experience of the author, Chap. 2 requires approximately 9 hours to do properly, due to the fact that this level of student has little previous exposure to covariance, covariance matrices, and eigen-analysis, material which the students usually find difficult. Chapter 6 on the QR decomposition is usually treated lightly to allow more time for other material.

Hamilton, ON, Canada James Reilly

Acknowledgements

I wish to express sincere appreciation to all the members of my family—my wife Beth and my offspring Sarah, Emma, and Liam for their encouragement, support, and patience during the preparation of this work. I also wish to express my indebtedness to the main mentors in my life: my father Prof. Park Reilly and my Ph.D. supervisor Prof. Simon Haykin. This book has evolved over the period of many years from teaching a graduate course in linear algebra. I am extremely grateful to the many students who have taken the course, which through the period of many years has given rise to this book. They have provided me with very useful feedback and helpful suggestions and have also provided me with the incentive to complete this task. I sincerely appreciate the contributions of Zhongda Zhang who helped me with the presentation of Chap. 10 and Mark Knez, Department of Electrical and Computer Engineering at McMaster, who provided the proof for Lemma 7.1 in Sect. 7.7. Finally, I am indebted to my colleagues Prof. Narges Armanfard, Prof. Tim Davidson, and Prof. Michael Noseworthy who have reviewed chapters from this book and whose comments have improved the quality of the manuscript immensely.

I am also indebted to the gracious help provided by the Springer staff in bringing the manuscript into a publishable state. In particular, my thanks go out to Bakiyalakshmi RM, Alicia Richard, Michael McCabe and Brian Halm from Springer who helped guide this book into fruition.

Contents

About the Author

James Reilly, Ph.D. is a Professor Emeritus in the Department of Electrical and Computer Engineering and the School of Biomedical Engineering at McMaster University. He works at the interface of machine learning and signal processing applied to health-related problems, and he has made pioneering contributions to treating and diagnosing psychiatric illnesses and disorders of consciousness. He has 70 refereed publications in top-tier journals, 9 patents, and over 90 reviewed conference publications. In his spare time, Prof. Reilly enjoys photography, flying model airplanes, and hiking/biking on the trails near his home in Hamilton, ON.

Chapter 1
Fundamental Concepts

1.1 Notation

Throughout this book, we indicate that a matrix $\boldsymbol{A}$ is of dimension $m \times n$ and whose elements are taken from the set of real numbers, by the notation $\boldsymbol{A} \in \mathbb{R}^{m \times n}$. This means that the matrix $\boldsymbol{A}$ belongs to the Cartesian product of the real numbers, taken $m \times n$ times, one for each element of $\boldsymbol{A}$. In a similar way, the notation $\boldsymbol{A} \in \mathbb{C}^{m \times n}$ means the matrix is of dimension $m \times n$, and the elements are taken from the set of complex numbers. By the matrix dimension "$m \times n$", we mean $\boldsymbol{A}$ consists of m rows and n columns. Matrices with $m > n$ (more rows than columns) are referred to as *tall* matrices, whereas matrices where $m < n$ are referred to as *short*. Matrices where $m = n$ are called *square*. The set of elements $a_{kk}, k = 1, \ldots, n$ constitute the *main diagonal*, a term which is most commonly used with respect to square matrices. A square matrix which has non-zero elements only on its main diagonal is referred to as a *diagonal* matrix.

Similarly, the notation $\boldsymbol{a} \in \mathbb{R}^m (\mathbb{C}^m)$ refers to a vector of m elements, which are taken from the set of real (complex) numbers. When referring to a single vector, we use the term *dimension* or *length* to denote the number of elements.

Also, we indicate that a scalar a is from the set of real (complex) numbers by the notation $a \in \mathbb{R}(\mathbb{C})$. Thus, an upper case bold character denotes a *matrix*, a lower case bold character denotes a vector, and a lower case non-bold character denotes a scalar.

By convention, a vector by default is taken to be a *column* vector. Further, for a matrix $\boldsymbol{A}$, we denote its ith column as $\boldsymbol{a}_i$. We also imply that its jth row is $\boldsymbol{a}_j^T$, even though this notation may be ambiguous, since it may also be taken to mean the

---

Supplementary Information The online version contains supplementary material available at (https://doi.org/10.1007/978-3-031-68915-4_1).

J. Reilly, *Fundamentals of Linear Algebra for Signal Processing*,
https://doi.org/10.1007/978-3-031-68915-4_1

transpose of the jth column. The context of the discussion will help to resolve the ambiguity. In the hand-written case, vectors and matrices are denoted by underlining the respective character.

1.2 Fundamental Linear Algebra

In this section, we introduce fundamental concepts such as vector spaces, linear independence, bases, rank, and related ideas. These concepts form the foundation for most of what follows in this book. First we introduce a few preliminaries:

The Matrix Transpose Consider an $m \times n$ matrix $\boldsymbol{A}$. Its transpose, which is $n \times m$ and is denoted $\boldsymbol{A}^T$, is formed by converting each column of $\boldsymbol{A}$ into the corresponding row of $\boldsymbol{A}^T$. The transpose operation can also apply to vectors.

The Identity Matrix We define a set of *elementary* vectors $\boldsymbol{e}_i, i = 1, \ldots, m$ as a vector of m elements having all zeros, except for a 1 in the ith position. The $m \times m$ *identity* matrix $\boldsymbol{I}$ is defined as $\boldsymbol{I} = \left[\boldsymbol{e}_1, \boldsymbol{e}_2, \ldots \boldsymbol{e}_m\right]$. As may be seen, it is a diagonal matrix consisting of 1's along the main diagonal with zeros elsewhere. This matrix is called the identity because of its property $\boldsymbol{I}\boldsymbol{x} = \boldsymbol{x}$ for any $\boldsymbol{x}$. The inverse of $\boldsymbol{I}$ is $\boldsymbol{I}$ itself.

The Matrix Inverse The inverse $\boldsymbol{A}^{-1}$ of a matrix $\boldsymbol{A}$ is defined such that $\boldsymbol{A}\boldsymbol{A}^{-1} = \boldsymbol{A}^{-1}\boldsymbol{A} = \boldsymbol{I}$. To be "invertible", $\boldsymbol{A}$ must be square and full rank. More on this later.

Trace The trace of a matrix $\boldsymbol{A}$, denoted as $\mathrm{tr}(\boldsymbol{A})$, is the sum of its diagonal elements.

Inner and Outer Products Given two vectors $\boldsymbol{a}, \boldsymbol{b} \in \mathbb{R}^n$, then their *inner product* p is a scalar defined as $p = \sum_{i=1}^{n} a_i b_i$. This is equivalent to the expression $p = \boldsymbol{a}^T \boldsymbol{b}$. Recall that if $\boldsymbol{a}, \boldsymbol{b}$ are orthogonal, then $p = 0$. If $\boldsymbol{a} \in \mathbb{R}^m$ and $\boldsymbol{b} \in \mathbb{R}^n$, then the *outer product* $\boldsymbol{P}$ is the $m \times n$ matrix defined as $\boldsymbol{P} = \boldsymbol{a}\boldsymbol{b}^T$.

1.2.1 Linear Vector Spaces

In this section, vectors are assumed to be length m. A *linear* vector space $\mathcal{S}$ is a set of vectors, which satisfy two properties:

1. If $\boldsymbol{x}, \boldsymbol{y} \in \mathcal{S}$, then $\boldsymbol{x} + \boldsymbol{y} \in \mathcal{S}$, where "+" denotes vector addition.
2. If $\boldsymbol{x} \in \mathcal{S}$, then $c\boldsymbol{x} \in \mathcal{S}$, where $c \in \mathbb{R}$, where vector multiplication by a scalar has been indicated.

A more mathematically rigorous presentation is given, e.g. in [2].

Given a set of vectors $[\boldsymbol{a}_1, \ldots, \boldsymbol{a}_n]$, where $\boldsymbol{a}_i \in \mathbb{R}^m$, $i = 1, \ldots, n$, and a set of n scalars $c_i \in \mathbb{R}$, then the vector $\boldsymbol{y} \in \mathbb{R}^m$ defined by

$$\boldsymbol{y} = \sum_{i=1}^{n} c_i \boldsymbol{a}_i \tag{1.1}$$

is referred to as a *linear combination* of the vectors $\boldsymbol{a}_i$. Then according to the properties above, if the $\boldsymbol{a}_i \in \mathcal{S}$, then the linear combination $\boldsymbol{y} \in \mathcal{S}$.

Equation (1.1) can be represented more compactly as the matrix-vector product given as

$$\boldsymbol{y} = \boldsymbol{A}\boldsymbol{c} \tag{1.2}$$

where $\boldsymbol{A} \in \mathbb{R}^{m \times n} = [\boldsymbol{a}_1, \ldots, \boldsymbol{a}_n]$. To see this, we can depict the product $\boldsymbol{A}\boldsymbol{c}$ in the following form for the 3×3 case:

$$\boldsymbol{y} = \begin{bmatrix} a & d & g \\ b & e & h \\ c & f & i \end{bmatrix} \begin{bmatrix} 1 \\ 2 \\ 3 \end{bmatrix}. \tag{1.3}$$

Then, from the conventional rules of matrix-vector multiplication, we have

$$\boldsymbol{y} = \begin{bmatrix} 1a + 2d + 3g \\ 1b + 2e + 3h \\ 1c + 2f + 3i \end{bmatrix}.$$

Note, in this example, that all elements of the first column of $\boldsymbol{A}$ are multiplied only by the coefficient $c_1 = 1$, that the entire second column is multiplied only by $c_2 = 2$, and that all elements of the third column of $\boldsymbol{A}$ are multiplied only by $c_3 = 3$. Therefore (1.2) can be written in the form $\boldsymbol{y} = c_1\boldsymbol{a}_1 + c_2\boldsymbol{a}_2 + c_3\boldsymbol{a}_3$, which is identical to (1.1). If $\boldsymbol{C}$ were a matrix with, e.g. n columns instead of one, then the resulting $\boldsymbol{Y}$ would be a matrix with n columns. In this case, e.g. the second column of $\boldsymbol{Y}$ would also be a linear combination of the columns of $\boldsymbol{A}$, but in this case the coefficients are from the second column of $\boldsymbol{C}$.

It is very important that the reader understand the concept that $\boldsymbol{y}$ in (1.2) is a linear combination of the columns of $\boldsymbol{A}$, since it is one of the fundamental ideas of linear algebra. In this vein, it helps greatly to visualise an entire column as being a single entity, rather than treating each element individually. We present an additional example to illustrate the concept further. Consider the following depiction of matrix-vector multiplication:

$$
y = \begin{bmatrix} \Big| & \Big| & \cdots & \Big| \end{bmatrix} \begin{bmatrix} c_1 \\ c_2 \\ \vdots \\ c_n \end{bmatrix}
$$

$$\quad A \qquad\qquad c$$

Each vertical line in the long brackets represents an entire column $\boldsymbol{a}_i$ of $\boldsymbol{A}, i = 1, \ldots, n$. In a manner similar to (1.1), from the rules of matrix-vector multiplication, we see from the above diagram that each element c_i of $\boldsymbol{c}$ multiples only the corresponding column $\boldsymbol{a}_i$; i.e. coefficient c_i interacts only with the column $\boldsymbol{a}_i$. Thus, $\boldsymbol{y} = \sum_i \boldsymbol{a}_i c_i$ which is the same result as (1.2).

We now transpose (1.3) to obtain

$$
\begin{aligned}
\boldsymbol{y}^T &= \boldsymbol{c}^T \boldsymbol{A}^T \\
&= \begin{bmatrix} 1 \; 2 \; 3 \end{bmatrix} \begin{bmatrix} a & b & c \\ d & e & f \\ g & h & i \end{bmatrix} \\
&= \begin{bmatrix} 1a & 1b & 1c \\ +2d & +2e & +2f \\ +3g & +3h & +3i \end{bmatrix} \\
&= \sum_{i=1}^{n} c_i \boldsymbol{a}_i^T
\end{aligned} \tag{1.4}
$$

where $\boldsymbol{a}_i^T$ is the ith row of $\boldsymbol{A}^T$ in this case. With respect to the third equation above, note that $\boldsymbol{y}^T$ is a 1×3 row vector, where the summation corresponding to each element is represented in a column format for clarity of presentation. In a transposed manner corresponding to (1.3), note the ith element of $\boldsymbol{c}$ interacts only with the ith row $\boldsymbol{a}_i^T$ of $\boldsymbol{A}^T, i = 1, \ldots, 3$. Thus, using similar logic as the column case, we see that the row vector $\boldsymbol{y}^T$ in this case is a linear combination of the *rows* of $\boldsymbol{A}^T$, whose coefficients are the elements of $\boldsymbol{c}^T$.

Thus to summarise, if we pre-multiply a matrix $\boldsymbol{A}^T$ by a row vector $\boldsymbol{c}^T$, then each row of the product is a linear combination of the rows of $\boldsymbol{A}^T$. Likewise, when a matrix $\boldsymbol{A}$ is post-multiplied by a column vector $\boldsymbol{c}$, then each column of the product is a linear combination of the columns of $\boldsymbol{A}$.

Instead of using (1.2) to define a single vector $\boldsymbol{y}$, we can extend it to define a vector space $\mathcal{S}$:

$$\mathcal{S} = \left[\boldsymbol{y} \in \mathbb{R}^m \,|\, \boldsymbol{y} = \boldsymbol{A}\boldsymbol{c}, \boldsymbol{c} \in \mathbb{R}^n \right] \tag{1.5}$$

Here it is implied that $\boldsymbol{c}$ takes on the infinite set of all possible values within $\mathbb{R}^n$ and consequently $\{\boldsymbol{y}\}$ is the set of all possible linear combinations of the columns $\boldsymbol{a}_i$. The *dimension* of $\mathcal{S}$ (denoted as $\dim(\mathcal{S})$) is the number of independent directions that span the space; e.g. the dimension of the universe we live in is 3. In the case where $n = 2$ and if $\boldsymbol{a}_1$ and $\boldsymbol{a}_2$ are linearly independent (to be defined), then $\mathcal{S}$ is a two-dimensional plane which contains $\boldsymbol{a}_1$ and $\boldsymbol{a}_2$.

Dim$(\mathcal{S})$ is not necessarily n, the number of vectors or columns of $\boldsymbol{A}$. In fact, $\dim(\mathcal{S}) \leq n$. The quantity $\dim(\mathcal{S})$ depends on the characteristics of the vectors $\boldsymbol{a}_i$. For example, the vector space defined by the vectors $\boldsymbol{a}_1$ and $\boldsymbol{a}_2$ in Fig. 1.1 is infinite extension of the plane of the paper. The dimension of this vector space is 2: if a third vector $\boldsymbol{a}_3$ which is orthogonal to the plane of the paper were added to the set, then the resulting vector space would be the three-dimensional universe.

A third example is shown in Fig. 1.2. Here, since none of the vectors $\boldsymbol{a}_1 \ldots, \boldsymbol{a}_3$ have a component which is orthogonal to the plane of the paper, all linear combinations of this vector set, and hence the corresponding vector space, lies in the plane of the paper. Thus, in this case, $\dim(\mathcal{S}) = 2$, even though there are three vectors in the set.

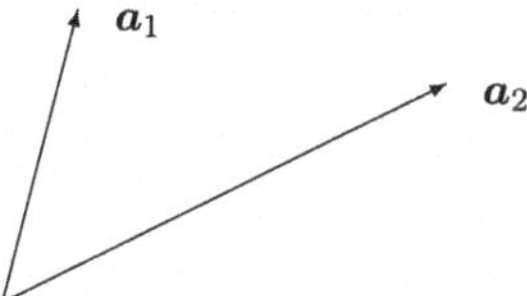

Fig. 1.1 A vector set containing two linearly independent vectors. The dimension of the corresponding vector space $\mathcal{S}$ is 2

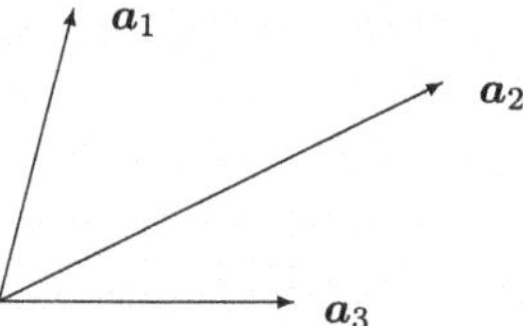

Fig. 1.2 A vector set containing three linearly dependent vectors. The dimension of the corresponding vector space $\mathcal{S}$ is 2, even though there are three vectors in the set. This is because the vectors all lie in the plane of the paper

1.2.2 Linear Independence

A vector set $[\boldsymbol{a}_1, \ldots, \boldsymbol{a}_n]$, $\boldsymbol{a}_i \in \mathbb{R}^m$, is linearly independent under the condition

$$\boldsymbol{y} = \sum_{j=1}^{n} c_j \boldsymbol{a}_j = \boldsymbol{A}\boldsymbol{c} = \boldsymbol{0} \quad \text{if and only if} \quad c_1, \ldots, c_n = 0, \tag{1.6}$$

i.e. the only way to make a linear combination of a set of linearly independent vectors to be zero is to make all the coefficients $[c_1, \ldots, c_n] = 0$. We can also say that the set $[\boldsymbol{a}_1, \ldots, \boldsymbol{a}_n]$ is linearly independent if and only if $\dim(\mathcal{S}) = n$, where $\mathcal{S}$ is the vector space corresponding to the set. Loosely speaking, a vector set consisting of n vectors (of length $m \geq n$) is linearly independent if and only if (iff) the corresponding vector space "fills up" n dimensions. Notice that if $m < n$, then the vectors must be linearly dependent, since a set of vectors of length m can only span at most an m-dimensional space. Further, a linearly dependent vector set can be made independent by removing appropriate vectors from the set.

Example 1.1

$$\boldsymbol{A} = [\boldsymbol{a}_1 \ \boldsymbol{a}_2 \ \boldsymbol{a}_3] = \begin{bmatrix} 1 & 2 & 1 \\ 0 & 3 & -1 \\ 0 & 0 & 1 \end{bmatrix} \tag{1.7}$$

This set is linearly independent. On the other hand, the set

$$\boldsymbol{B} = [\boldsymbol{b}_1 \ \boldsymbol{b}_2 \ \boldsymbol{b}_3] = \begin{bmatrix} 1 & 2 & -3 \\ 0 & 3 & -3 \\ 1 & 1 & -2 \end{bmatrix} \tag{1.8}$$

is not. This follows because the third column is a linear combination of the first two. (-1 times the first column plus -1 times the second equals the third column.) Thus, the coefficients of the vector $\boldsymbol{c}$ in (1.6) which results in zero are any scalar multiple of $(1, 1, 1)$. We will see later that this vector defines the *null space* of $\boldsymbol{B}$.

1.2.3 Span, Subspaces, and Range

In this section, we explore these three closely related ideas. In fact, their mathematical definitions are similar, but the interpretation is different for each case.

Span The *span* of a vector set $[\boldsymbol{a}_1, \ldots, \boldsymbol{a}_n]$, written as span$[\boldsymbol{a}_1, \ldots, \boldsymbol{a}_n]$, is the vector space $\mathcal{S}$ corresponding to this set; i.e.

$$\mathcal{S} = \text{span}\,[\boldsymbol{a}_1, \ldots, \boldsymbol{a}_n] = \left\{ \boldsymbol{y} \in \mathbb{R}^m \;\middle|\; \boldsymbol{y} = \sum_{j=1}^{n} c_j \boldsymbol{a}_j, \quad c_j \in \mathbb{R} \right\},$$

where the notation $c_j \in \mathbb{R}$ implies that the coefficients c_j are each assumed to take on the infinite range of values in the set of real numbers. In the above, we have $\dim(\mathcal{S}) \leq n$, where the equality is satisfied iff the vectors $\boldsymbol{a}_i$ are linearly independent. Note that the argument of span$(\cdot)$ is a vector set.

Subspaces A subspace is a subset of a vector space. Formally speaking, a subspace $\mathcal{U}$ of $\mathcal{S} = \text{span}[\boldsymbol{a}_1, \ldots, \boldsymbol{a}_n]$ is determined by $\mathcal{U} = \text{span}[\boldsymbol{a}_{i_1}, \ldots \boldsymbol{a}_{i_k}]$, where the indices satisfy $\{i_1, \ldots, i_k\} \subset \{1, \ldots, n\}$. In other words, a subspace is a vector space formed from a subset of the vectors $[\boldsymbol{a}_1 \ldots \boldsymbol{a}_n]$. The subspace is of dimension k if the $[\boldsymbol{a}_{i_1}, \ldots \boldsymbol{a}_{i_k}]$ are linearly independent.

Note that $[\boldsymbol{a}_{i1}, \ldots \boldsymbol{a}_{ik}]$ is not necessarily *a basis* for the subspace S. This set is a basis iff it is a maximally independent set. This idea is discussed shortly. The set $\{\boldsymbol{a}_i\}$ need not be linearly independent to define the span or subspace.

Example 1.2 The vectors $[\boldsymbol{a}_1, \boldsymbol{a}_2]$ in Fig. 1.1 define a subspace (the plane of the paper) which is a subset of the three–dimensional universe $\mathbb{R}^3$.

$*$ What is the span of the vectors $[\mathbf{b}_1, \ldots, \mathbf{b}_3]$ in Example 1.1?

Range The *range* of a matrix $\boldsymbol{A} \in \mathbb{R}^{m \times n}$, denoted $R(\boldsymbol{A})$, is the vector space satisfying

$$R(\boldsymbol{A}) = \left\{ \boldsymbol{y} \in \mathbb{R}^m \mid \boldsymbol{y} = \boldsymbol{A}\boldsymbol{x}, \ \text{for}\ \boldsymbol{x} \in \mathbb{R}^n \right\}.$$

Thus, we see that $R(\boldsymbol{A})$ is the vector space consisting of all linear combinations of the columns $\boldsymbol{a}_i$ of $\boldsymbol{A}$, whose coefficients are the elements x_i of $\boldsymbol{x}$. Therefore, $R(\boldsymbol{A}) \equiv \text{span}[\boldsymbol{a}_1, \ldots, \boldsymbol{a}_n]$. The distinction between *range* and *span* is that the argument of *range* is a matrix, whereas we have seen that the argument of *span* is a vector set. We have $\dim[R(\boldsymbol{A})] \leq n$, where the equality is satisfied iff the columns are linearly independent. Any vector $\boldsymbol{y} \in R(\boldsymbol{A})$ is of dimension (length) m.

In the case when $m > n$ (i.e. $\boldsymbol{A}$ is a *tall* matrix), it is important to note that $R(\boldsymbol{A})$ is indeed a subspace of the m-dimensional "universe" $\mathbb{R}^m$. In this case, the dimension of $R(\boldsymbol{A})$ is less than or equal to n. Thus, $R(\boldsymbol{A})$ does not span the whole m-dimensional universe and therefore is a subspace of it.

An example illustrating the concept of range is given by Fig. 1.3. For this example, we have a 3×2 matrix $\boldsymbol{A} = [\boldsymbol{a}_1 \quad \boldsymbol{a}_2]$ where $\boldsymbol{a}_1$ and $\boldsymbol{a}_2$, as seen from the figure, are linearly independent. The darker extensions of these vectors indicate the spans of $\boldsymbol{a}_1$ and $\boldsymbol{a}_2$, respectively. The infinite extension of the red cross-hatched region is $R(\boldsymbol{A})$, since it represents the set of all possible linear combinations of the vectors $\boldsymbol{a}_1$ and $\boldsymbol{a}_2$. The vectors $\boldsymbol{e}_i$ are the coordinate axes.

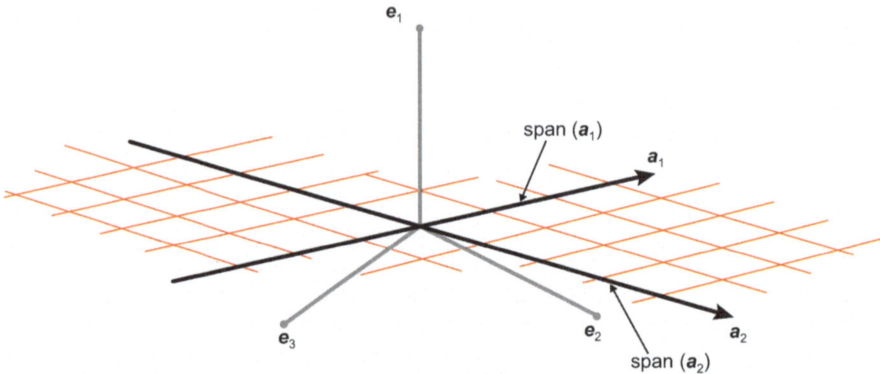

Fig. 1.3 An example of range for the 3×2 case. Two linearly independent vectors $\boldsymbol{a}_1$ and $\boldsymbol{a}_2$ are shown. The infinite extension of the darker black lines represent the spans of $\boldsymbol{a}_1$ and $\boldsymbol{a}_2$, respectively. The infinite extension of the red cross–hatched region is $R(\boldsymbol{A})$

1.2.4 The Orthogonal Complement of a Subspace

Recall for any pair of vectors $\boldsymbol{x}, \boldsymbol{y} \in \mathbb{R}^m$, their dot product, or *inner product* c is defined as $c = \sum_{i=1}^{m} x_i y_i = \boldsymbol{x}^T \boldsymbol{y}$, where $(\cdot)^T$ denotes transpose. Further, recall that two non-zero vectors are orthogonal iff their inner product is zero. Now suppose we have a subspace $\mathcal{S}$ of dimension r corresponding to the vectors $[\boldsymbol{a}_1, \ldots, \boldsymbol{a}_n]$, for $r \leq n \leq m$; i.e. the respective matrix $\boldsymbol{A}$ is tall, and the $\boldsymbol{a}_i$'s are not necessarily linearly independent. With this background, the orthogonal complement subspace $\mathcal{S}_\perp$ of $\mathcal{S}$ of dimension $m - r$ is defined as

$$\mathcal{S}_\perp = \left\{ \boldsymbol{y} \in \mathbb{R}^m | \boldsymbol{y}^T \boldsymbol{x} = 0 \text{ for all } \boldsymbol{x} \in \mathcal{S} \right\} \tag{1.9}$$

i.e. any vector in $\mathcal{S}_\perp$ is orthogonal to any vector in $\mathcal{S}$. The quantity $\mathcal{S}_\perp$ is pronounced "$\mathcal{S}$–perp".

Example 1.3 A vector set defining $\mathcal{S}$ is given as

$$\mathcal{S} \equiv \begin{bmatrix} 1 & 2 \\ 0 & 3 \\ 0 & 0 \end{bmatrix}.$$

Then, a basis for $\mathcal{S}_\perp$ is

$$\begin{bmatrix} 0 \\ 0 \\ 1 \end{bmatrix}$$

1.2.5 Bases

A *maximally independent set* is a vector set which cannot be made larger without losing independence and smaller without remaining maximal; i.e. it is a set containing the maximum number of linearly independent vectors that span the space. A *basis* for a subspace is any maximally independent set within the subspace. It is not unique.

A commonly used basis is $\boldsymbol{I}$. Another commonly used form of bases are columns from an orthonormal matrix; these matrices have mutually orthogonal columns with unit norm. More on this later in Chap. 2. Given a basis, any vector in the corresponding subspace is uniquely represented by that basis.

Example 1.4 A basis for the subspace $\mathcal{U}$ spanning the first two columns of

$$\boldsymbol{A} = \begin{bmatrix} 1 & 2 & 3 \\ & 3 & -3 \\ & & 3 \end{bmatrix}$$

is

$$\mathbf{e}_1 = (1, 0, 0)^T$$
$$\mathbf{e}_2 = (0, 1, 0)^T.$$

Note that any linearly independent set in span$[\boldsymbol{e}_1, \boldsymbol{e}_2]$ is also a basis.

Change of Basis We are given a basis formed by $\boldsymbol{A}$ which we use to express a vector $\boldsymbol{y}$ as $\boldsymbol{y} = \boldsymbol{A}\boldsymbol{c}_1$, where $\boldsymbol{c}_1$ are the respective coefficients. Suppose we wish to represent $\boldsymbol{y}$ using the basis $\boldsymbol{B}$ instead of $\boldsymbol{A}$. To determine the coefficients $\boldsymbol{c}_2$ to represent $\boldsymbol{y}$ in the basis $\boldsymbol{B}$, we solve the system of linear equations

$$\boldsymbol{y} = \boldsymbol{B}\boldsymbol{c}_2; \tag{1.10}$$

i.e. $\boldsymbol{c}_2 = \boldsymbol{B}^{-1}\boldsymbol{y}$. (The reader is invited to explain why this inverse always exists in this case.) It is interesting to note that if the basis $\boldsymbol{B} = \boldsymbol{I}$, then the respective coefficients $\boldsymbol{c}_2 = \boldsymbol{y}$. This fact greatly simplifies the underlying algebra and explains why the basis $\boldsymbol{I} = [\boldsymbol{e}_1, \ldots, \boldsymbol{e}_n]$ is so commonly used.

1.2.6 Rank

Rank is a fundamental concept of linear algebra. Its definition and properties follow:

1. The rank of a matrix $\boldsymbol{A}$ (denoted rank$(\boldsymbol{A})$) is the number of linearly independent rows or columns in $\boldsymbol{A}$. Thus, it is the dimension of $R(\boldsymbol{A})$, which is the subspace

formed from the columns of $\boldsymbol{A}$. The symbol r is commonly used to denote rank; i.e. $r = \text{rank}(\boldsymbol{A})$.
2. A matrix $\boldsymbol{A} \in \mathbb{R}^{m\times n}$ is said to be *rank deficient* if $r < \min(m, n)$. Otherwise, it is said to be *full rank*. The columns (if $\boldsymbol{A}$ is tall) or rows (if $\boldsymbol{A}$ is short) of a full rank matrix must be linearly independent. This follows directly from the definition of rank. For tall matrices, *full column rank* is sometimes used in place of *full rank*, and for short matrices, the term *full row rank* applies.
3. As indicated earlier, a matrix is invertible if and only if it is square and full rank.
4. The columns (rows) of a rank-deficient matrix are linearly dependent.
5. If $\boldsymbol{A}$ is square and rank deficient, then $\det(\boldsymbol{A}) = 0$.
6. If $\boldsymbol{A} = \boldsymbol{BC}$, and $r_1 = \text{rank}(\boldsymbol{B})$, $r_2 = \text{rank}(\boldsymbol{C})$, then $\text{rank}(\boldsymbol{A}) \leq \min(r_1, r_2)$.
7. It can be shown that $\text{rank}(\boldsymbol{A}) = \text{rank}(\boldsymbol{A}^T)$. More is said on this point later.

Example 1.5 The rank of $\boldsymbol{A}$ in Example 1.1 is 3, whereas the rank of $\boldsymbol{B}$ in Example 1.1 is 2.

Example 1.6 Consider the matrix multiplication $\boldsymbol{C} \in \mathbb{R}^{m\times n} = \boldsymbol{AB}$, where $\boldsymbol{A} \in \mathbb{R}^{m\times 2}$ and $\boldsymbol{B} \in \mathbb{R}^{2\times n}$, depicted by the following diagram:

$$\underset{\boldsymbol{C}\in\mathbb{R}^{m\times n}}{\Big[\;|\;|\;|\;|\;\Big]} = m\;\underset{\boldsymbol{A}}{\Big[\;|\;|\;\Big]}\;\underset{\boldsymbol{B}}{\overset{n}{\begin{bmatrix}\times&\times&\times&\times\\ \times&\times&\times&\times\end{bmatrix}}}.$$

where the symbol $\times$ represents a respective element of $\boldsymbol{B}$. Then, the rank of $\boldsymbol{C}$ is at most two. To see this, we realise from our discussion on representing a linear combination of vectors by matrix multiplication that the ith column of $\boldsymbol{C}$ is a linear combination of the two columns of $\boldsymbol{A}$ whose coefficients are the ith column of $\boldsymbol{B}$. Thus, all columns of $\boldsymbol{C}$ reside in the vector space $R(\boldsymbol{A})$. If the columns of $\boldsymbol{A}$ and the rows of $\boldsymbol{B}$ are linearly independent, then the dimension of this vector space is two, and hence $\text{rank}(\boldsymbol{C}) = 2$. If the columns of $\boldsymbol{A}$ or the rows of $\boldsymbol{B}$ are linearly *dependent* and non-zero, then $\text{rank}(\boldsymbol{C}) = 1$. This example can be extended in an obvious way to matrices of arbitrary size. It follows that the rank of an outer product is one.

1.2.7 Null Space of A

The null space $\mathcal{N}(\boldsymbol{A})$ of $\boldsymbol{A}$ is defined as

$$\mathcal{N}(\boldsymbol{A}) = \{\boldsymbol{x} \in \mathbb{R}^n \mid \boldsymbol{Ax} = \boldsymbol{0}\}, \tag{1.11}$$

where the trivial value $\boldsymbol{x} = \mathbf{0}$ is normally excluded from the space. From previous discussions, the product $\boldsymbol{Ax}$ is a linear combination of the columns $\boldsymbol{a}_i$ of $\boldsymbol{A}$, where the elements x_i of $\boldsymbol{x}$ are the corresponding coefficients. Thus, from (1.11), $\mathcal{N}(\boldsymbol{A})$ is the set of non-zero coefficients of all zero linear combinations of the columns of $\boldsymbol{A}$. Therefore, if $\mathcal{N}(\boldsymbol{A})$ is non-empty, then $\boldsymbol{A}$ must have linearly dependent columns and thus be column rank deficient.

If the columns of $\boldsymbol{A}$ are linearly independent, then $\mathcal{N}(\boldsymbol{A}) = \emptyset$ by definition (where $\emptyset$ denotes the empty set) because there can be no coefficients except zero which result in a zero linear combination. In this case, the dimension of the null space is zero (i.e. $\mathcal{N}(\boldsymbol{A})$ is empty), and $\boldsymbol{A}$ is full column rank. Note that any vector in $\mathcal{N}(\boldsymbol{A})$ is of dimension n. Any vector in $\mathcal{N}(\boldsymbol{A})$ is orthogonal to the rows of $\boldsymbol{A}$ and is thus in the orthogonal complement subspace of the rows of $\boldsymbol{A}$.

Example 1.7 Let $\boldsymbol{B}$ be as in Example 1.1. Then $\mathcal{N}(\boldsymbol{B}) = c(1, 1, -2)^T$, where $c \in \mathbb{R}$.

Example 1.8 Consider a matrix $\boldsymbol{A} \in \mathbb{R}^{3\times 3}$ whose columns are constrained to lie in a two-dimensional plane. Then there exists a zero linear combination of these vectors. The coefficients of this linear combination define a vector $\boldsymbol{x}$ which is in the nullspace of $\boldsymbol{A}$. In this case, $\boldsymbol{A}$ must be rank deficient.

Another important characterisation of a matrix is its *nullity*. The nullity of $\boldsymbol{A} \in \mathbb{R}^{m\times n}$ is the dimension of the nullspace of $\boldsymbol{A}$. In Example 1.7 above, the nullity of $\boldsymbol{A}$ is one. We then have the following interesting property, which is proved later in Chap. 2:

$$\text{rank}(\boldsymbol{A}) + \text{nullity}(\boldsymbol{A}) = n. \tag{1.12}$$

1.3 The Four Fundamental Subspaces of a Matrix

The four matrix subspaces of concern are *the column space, the row space*, and their respective *orthogonal complements*. The development of these four subspaces is closely linked to $\mathcal{N}(\boldsymbol{A})$ and $R(\boldsymbol{A})$. We assume for this section that $\boldsymbol{A} \in \mathbb{R}^{m\times n}$, $r \leq \min(m, n)$, where $r = \text{rank}(\boldsymbol{A})$.

The Column Space This is simply $R(\boldsymbol{A})$. Its dimension is r. It is the set of all linear combinations of the columns of $\boldsymbol{A}$. Any vector in $R(\boldsymbol{A})$ is of dimension m.

The Orthogonal Complement of the Column Space This may be expressed as $R(\boldsymbol{A})_\perp$, with dimension $m - r$. It may be shown to be equivalent to $\mathcal{N}(\boldsymbol{A}^T)$, as follows: By definition, $\mathcal{N}(\boldsymbol{A}^T)$ is the set $\boldsymbol{x}$ satisfying:

$$\underbrace{\begin{bmatrix} \text{———} \\ \text{———} \\ \text{———} \\ \text{———} \end{bmatrix}}_{\boldsymbol{A}^T} \begin{bmatrix} x_1 \\ \vdots \\ x_m \end{bmatrix} = \mathbf{0},$$

where columns of $\boldsymbol{A}$ are the rows of $\boldsymbol{A}^T$. From above, we see that $\mathcal{N}(\boldsymbol{A}^T)$ is the set of $\boldsymbol{x} \in \mathbb{R}^m$ which is orthogonal to all columns of $\boldsymbol{A}$ (rows of $\boldsymbol{A}^T$). This by definition is the orthogonal complement of $R(\boldsymbol{A})$. Any vector in $R(\boldsymbol{A})_\perp$ is of dimension m.

The Row Space The row space is defined simply as $R(\boldsymbol{A}^T)$, with dimension r. The row space is the span of the rows of $\boldsymbol{A}$. Any vector in $R(\boldsymbol{A}^T)$ is of dimension n.

The Orthogonal Complement of the Row Space This may be denoted as $R(\boldsymbol{A}^T)_\perp$. Its dimension is $n - r$. This set must be that which is orthogonal to all rows of $\boldsymbol{A}$: i.e. for $\boldsymbol{x}$ to be in this space, $\boldsymbol{x}$ must satisfy

$$\begin{matrix} \text{rows} \\ \text{of} \\ \boldsymbol{A} \rightarrow \end{matrix} \begin{bmatrix} \text{---} \\ \text{---} \\ \text{---} \\ \vdots \\ \text{---} \end{bmatrix} \begin{bmatrix} x_1 \\ \vdots \\ x_n \end{bmatrix} = \mathbf{0}.$$

Thus it is apparent that the set $\boldsymbol{x}$ satisfying the above is $\mathcal{N}(\boldsymbol{A})$. Any vector in $R(\boldsymbol{A}^T)_\perp$ is of dimension n.

We have noted before that $\text{rank}(\boldsymbol{A}) = \text{rank}(\boldsymbol{A}^T)$. Thus, the dimension of the row and column subspaces are equal. This is surprising, because it implies the number of linearly independent rows of a matrix is the same as the number of linearly independent columns. This holds regardless of the size or rank of the matrix. It is not an intuitively obvious fact and there is no immediately obvious reason why this should be so. Nevertheless, the rank of a matrix is the number of independent rows *or* columns.

1.4 Further Interpretations of Matrix Multiplication

1.4.1 Bigger-Block Interpretations of Matrix Multiplication

In this section, we generalise on the discussion of Sect. 1.2.1 for matrix-vector multiplication. To standardise our discussion, we define the matrix product $\boldsymbol{C}$ as

$$\underset{m \times n}{\mathbf{C}} = \underset{m \times k}{\mathbf{A}} \; \underset{k \times n}{\mathbf{B}} \tag{1.13}$$

Note that the matrix multiplication operation requires the inner dimensions k of $\boldsymbol{A}$ and $\boldsymbol{B}$ to be equal. Such matrices are said to have *conformable* dimensions. Four interpretations of the matrix multiplication operation follow:

1. Inner-Product Representation

If we define $\mathbf{a}_i^T \in \mathbb{R}^k$ as the ith row of $\boldsymbol{A}$ and $\mathbf{b}_j \in \mathbb{R}^k$ as the jth column of $\boldsymbol{B}$, then the element c_{ij} of $\boldsymbol{C}$ is defined as the inner product $\mathbf{a}_i^T \mathbf{b}_j$. This is the conventional small-block representation of matrix multiplication.

2. Column Representation

This is the next bigger-block view of matrix multiplication. Here we form the product one column at a time. We have seen this idea before in Sect. 1.2.1—the jth column $\mathbf{c}_j$ of $\boldsymbol{C}$ may be expressed as a linear combination of columns $\mathbf{a}_i$ of $\boldsymbol{A}$ with coefficients which are the elements of the jth column of $\boldsymbol{B}$. Thus,

$$\boldsymbol{c}_j = \sum_{i=1}^{k} \boldsymbol{a}_i b_{ij} = \boldsymbol{A}\boldsymbol{b}_j \qquad j = 1, \ldots, n. \tag{1.14}$$

For example, if we evaluate only the pth element of the jth column $\mathbf{c}_j$ (i.e. element c_{pj}), we see that (1.14) degenerates into $c_{pj} = \sum_{i=1}^{k} a_{pi} b_{ij}$. This is the inner product of the pth row and jth column of $\boldsymbol{A}$ and $\boldsymbol{B}$, respectively. Even though the column representation is more compact from the algebraic viewpoint, its computer execution is identical to that of the inner-product representation. From this column representation, it is straightforward to show that $R(\boldsymbol{C}) = R(\boldsymbol{A})$ (see Problem 2 at the end of this chapter).

3. Row Representation

This is the transpose operation of the column representation above. The ith row $\boldsymbol{c}_i^T$ of $\boldsymbol{C}$ can be written as a linear combination of the rows $\boldsymbol{b}_j^T$ of $\boldsymbol{B}$, whose coefficients are given as the ith row of $\boldsymbol{A}$, i.e.

$$\boldsymbol{c}_i^T = \sum_{j=1}^{k} a_{ij} \boldsymbol{b}_j^T = \boldsymbol{a}_i^T \boldsymbol{B} \qquad i = 1, \ldots, m.$$

It is possible to show that $R(\boldsymbol{C}^T) = R(\boldsymbol{B}^T)$ (Problem 2).

4. Outer-Product Representation

This is the largest-block representation. Let $\mathbf{a}_i$ and $\mathbf{b}_i^T$ be the ith column and row of $\boldsymbol{A}$ and $\boldsymbol{B}$ respectively. Then the product $\boldsymbol{C}$ may also be expressed as the sum of outer products of $\boldsymbol{a}_i$ and $\boldsymbol{b}_i^T$, respectively:

$$C = \sum_{i=1}^{k} a_i b_i^T. \tag{1.15}$$

By looking at this operation one column at a time, we see this form of matrix multiplication performs exactly the same operations as the column representation above. For example, the jth column c_j of the product is determined from (1.15) to be $c_j = \sum_{i=1}^{k} a_i b_{ij}$, which is identical to (1.14) above.

An extension to the outer-product rule is that the product of the form $C = ADB$, where D is diagonal, is given by

$$C = \sum_{i=1}^{k} d_{ii} a_i b_i^T, \tag{1.16}$$

where d_{ii} are the diagonal elements of D. The proof is left to Problem 7.

Multiplication by Diagonal Matrices We consider a matrix A premultiplied by D of conformable dimensions, i.e. $C = DA$, where D is diagonal. In this case, the ith *row* c_i^T of C can be written as $c_i^T = d_{ii} a_i^T$; i.e. pre-multiplication by a diagonal matrix has the effect of scaling each row of the product by the corresponding diagonal element of D. Similarly, if we take a matrix B postmultiplied by a diagonal D, i.e. $C = BD$ then each *column* $c_i = d_{ii} b_i$; i.e. post-multiplication by a diagonal matrix scales each column by the corresponding diagonal element of D.

Multiplication with Block Matrices In many cases, matrix analysis becomes much easier if we partition a matrix into blocks. This happens extensively in Chap. 2, for example. Manipulating block matrices is a very straightforward process, since the blocks can be treated just as regular elements are when manipulating ordinary matrices, provided the dimensions of the blocks of the two matrices being operated upon are conformable. We partition two matrices A and B into blocks as shown, where the dimensions of each block are as indicated [1]:

$$A = \begin{bmatrix} A_{11} & \dots & A_{1p} \\ \vdots & & \vdots \\ A_{q1} & \dots & A_{qp} \end{bmatrix} \begin{matrix} m_1 \\ \vdots \\ m_q \end{matrix}$$

$$\begin{matrix} s_1 & \dots & s_p \end{matrix}$$

$$B = \begin{bmatrix} B_{11} & \dots & B_{1r} \\ \vdots & & \vdots \\ B_{p1} & \dots & B_{pr} \end{bmatrix} \begin{matrix} s_1 \\ \vdots \\ s_p \end{matrix}$$

$$\begin{matrix} n_1 & \dots & n_r \end{matrix}$$

Then the product $\boldsymbol{C} = \boldsymbol{AB}$ can be formed by treating each block as a regular element when performing ordinary matrix multiplication. For example, block $\boldsymbol{C}_{ij}$ can be written following a block form of the inner product rule for matrix multiplication as:

$$\boldsymbol{C}_{ij} = \sum_{k=1}^{p} \boldsymbol{A}_{ik}\boldsymbol{B}_{kj}. \tag{1.17}$$

Notice that for each term in the above, the number of columns of the kth $\boldsymbol{A}$–block is equal to the number of rows in the kth $\boldsymbol{B}$–block, which is the dimension s_k, $k = 1, \ldots, p$ in the above equations. This way, the blocks have conformable dimensions with regard to matrix multiplication. Also the number of blocks p in a row of $\boldsymbol{A}$ must equal the number of blocks in a column of $\boldsymbol{B}$. Equation (1.17) can be proved by verifying that for any element $\boldsymbol{c}_{ij}$, (1.17) performs exactly the same operations as does ordinary matrix multiplication.

A relevant example of block matrix multiplication, which is related to the eigendecomposition in the next chapter, is the following, where all the blocks are assumed to have conformable dimensions:

$$\begin{aligned}\begin{bmatrix} \boldsymbol{V}_1 & \boldsymbol{V}_2 \end{bmatrix} \begin{bmatrix} \boldsymbol{\Lambda} & \boldsymbol{0} \\ \boldsymbol{0} & \boldsymbol{0} \end{bmatrix} \begin{bmatrix} \boldsymbol{V}_1^T \\ \boldsymbol{V}_2^T \end{bmatrix} &= \begin{bmatrix} \boldsymbol{V}_1 & \boldsymbol{V}_2 \end{bmatrix} \begin{bmatrix} \boldsymbol{\Lambda}\boldsymbol{V}_1^T \\ \boldsymbol{0} \end{bmatrix} \\ &= \boldsymbol{V}_1\boldsymbol{\Lambda}\boldsymbol{V}_1^T.\end{aligned}$$

1.5 Vector Norms

A *vector norm* is a means of expressing the length or distance associated with a vector. A norm on a vector space $\mathbb{R}^m$ is a *function* f, which maps a point in $\mathbb{R}^m$ into a point in $\mathbb{R}$. Formally, this is stated mathematically as $f : \mathbb{R}^m \rightarrow \mathbb{R}$. The norm has the following properties:

1. $f(\boldsymbol{x}) \geq 0$ for all $\mathbf{x} \in \mathbb{R}^n$.
2. $f(x) = 0$ if and only if $\boldsymbol{x} = 0$.
3. $f(\boldsymbol{x} + \mathbf{y}) \leq f(\boldsymbol{x}) + f(\mathbf{y})$ for $\mathbf{x}, \mathbf{y} \in \mathbb{R}^m$.
4. $f(a\boldsymbol{x}) = |a| f(\boldsymbol{x})$ for $a \in \mathbb{R}, \boldsymbol{x} \in \mathbb{R}^m$.

We denote the function $f(\boldsymbol{x})$ as $||\boldsymbol{x}||$.

The p-Norms This is a useful class of norms, generalising on the idea of the Euclidean norm. They are defined by

$$||\boldsymbol{x}||_p = (|x_1|^p + |x_2|^p + \ldots + |x_m|^p)^{1/p}, \tag{1.18}$$

where p can assume any positive value. Below we discuss commonly used values for p:

$p = 1$:

$$||\boldsymbol{x}||_1 = \sum_i |x_i|$$

which is simply the sum of absolute values of the elements.

$p = 2$:

$$||\boldsymbol{x}||_2 = \left(\sum_i {x_i}^2\right)^{\frac{1}{2}} = (\boldsymbol{x}^T\boldsymbol{x})^{\frac{1}{2}}$$

which is the familiar Euclidean norm. As implied from the above, we have the important identity $||\boldsymbol{x}||_2^2 = \boldsymbol{x}^T\boldsymbol{x}$.

$p = \infty$:

$$||\boldsymbol{x}||_\infty = \max_i |x_i|,$$

which is the element of $\boldsymbol{x}$ with the largest magnitude. This may be shown in the following way. As $p \to \infty$, the largest term within the round brackets in (1.18) dominates all others in the summation. Therefore (1.18) may be written as

$$\begin{aligned} ||\boldsymbol{x}||_\infty = \lim_{p\to\infty}\left[\sum_{i=1}^m |x|_i^p\right]^{\frac{1}{p}} &\to \left[|x_k|^p\right]^{\frac{1}{p}} \\ &= |x_k| \end{aligned}$$

where k is the index corresponding to the element of $\boldsymbol{x}$ with the largest absolute value.

Note that the $p = 2$ norm has many useful properties, but is expensive to compute. The 1- and ∞-norms are easier to compute, but are more difficult to deal with algebraically. All the p–norms obey all the properties of a vector norm.

Figure 1.4 shows the locus of points of the set $\{\boldsymbol{x} \mid ||\boldsymbol{x}||_p = 1\}$ for $p = 1, 2, \infty$. We now consider the relation between $||\boldsymbol{x}||_1$ and $||\boldsymbol{x}||_2$ for some point $\boldsymbol{x}$ (assumed not to be on a coordinate axis, for the sake of argument). Let $\boldsymbol{x}$ be a point which lies on the $||\boldsymbol{x}||_2 = 1$ locus. Because the $p = 1$ locus lies inside the $p = 2$ locus, the $p = 1$ locus must expand outwards (i.e. $||\boldsymbol{x}||_1$ must assume a larger value), to intersect the $p = 2$ locus at the point $\boldsymbol{x}$. Therefore we have $||\boldsymbol{x}||_1 \geq ||\boldsymbol{x}||_2$. The same reasoning can be used to show the same relation holds for $||\boldsymbol{x}||_1$ and $||\boldsymbol{x}||_2$ vs. $||\boldsymbol{x}||_\infty$. Even though we have considered only the two-dimensional case, the same argument is readily extended to vectors of arbitrary dimension. Therefore, we have the following generalisation: for any vector $\boldsymbol{x}$, we have

$$||\boldsymbol{x}||_1 \geq ||\boldsymbol{x}||_2 \geq ||\boldsymbol{x}||_\infty.$$

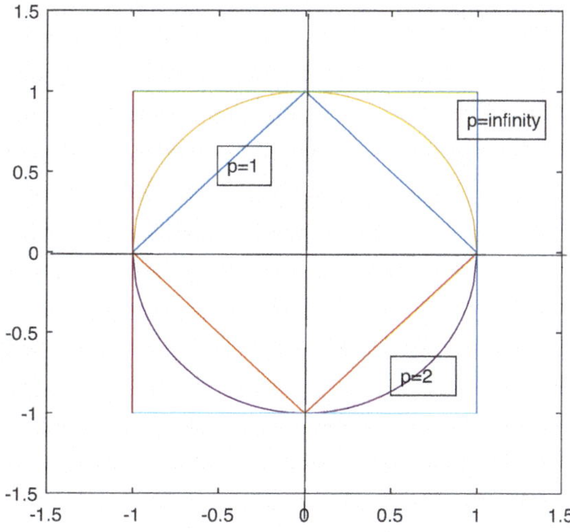

Fig. 1.4 Locus of points of the set $\{\boldsymbol{x} \mid ||\boldsymbol{x}||_p = 1\}$ for various values of p

1.6 Determinants

Consider a square matrix $\boldsymbol{A} \in \mathbb{R}^{m\times m}$. We can define the matrix $\boldsymbol{A}_{ij}$ as the submatrix obtained from $\boldsymbol{A}$ by deleting the ith row and jth column of $\boldsymbol{A}$. The scalar number $\det(\boldsymbol{A}_{ij})$ (where $\det(\cdot)$ denotes *determinant*) is called the *minor* associated with the element a_{ij} of $\boldsymbol{A}$. The signed minor $c_{ij} \triangleq (-1)^{j+i}\det(\boldsymbol{A}_{ij})$ is called the *cofactor* of a_{ij}.

The determinant of $\boldsymbol{A}$ is the m-dimensional volume contained within the columns (rows) of $\boldsymbol{A}$. This interpretation of determinant is very useful as we see shortly. The determinant of a matrix may be evaluated by the expression

$$\det(\boldsymbol{A}) = \sum_{j=1}^{m} a_{ij}c_{ij}, \qquad i \in (1\dots m). \tag{1.19}$$

or

$$\det(\boldsymbol{A}) = \sum_{i=1}^{m} a_{ij}c_{ij}, \qquad j \in (1\dots m). \tag{1.20}$$

Both the above are referred to as the *cofactor expansion* of the determinant. Equation (1.19) is along the ith *row* of $\boldsymbol{A}$, whereas (1.20) is along the jth *column*. It is indeed interesting to note that both versions above give exactly the same value, regardless of the value of i or j.

Equations (1.19) and (1.20) express the $m \times m$ determinant $\det\boldsymbol{A}$ in terms of the cofactors c_{ij} of $\boldsymbol{A}$, which are themselves $(m-1)\times(m-1)$ determinants. Thus,

$m - 1$ recursions of (1.19) or (1.20) will finally yield the determinant of the $m \times m$ matrix $\boldsymbol{A}$.

From (1.19) it is evident that if $\boldsymbol{A}$ is triangular, then $\det(\boldsymbol{A})$ is the product of the main diagonal elements. Since diagonal matrices are in the upper triangular set, then the determinant of a diagonal matrix is also the product of its diagonal elements.

Properties of Determinants

Before we begin this discussion, let us define the volume of a parallelepiped defined by the set of column vectors comprising a matrix as the *principal volume* of that matrix.

We have the following properties of determinants, which are stated without proof:

1. $\det(\boldsymbol{AB}) = \det(\boldsymbol{A})\det(\boldsymbol{B}) \qquad \boldsymbol{A}, \boldsymbol{B} \in \mathbb{R}^{m\times m}$.
   The principal volume of the product of matrices is the product of principal volumes of each matrix.
2. $\det(\boldsymbol{A}) = \det(\boldsymbol{A}^T)$
   This property shows that the characteristic polynomials[1] of $\boldsymbol{A}$ and $\boldsymbol{A}^T$ are identical. Consequently, as we see later, eigenvalues of $\boldsymbol{A}^T$ and $\boldsymbol{A}$ are identical.
3. $\det(c\boldsymbol{A}) = c^m \det(\boldsymbol{A}) \qquad c \in \mathbb{R}, \boldsymbol{A} \in \mathbb{R}^{m\times m}$.
   This is a reflection of the fact that if each vector defining the principal volume is multiplied by c, then the resulting volume is multiplied by c^m.
4. $\det(\boldsymbol{A}) = 0 \rightleftharpoons \boldsymbol{A}$ is singular.
   This implies that at least one dimension of the principal volume of the corresponding matrix has collapsed to zero length.
5. $\det(\boldsymbol{A}) = \prod_{i=1}^{m} \lambda_i$, where λ_i are the eigen (singular) values of $\boldsymbol{A}$.
   This means the parallelepiped defined by the column or row vectors of a matrix may be transformed into a regular rectangular solid of the same m-dimensional volume whose edges have lengths corresponding to the eigen (singular) values of the matrix.
6. The determinant of an orthonormal[2] matrix is ± 1.
   This is easy to see, because the vectors of an orthonormal matrix are all unit length and mutually orthogonal. Therefore the corresponding principal volume is ± 1.
7. If $\boldsymbol{A}$ is nonsingular, then $\det(\boldsymbol{A}^{-1}) = [\det(\boldsymbol{A})]^{-1}$.
8. If $\boldsymbol{B}$ is nonsingular, then $\det(\boldsymbol{B}^{-1}\boldsymbol{AB}) = \det(\boldsymbol{A})$.
9. If $\boldsymbol{B}$ is obtained from $\boldsymbol{A}$ by interchanging any two rows (or columns), then $\det(\boldsymbol{B}) = -\det(\boldsymbol{A})$.
10. If $\boldsymbol{B}$ is obtained from $\boldsymbol{A}$ by adding a scalar multiple of one row to another (or a scalar multiple of one column to another), then $\det(\boldsymbol{B}) = \det(\boldsymbol{A})$.

[1] The characteristic polynomial of a matrix is defined in Chap. 2.

[2] An *orthonormal* matrix is defined in Chap. 2.

A further property of determinants allows us to compute the *inverse* of $\boldsymbol{A}$. Define the matrix $\tilde{\boldsymbol{A}}$ as the *adjoint* of $\boldsymbol{A}$

$$\tilde{\boldsymbol{A}} = \begin{bmatrix} c_{11} & \dots & c_{1m} \\ \vdots & & \vdots \\ c_{m1} & \dots & c_{mm} \end{bmatrix}^T \tag{1.21}$$

where the c_{ij} are the cofactors of $\boldsymbol{A}$. According to (1.19) or (1.20), the ith row $\tilde{\boldsymbol{a}}_i^T$ of $\tilde{\boldsymbol{A}}$ times the ith column $\boldsymbol{a}_i$ is $\det(\boldsymbol{A})$; i.e.

$$\tilde{\boldsymbol{a}}_i^T \boldsymbol{a}_i = \det(\boldsymbol{A}), \qquad i = 1, \dots, m. \tag{1.22}$$

It can also be shown that

$$\tilde{\boldsymbol{a}}_i^T \boldsymbol{a}_j = 0, \qquad i \neq j. \tag{1.23}$$

Then, combining (1.22) and (1.23) for $i, j \in \{1, \dots, m\}$, we have the following interesting property:

$$\tilde{\boldsymbol{A}}\boldsymbol{A} = \det(\boldsymbol{A})\boldsymbol{I}, \tag{1.24}$$

where $\boldsymbol{I}$ is the $m \times m$ identity matrix. It then follows from (1.24) that the inverse $\boldsymbol{A}^{-1}$ of $\boldsymbol{A}$ is given as

$$\boldsymbol{A}^{-1} = [\det(\boldsymbol{A})]^{-1}\tilde{\boldsymbol{A}}. \tag{1.25}$$

Neither (1.20) nor (1.25) are computationally efficient ways of calculating a determinant or an inverse, respectively. Better methods which exploit the properties of various matrix decompositions are made evident later in the course.

Problems

1. Explain how to construct a short $m \times n$ matrix with rank $r < m$.
2. We are given a matrix $\boldsymbol{A} \in \mathbb{R}^{10\times 2}$ and a matrix $\boldsymbol{B} \in \mathbb{R}^{2\times 5}$. Consider the product $\boldsymbol{C} = \boldsymbol{A}\boldsymbol{B}$. The columns and rows of $\boldsymbol{A}$ and $\boldsymbol{B}$, respectively, are linearly independent.

    (a) What is $\text{rank}(\boldsymbol{C})$?
    (b) Express a basis for the row space of $\boldsymbol{C}$ that does involve $\boldsymbol{C}$ itself. Justify your response.
    (c) Likewise, for the column space of $\boldsymbol{C}$.

3. The supplementary material described in the opening pages of this chapter contains a link to a matlab® data file Ch1Q3.mat. This file contains a short

matrix $\boldsymbol{A}$. Determine the orthogonal complement of the row space, using matlab®. What can you infer about the rank of $\boldsymbol{A}$ from your results? Also find, using matlab®, an orthonormal basis for $R(\boldsymbol{A})$ as well as its orthogonal complement subspace. Verify using matlab® that the respective subspaces are indeed orthogonal to each other.

4. You will also find the file Ch1Q4.mat in the supplementary material described in the opening pages of this chapter. This file contains a 6×3 matrix $\boldsymbol{A}_1$ along with vectors $\boldsymbol{b}_1$ and $\boldsymbol{b}_2$:

   (a) Given that $\boldsymbol{A}_1$ is tall, explain how to determine whether $\boldsymbol{b}_1 \in R(\boldsymbol{A}_1)$. *Hint:* Consider the system of equations $\boldsymbol{Ax} = \boldsymbol{b}$. But be careful, because the matlab® command that solves $\boldsymbol{Ax} = \boldsymbol{b}$ for $\boldsymbol{x}$ yields a solution such that the quantity $||\boldsymbol{Ax} - \boldsymbol{b}||_2^2$ is minimised, regardless of the properties of $\boldsymbol{b}_1$.
   (b) Find the coefficients of the linear combination of the columns of $\boldsymbol{A}_1$ that give the vector $\boldsymbol{b}_1$.
   (c) Now repeat using the vector $\boldsymbol{b}_2$. Does a set of coefficients exist that yield exactly $\boldsymbol{b}_2$? If not, why not? Explain the difference for the cases $\boldsymbol{b}_1$ and $\boldsymbol{b}_2$.
   (d) This problem suggests we can determine whether or not a vector $\boldsymbol{b} \in R(\boldsymbol{A})$ by solving the system of equations $\boldsymbol{Ax} = \boldsymbol{b}$. Suggest an alternative approach.

5. This question has to do with blind recovery of a source signal that has been transmitted through a pair of linear finite impulse response (FIR) systems, as shown in Fig. 1.5a. FIR filters are discussed in the Appendix of this volume. This problem is relevant to the situation where speech is recorded within a reverberant room using a pair of microphones. The impulse responses f_1 and f_2 model the reverberation effect of the room from the source to the individual microphones. Under certain conditions, the source signal $x[n]$ can be recovered without error and without knowledge of f_1 and f_2, using the concepts developed in this chapter. The sequence $x[n]$ is of length m and $f_1[n]$ and $f_2[n]$ are FIR sequences of length $n \ll m$. The outputs $y_1[n]$ and $y_2[n]$ are the convolution of $x[n]$ with $f_1[n]$ and $f_2[n]$, respectively, i.e.

$$y_i[n] = \sum_k f_i[k]x[n-k], \quad i \in [1, 2],$$

as shown in Fig. 1.5a. We observe only the sequences $y_1[n]$ and $y_2[n]$.

   (a) Show how to express the convolution of two sequences as a matrix-vector multiplication.
   (b) Show how $f_1[n]$, $f_2[n]$ can be determined, based ONLY on the observations $y_1[n]$ and $y_2[n]$. *Hint: Consider the configuration of Fig. 1.5b. What are g_1 and g_2 so that the output $z[n] = 0$? This condition can be used to identify f_1 and f_2. Note that the sequences g_2 and g_1 are the recovered versions of f_1 and f_2, respectively. The g_2 and g_1 can be computed to represent f_1 and f_2 with insignificant error.*

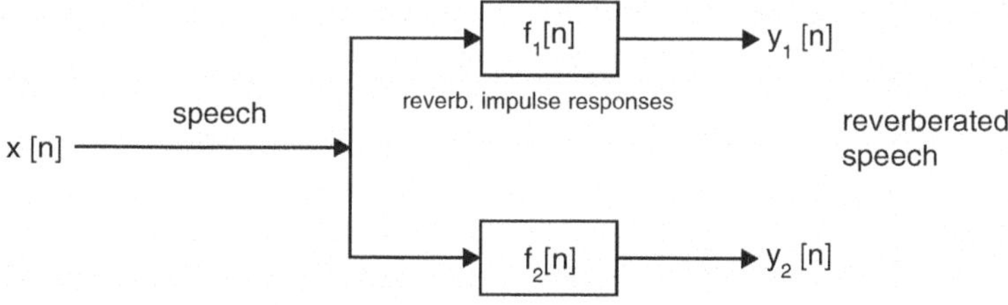

(a) Generation of reverberated speech

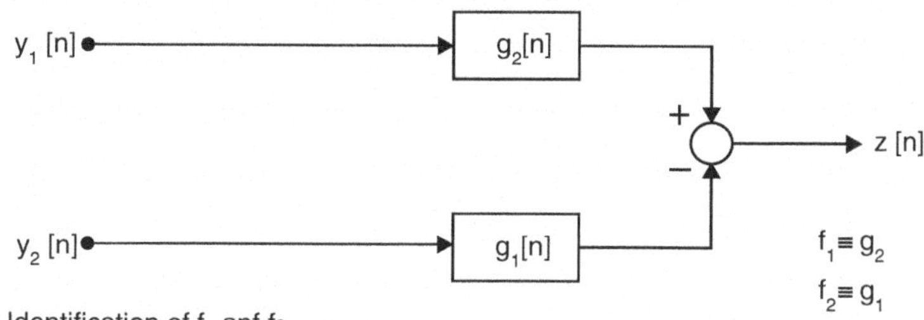

(b) Identification of f_1 anf f2

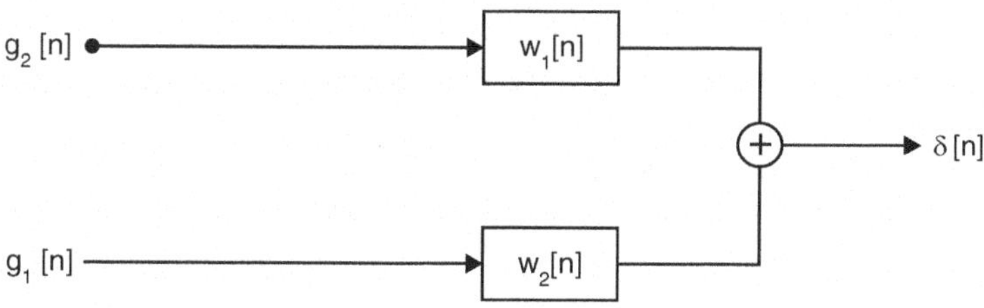

(c) w_1 and w_2 are chosen to that the output is a δ-function

Fig. 1.5 A configuration for blind deconvolution

(c) Explain how to determine g_1 and g_2. Are there any required conditions on $x[n]$, or on $f_1[n]$, $f_2[n]$? The true versions of $f_1[n]$ and $f_2[n]$ are available from the file Ch1Q6.mat in the supplementary material described in the opening pages of this chapter, for the purposes of comparison.

6. The speech signal used to generate the sequences $y_1[n]$ and $y_2[n]$ in Problem 5 is the variable "speech", which is one of the variables contained in the file Ch1Q6, as decribed in Problem 5(c) above. Also available in the file are the signals $y_1[n]$ and $y_2[n]$. It is possible to recover the source signal $x[n]$ from the observations $y_1[n]$ and $y_2[n]$ knowing f_1 and f_2 or equivalently g_1 and g_2.

If the sequences f are of length n, then there exist sequences $w_1[n]$, $w_2[n]$ of length $n - 1$ that satisfy the expression

$$f_1[n] \star w_1[n] + f_2[n] \star w_2[n] = \delta[n]$$

where $\delta[n]$ is the delta-function sequence and $\star$ denotes the convolution operation. (The Appendix of this volume presents a treatment of the digital convolution operation for those who may find it useful.) The configuration corresponding to this equation is shown in Fig. 1.5c. See [3] for more details. Note that if the structure of Fig. 1.5a is cascaded with that of Fig. 1.5c, then the impulse response of the overall structure becomes a zero-delay impulse function, and thus the original speech $x[n]$ is recovered. Find the sequences $w_1[n]$, $w_2[n]$ and recover the source $x[n]$. The true impulse responses f_1 and f_2 may be found in file Ch1Q6.mat so you can compare your responses with the true values. You can also play the recovered speech file through your computer sound system by issuing the command "soundsc(*vector*)" within matlab®.

7. Prove Eq. (1.16).
8. A matrix $\boldsymbol{A} \in \mathbb{R}^{m \times n}$ may be decomposed according to the *singular value decomposition*, to be discussed in Chap. 3, as $\boldsymbol{A} = \boldsymbol{U}\boldsymbol{\Sigma}\boldsymbol{V}^T$, where $\boldsymbol{U}$ is $m \times m$, $\boldsymbol{\Sigma}$ is $m \times n$ diagonal (i.e. a zero block either below or to the right of the diagonal block is appended in order to maintain dimensional consistency), and $\boldsymbol{V}$ is $n \times n$. Express $\boldsymbol{A}$ using a modified form of the outer product rule for matrix multiplication.
9. We have a matrix $\boldsymbol{A} \in \mathbb{R}^{m \times n}$, $m \geq n$. Under what conditions is $\boldsymbol{A}^T\boldsymbol{A}$ diagonal? Explain fully.
10. Consider the matrix product $\boldsymbol{C} = \boldsymbol{A}\boldsymbol{\Lambda}^{-1}\boldsymbol{B}$, where all matrices are $n \times n$ and $\boldsymbol{\Lambda}$ is diagonal. Express $\boldsymbol{C}$ using a modified form of the outer product rule.

References

1. G.H. Golub, C.F. Van Loan, *Matrix Computations*, 3rd edn. (The Johns Hopkins University, Baltimore, 1996)
2. A. J. Laub, *Matrix Analysis for Scientists and Engineers*, vol. 91 (Siam, Philadelphia, 2005)
3. M. Miyoshi, Y. Kaneda, Inverse filtering of room acoustics. IEEE Trans. Acoust. Speech Signal process. **36**(2), 145–152 (1988)

Chapter 2
Eigenvalues, Eigenvectors, and Correlation

2.1 Eigenvalues and Eigenvectors

We first discuss this subject from the classical mathematical viewpoint, and then when the requisite background is in place, we will apply eigenvalue and eigenvectors in a signal processing context. We investigate the underlying ideas of this topic using the matrix $\boldsymbol{A}$ as an example:

$$\boldsymbol{A} = \begin{bmatrix} 4 & 1 \\ 1 & 4 \end{bmatrix} \tag{2.1}$$

The product $\boldsymbol{A}\boldsymbol{x}_1$, where $\boldsymbol{x}_1 = [1, 0]^T$, is shown in Fig. 2.1. Then,

$$\boldsymbol{A}\boldsymbol{x}_1 = \begin{bmatrix} 4 \\ 1 \end{bmatrix}. \tag{2.2}$$

By comparing the vectors $\boldsymbol{x}_1$ and $\boldsymbol{A}\boldsymbol{x}_1$, we see that the product vector is scaled and rotated *counter-clockwise* with respect to $\boldsymbol{x}_1$. Now consider the case where $\boldsymbol{x}_2 = [0, 1]^T$. Then $\mathbf{Ax}_2 = [1, 4]^T$. Here, we note a *clockwise* rotation of $\mathbf{Ax}_2$ with respect to $\boldsymbol{x}_2$. Now let $\boldsymbol{x}_3 = [1, 1]^T$. Then $\boldsymbol{A}\,\boldsymbol{x}_3 = [5, 5]^T$ and the product vector points in the *same* direction as $\boldsymbol{x}_3$; i.e. $\boldsymbol{A}\boldsymbol{x}_3 \in \text{span}(\boldsymbol{x}_3)$ and $\boldsymbol{A}\boldsymbol{x}_3 = \lambda\boldsymbol{x}_3$. Because of this property, $\boldsymbol{x}_3 = [1, 1]^T$ is an *eigenvector* of $\boldsymbol{A}$. The scale factor (which in this case is 5) is given the symbol λ and is referred to as an *eigenvalue*.

Note that $\boldsymbol{x} = [1, -1]^T$ is also an eigenvector, because in this case, $\boldsymbol{A}\boldsymbol{x} = [3, -3]^T = 3\boldsymbol{x}$. The corresponding eigenvalue is 3.

Supplementary Information The online version contains supplementary material available at (https://doi.org/10.1007/978-3-031-68915-4_2).

J. Reilly, *Fundamentals of Linear Algebra for Signal Processing*,
https://doi.org/10.1007/978-3-031-68915-4_2

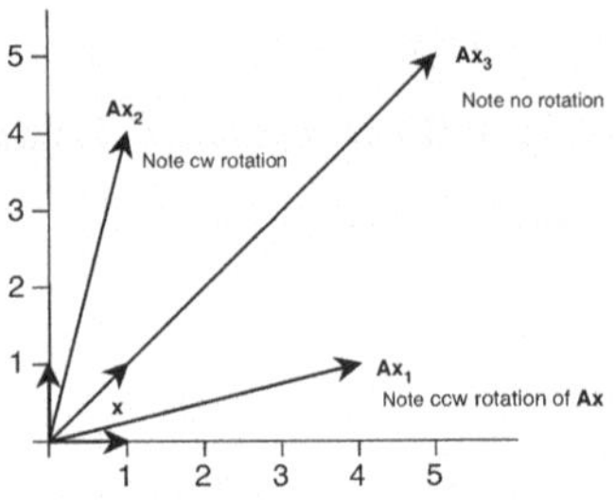

Fig. 2.1 Matrix-vector multiplication for various vectors

Thus if $\boldsymbol{x}$ is an eigenvector of $\boldsymbol{A} \in \mathbb{R}^{n \times n}$, we have

$$\boldsymbol{A}\boldsymbol{x} = \lambda \boldsymbol{x} \tag{2.3}$$

i.e. the vector $\boldsymbol{A}\boldsymbol{x}$ is in the same direction as $\boldsymbol{x}$ but scaled by a factor λ.

Equation (2.3) may be written in the form $\boldsymbol{A}\boldsymbol{x} = \lambda \boldsymbol{I}\boldsymbol{x}$ or

$$(\boldsymbol{A} - \lambda \boldsymbol{I})\boldsymbol{x} = \boldsymbol{0} \tag{2.4}$$

where $\boldsymbol{I}$ is the $n \times n$ identity matrix. Thus to be an eigenvector, $\boldsymbol{x}$ must lie in the nullspace of $\boldsymbol{A} - \lambda \boldsymbol{I}$. We know that a nontrivial solution to (2.4) exists if and only if $\mathcal{N}(\boldsymbol{A} - \lambda \boldsymbol{I})$ is non-empty, which implies that

$$\det(\boldsymbol{A} - \lambda \boldsymbol{I}) = 0 \tag{2.5}$$

where $\det(\cdot)$ denotes determinant.[1] Equation (2.5), when evaluated, becomes a polynomial in λ of degree n whose roots are the eigenvalues. For example, for the matrix $\boldsymbol{A}$ above we have

$$\det\left[\begin{pmatrix} 4 & 1 \\ 1 & 4 \end{pmatrix} - \lambda \begin{pmatrix} 1 & 0 \\ 0 & 1 \end{pmatrix}\right] = 0$$

$$\det\begin{bmatrix} 4-\lambda & 1 \\ 1 & 4-\lambda \end{bmatrix} = (4-\lambda)^2 - 1$$

$$= \lambda^2 - 8\lambda + 15 = 0. \tag{2.6}$$

It is easily verified that the roots of this polynomial are (5,3), which correspond to the eigenvalues indicated above.

[1] It turns out that determining the nullspace by evaluating the determinant is *very* inefficient. Nevertheless, the method presented here is what appears in most undergraduate texts and leads to the useful concept of the *characteristic equation* relating to a matrix.

Equation (2.6) is referred to as the *characteristic equation* of $\boldsymbol{A}$, and the corresponding polynomial is the *characteristic polynomial*. The characteristic polynomial is of degree n.

More generally, if $\boldsymbol{A}$ is $n \times n$, then there are n solutions of (2.5), or n roots of the characteristic polynomial. Thus there are n eigenvalues of $\boldsymbol{A}$ satisfying (2.3), i.e.

$$\boldsymbol{A}\boldsymbol{x}_i = \lambda_i \boldsymbol{x}_i, \quad i = 1, \ldots, n. \tag{2.7}$$

Note that the eigenvalues of a diagonal matrix are the diagonal elements themselves, with corresponding eigenvector given by the respective elementary vector $\boldsymbol{e}_1, \ldots, \boldsymbol{e}_n$. If the matrix is triangular, it is straightforward to show that the eigenvalues are also the diagonal elements, but the eigenvectors are no longer the elementary vectors.

If the eigenvalues are all *distinct*, there are n associated linearly independent eigenvectors, whose directions are unique, which span an n-dimensional Euclidean space.

In the case where there are $r \leq n$ *repeated* eigenvalues, then a linearly independent set of n eigenvectors still exist.[2] However, their directions are not unique in this case. In fact, if $[\boldsymbol{v}_1 \ldots \boldsymbol{v}_r]$ are a set of r linearly independent eigenvectors associated with a repeated eigenvalue, then any vector in span$[\boldsymbol{v}_1 \ldots \boldsymbol{v}_r]$ is also an eigenvector. The proof is left as an exercise.

Example 2.1 Consider the matrix given by

$$\begin{bmatrix} 1 & 0 & 0 \\ 0 & 0 & 0 \\ 0 & 0 & 0 \end{bmatrix}.$$

Since the eigenvalues of a diagonal matrix are the diagonal elements themselves, the eigenvalues are $[1, 0, 0]^T$. We have a repeated zero eigenvalue, and a corresponding linearly independent eigenvector set is $[\boldsymbol{e}_1, \boldsymbol{e}_2, \boldsymbol{e}_3]$. Then it may be verified that any vector in span$[\boldsymbol{e}_2, \boldsymbol{e}_3]$ is also an eigenvector associated with the zero repeated eigenvalue.

Example 2.2 Consider the $n \times n$ identity matrix $\boldsymbol{I}$. It has n repeated eigenvalues equal to one. In this case, any n-dimensional vector is an eigenvector with a corresponding eigenvalue of 1.

Equation (2.5) gives us a clue how to compute eigenvalues. We can formulate the characteristic polynomial and evaluate its roots to give the λ_i. Once the eigenvalues are available, it is possible to compute the corresponding eigenvectors $\boldsymbol{v}_i$ by evaluating the nullspace of the quantity $\boldsymbol{A} - \lambda_i \boldsymbol{I}$, for $i = 1, \ldots, n$. This approach is adequate for small systems, but for those of appreciable size, this method is slow

[2] This condition holds provided rank$(\boldsymbol{A} - \lambda \boldsymbol{I}) = n - r$ [11].

and prone to appreciable numerical error. Later, we consider various orthogonal transformations which lead to much more effective techniques for finding the eigenvalues.

We now present some useful properties of eigenvalues and eigenvectors, to aid in our understanding.

Property 1 *If the eigenvalues of a (Hermitian)*[3] *symmetric matrix are distinct, then the eigenvectors are orthogonal.*

Proof Let $\{\boldsymbol{v}_i\}$ and $\{\lambda_i\}, i = 1, \ldots, n$ be the eigenvectors and corresponding eigenvalues of $\boldsymbol{A} \in \mathbb{R}^{n \times n}$. Choose any $i, j \in [1, \ldots, n], i \neq j$. Then

$$\boldsymbol{A}\boldsymbol{v}_i = \lambda_i \boldsymbol{v}_i \tag{2.8}$$

and

$$\boldsymbol{A}\boldsymbol{v}_j = \lambda_j \boldsymbol{v}_j. \tag{2.9}$$

Premultiply (2.8) by $\boldsymbol{v}_j^T$ and (2.9) by $\boldsymbol{v}_i^T$:

$$\boldsymbol{v}_j^T \boldsymbol{A}\boldsymbol{v}_i = \lambda_i \boldsymbol{v}_j^T \boldsymbol{v}_i \tag{2.10}$$

$$\boldsymbol{v}_i^T \boldsymbol{A}\boldsymbol{v}_j = \lambda_j \boldsymbol{v}_i^T \boldsymbol{v}_j \tag{2.11}$$

The quantities on the left are equal when $\boldsymbol{A}$ is symmetric. We show this as follows. Since the left-hand side of (2.10) is a scalar, its transpose is equal to itself. Therefore, we get $\boldsymbol{v}_j^T \boldsymbol{A}\boldsymbol{v}_i = \boldsymbol{v}_i^T \boldsymbol{A}^T \boldsymbol{v}_j$.[4] But, since $\boldsymbol{A}$ is symmetric, $\boldsymbol{A}^T = \boldsymbol{A}$. Thus, $\boldsymbol{v}_j^T \boldsymbol{A}\boldsymbol{v}_i = \boldsymbol{v}_i^T \boldsymbol{A}^T \boldsymbol{v}_j = \boldsymbol{v}_i^T \boldsymbol{A}\boldsymbol{v}_j$.

Subtracting (2.10) from (2.11), we have

$$(\lambda_i - \lambda_j)\boldsymbol{v}_j^T \boldsymbol{v}_i = 0 \tag{2.12}$$

where we have used the fact $\boldsymbol{v}_j^T \boldsymbol{v}_i = \boldsymbol{v}_i^T \boldsymbol{v}_j$. But by hypothesis, $\lambda_i - \lambda_j \neq 0$. Therefore, (2.12) is satisfied only if $\boldsymbol{v}_j^T \boldsymbol{v}_i = 0$, which means the vectors are orthogonal. □

[3] Recall a symmetric matrix is one where $\boldsymbol{A} = \boldsymbol{A}^T$, where the superscript ^{T} means *transpose*, i.e. for a symmetric matrix, an element $a_{ij} = a_{ji}$. The Hermitian transpose (denoted $\boldsymbol{A}^H$) is relevant only for the complex case and is where the matrix is transposed *and* complex conjugated. The Hermitian symmetric matrix has the property $a_{ij} = a_{ji}^*$.

In this book we generally consider only real matrices. However, when complex matrices are considered, *Hermitian symmetric* is implied instead of symmetric.

[4] Here, we have used the property that for matrices or vectors $\boldsymbol{A}$ and $\mathbf{B}$ of conformable size, $(\mathbf{AB})^T = \mathbf{B}^T \boldsymbol{A}^T$.

Here we have considered only the case where the eigenvalues are distinct. If an eigenvalue $\tilde{\lambda}$ is repeated r times, and $\text{rank}(\boldsymbol{A} - \tilde{\lambda}\boldsymbol{I}) = n - r$, then a mutually orthogonal set of n eigenvectors can still be found.

Another useful property of eigenvalues of symmetric matrices is as follows.

Property 2 *The eigenvalues of a (Hermitian) symmetric matrix are real.*

Proof From [11]. (By contradiction): First, we consider the case where $\boldsymbol{A}$ is real. Let λ be a non-zero complex eigenvalue of a symmetric matrix $\boldsymbol{A}$. Then, since the elements of $\boldsymbol{A}$ are real, λ^*, the complex conjugate of λ must also be an eigenvalue of $\boldsymbol{A}$, because the roots of the characteristic polynomial must occur in complex conjugate pairs. Also, if $\boldsymbol{v}$ is a non-zero eigenvector corresponding to λ, then an eigenvector corresponding λ^* must be $\boldsymbol{v}^*$, the complex conjugate of $\boldsymbol{v}$. But **Property 1** requires that the eigenvectors be orthogonal; therefore, $\boldsymbol{v}^T\boldsymbol{v}^* = 0$. But $\boldsymbol{v}^T\boldsymbol{v}^* = (\boldsymbol{v}^H\boldsymbol{v})^*$, which is by definition the complex conjugate of the norm of $\boldsymbol{v}$. But the norm of a vector is a pure real number; hence, $\boldsymbol{v}^T\boldsymbol{v}^*$ must be greater than zero, since $\boldsymbol{v}$ is by hypothesis non-zero. We therefore have a contradiction. It follows that the eigenvalues of a symmetric matrix cannot be complex; i.e. they are *real*.

While this proof considers only the real symmetric case, it is easily extended to the case where $\boldsymbol{A}$ is Hermitian symmetric. The proof is left as an exercise. □

Property 3 *Let $\boldsymbol{A}$ be a matrix with eigenvalues $\lambda_i, i = 1, \ldots, n$ and eigenvectors $\boldsymbol{v}_i$. Then the eigenvalues of the matrix $\mathbf{A} + s\mathbf{I}$ are $\lambda_i + s$, with corresponding eigenvectors $\boldsymbol{v}_i$, where s is any real number.*

Proof From the definition of an eigenvector, we have $\mathbf{A}\mathbf{v} = \lambda\boldsymbol{v}$. Further, we have $s\mathbf{I}\mathbf{v} = s\boldsymbol{v}$. Adding, we have $(\boldsymbol{A} + s\mathbf{I})\boldsymbol{v} = (\lambda + s)\mathbf{v}$. This new eigenvector relation on the matrix $(\boldsymbol{A} + s\mathbf{I})$ shows the eigenvectors are unchanged, while the eigenvalues are displaced by s. This property has very useful consequences with regard to regularisation of poorly conditioned systems of equations. We discuss this point further in Chap. 9. □

Property 4 *Let $\boldsymbol{A}$ be an $n \times n$ matrix with eigenvalues $\lambda_i, i = 1, \ldots, n$. Then*

- *The determinant $det(\boldsymbol{A}) = \prod_{i=1}^{n} \lambda_i$.*
- *The trace*[5] *$tr(\boldsymbol{A}) = \sum_{i=1}^{n} \lambda_i$.*

The proof is straightforward, but because it is easier using concepts presented later in the course, it is deferred until later.

Property 5 *If $\boldsymbol{v}$ is an eigenvector of a matrix $\mathbf{A}$, then $c\boldsymbol{v}$ is also an eigenvector, where c is any real or complex constant.*

The proof follows directly by substituting $c\boldsymbol{v}$ for $\boldsymbol{v}$ in $\boldsymbol{A}\boldsymbol{v} = \lambda\boldsymbol{v}$. This means that only the direction of an eigenvector can be unique; its norm is not unique.

[5] The trace denoted $\text{tr}(\cdot)$ of a *square* matrix is the sum of its diagonal elements.

Property 6 *If the eigenpairs of* $\boldsymbol{A}$ *are* $\{\lambda, \boldsymbol{v}\}$, *then the eigenpairs of* $c\boldsymbol{A}$ *are* $\{c\lambda, \boldsymbol{v}\}$, *where* $c \in \mathbb{R}$.

The proof follows from direct multiplication of $c\boldsymbol{A}$ with any $\boldsymbol{v}$.

2.1.1 Orthonormal Matrices

Before proceeding with the eigendecomposition of a matrix, we must develop the concept of an *orthonormal matrix*. This form of matrix has mutually orthogonal columns, each of unit 2-norm. This implies that

$$\boldsymbol{q}_i^T \boldsymbol{q}_j = \delta_{ij}, \tag{2.13}$$

where δ_{ij} is the Kronecker delta and $\boldsymbol{q}_i$ and $\boldsymbol{q}_j$ are columns of the orthonormal matrix $\boldsymbol{Q}$. When $i = j$, the quantity $\boldsymbol{q}_i^T \boldsymbol{q}_i$ defines the squared 2-norm of $\boldsymbol{q}_i$, which has been defined as unity. When $i \neq j$, $\boldsymbol{q}_i^T \boldsymbol{q}_j = 0$, due to the orthogonality of the $\boldsymbol{q}_i$. We therefore have

$$\boldsymbol{Q}^T \boldsymbol{Q} = \boldsymbol{I}. \tag{2.14}$$

Thus, for an orthonormal matrix, (2.14) implies $\boldsymbol{Q}^{-1} = \boldsymbol{Q}^T$. Thus the inverse may be computed simply by taking the transpose of the matrix, an operation which requires almost no computational effort.

Equation (2.14) follows directly from the fact $\boldsymbol{Q}$ has orthonormal columns. It is not so clear that the quantity $\boldsymbol{Q}\boldsymbol{Q}^T$ should also equal the identity. We can resolve this question in the following way. Suppose that $\boldsymbol{A}$ and $\boldsymbol{B}$ are any two *square invertible* matrices such that $\boldsymbol{AB} = \boldsymbol{I}$. Then, $\boldsymbol{BAB} = \boldsymbol{B}$. By parsing this last expression, we have

$$(\boldsymbol{BA}) \cdot \boldsymbol{B} = \boldsymbol{B}. \tag{2.15}$$

Clearly, if (2.15) is to hold, then the quantity $\boldsymbol{BA}$ must be the identity; hence, if $\boldsymbol{AB} = \boldsymbol{I}$, then $\boldsymbol{BA} = \boldsymbol{I}$. Therefore, if $\boldsymbol{Q}^T\boldsymbol{Q} = \boldsymbol{I}$, then also $\boldsymbol{Q}\boldsymbol{Q}^T = \boldsymbol{I}$. From this fact, it follows that if a matrix has orthonormal columns, then it also must have orthonormal rows. We now develop a further useful property of orthonormal matrices:

An orthonormal matrix is sometimes referred to as a *unitary* matrix. This follows because the determinant of an orthonormal matrix is ± 1.

Orthonormal Matrices as a Basis Set To represent an m-length vector $\boldsymbol{x}$ in an $m \times m$ orthonormal basis represented by $\boldsymbol{Q}$ (as per the discussion in Sect. 1.2.4), we form the coefficients $\boldsymbol{c}$ by taking $\boldsymbol{c} = \boldsymbol{Q}^T\boldsymbol{x}$. Then $\boldsymbol{x}$ is given as $\boldsymbol{x} = \boldsymbol{Qc}$. Because of the simplicity of these operations, orthonormal matrices are a convenient form of basis. Later in this chapter, we use the eigenvector set of a covariance matrix as a basis for principal component analysis.

Consider the case where we have a tall matrix $\boldsymbol{U} \in \mathbb{R}^{m \times n}$, where $m > n$, whose columns are orthonormal. $\boldsymbol{U}$ can be formed by extracting only the first n columns of an arbitrary orthonormal matrix. (We reserve the term *orthonormal matrix* to refer to a *complete* $m \times m$ matrix). Because $\boldsymbol{U}$ has orthonormal columns, it follows that the quantity $\boldsymbol{U}^T\boldsymbol{U} = \boldsymbol{I}_{n\times n}$. However, it is important to realise that the quantity $\boldsymbol{U}\boldsymbol{U}^T \neq \boldsymbol{I}_{m\times m}$ in this case, in contrast to the situation when $m \leq n$. This fact is easily verified, since rank$(\boldsymbol{U}\boldsymbol{U}^T) = n$, which less than m, and so cannot be the identity.

However, the matrix $\boldsymbol{U}\boldsymbol{U}^T$ has interesting consequences when we interpret the tall $\boldsymbol{U}$ as an (incomplete) orthonormal basis. For an m-length vector $\boldsymbol{x}$, we form the coefficients of $\boldsymbol{x}$ in the incomplete basis $\boldsymbol{U}$ as $\boldsymbol{c} = \boldsymbol{U}^T\boldsymbol{x}$, where $\boldsymbol{c}$ is length $n < m$. Then the representation $\tilde{\boldsymbol{x}}$ of $\boldsymbol{x}$ in the new incomplete basis is $\tilde{\boldsymbol{x}} = \boldsymbol{U}\boldsymbol{c} = \boldsymbol{U}\boldsymbol{U}^T\boldsymbol{x}$. Notice however that $R(\boldsymbol{U})$ is a subspace of dimension n of the m-dimensional universe $\mathbb{R}^m$ and that $\tilde{\boldsymbol{x}}$ is in this subspace, even though $\boldsymbol{x}$ itself is in the full m-dimensional universe. So the operation $\boldsymbol{U}\boldsymbol{U}^T\boldsymbol{x}$ projects $\boldsymbol{x}$ into the subspace $R(\boldsymbol{U})$. The matrix $\boldsymbol{U}\boldsymbol{U}^T$ is referred as a *projector*. We will study projector matrices in more detail later in Chap. 3.

Property 7 *The vector* 2*-norm is invariant under an orthonormal transformation.*

Let $\boldsymbol{Q}$ be a matrix with orthonormal columns, i.e. it can be tall or square. Then for any $\boldsymbol{x}$, we have

$$||\boldsymbol{Q}\boldsymbol{x}||_2^2 = \boldsymbol{x}^T\boldsymbol{Q}^T\boldsymbol{Q}\boldsymbol{x} = \boldsymbol{x}^T\boldsymbol{x} = ||\boldsymbol{x}||_2^2.$$

Thus, because the norm does not change, an orthonormal transformation performs a *rotation* operation on a vector. We use this norm-invariance property later in our study of the least-squares problem.

2.1.2 The Eigendecomposition (ED) of a Square Symmetric Matrix

The eigendecomposition is a very useful tool in many branches of engineering analysis. While it may only be applied to square symmetric matrices, that is not too limiting a restriction in practice, since many matrices that are of interest in signal processing and related disciplines fit into this category.

Let $\boldsymbol{A} \in \mathbb{R}^{n\times n}$ be symmetric. Then, for eigenvalues λ_i and eigenvectors $\boldsymbol{v}_i$, we have

$$\mathbf{A}\mathbf{v}_i = \lambda_i \boldsymbol{v}_i, \qquad i = 1, \ldots, n. \tag{2.16}$$

Let $\boldsymbol{v}_i$ be an eigenvector of $\boldsymbol{A}$ having arbitrary 2-norm. Then, using Property 5, we can normalise $\boldsymbol{v}_i$ to unit 2-norm by dividing both sides of (2.16) by the quantity $||\boldsymbol{v}_i||_2$. Then $||\boldsymbol{v}_i||_2 = 1$. We assume all the eigenvectors have been normalised in this manner.

Then these n equations can be combined or stacked side-by-side together and represented in the following compact form:

$$\mathbf{AV} = \boldsymbol{V}\boldsymbol{\Lambda} \tag{2.17}$$

where $\boldsymbol{V} = [\boldsymbol{v}_1, \boldsymbol{v}_2, \ldots, \boldsymbol{v}_n]$ (i.e. each column of $\boldsymbol{V}$ is an eigenvector), and

$$\boldsymbol{\Lambda} = \begin{bmatrix} \lambda_1 & & & 0 \\ & \lambda_2 & & \\ & & \ddots & \\ 0 & & & \lambda_n \end{bmatrix} = \text{diag}(\lambda_1 \ldots \lambda_n). \tag{2.18}$$

Equation (2.17) may be verified by realising that corresponding columns from each side of this equation correspond to one specific value of the index i in (2.16). Because we have assumed $\boldsymbol{A}$ is symmetric, from Property 1, the $\boldsymbol{v}_i$ are orthogonal. Furthermore, since we have chosen $||\boldsymbol{v}_i||_2 = 1$, $\boldsymbol{V}$ is an orthonormal matrix. Thus, post-multiplying both sides of (2.17) by $\boldsymbol{V}^T$ and using $\boldsymbol{V}\boldsymbol{V}^T = \boldsymbol{I}$, we get

$$\boldsymbol{A} = \boldsymbol{V}\boldsymbol{\Lambda}\boldsymbol{V}^T. \tag{2.19}$$

Equation (2.19) is called the *eigendecomposition* (ED) of $\boldsymbol{A}$. The columns of $\boldsymbol{V}$ are eigenvectors of $\boldsymbol{A}$, and the diagonal elements of $\boldsymbol{\Lambda}$ are the corresponding eigenvalues. Any square symmetric matrix may be decomposed in this way. This form of decomposition, with $\boldsymbol{\Lambda}$ being diagonal, is of extreme interest and has many interesting consequences. It is this decomposition which leads directly to the concepts behind principal component analysis, which we discuss shortly.

Note that from (2.19), knowledge of the eigenvalues and eigenvectors of $\boldsymbol{A}$ is sufficient to completely specify $\boldsymbol{A}$. Note further that if the eigenvalues are distinct, then the ED is *unique*.

Equation (2.19) can also be written as

$$\boldsymbol{V}^T\mathbf{AV} = \boldsymbol{\Lambda}.$$

Since $\boldsymbol{\Lambda}$ is diagonal, we say that the orthonormal matrix $\boldsymbol{V}$ of eigenvectors *diagonalises* $\boldsymbol{A}$. In the case of distinct eigenvalues, $\boldsymbol{V}$ is unique and no other orthonormal matrix can diagonalise $\boldsymbol{A}$. (In the case of repeated roots, $\boldsymbol{V}$ is not unique, but any such eigenvector matrix will still diagonalise.) The fact that only $\boldsymbol{V}$ diagonalises $\boldsymbol{A}$ is a very important property of eigenvectors.

2.1.3 Conventional Notation on Eigenvalue Indexing

Let $\boldsymbol{A} \in \mathbb{R}^{n \times n}$ be symmetric and have rank $r \leq n$. Then, we see in the next section we have r non-zero eigenvalues and $n - r$ zero eigenvalues. Life becomes simpler if we re-order the eigenvalues so that

$$\underbrace{|\lambda_1| \geq |\lambda_2| \geq \ldots \geq |\lambda_r|}_{r \text{ non-zero eigenvalues}} > \underbrace{\lambda_{r+1} = \ldots = \lambda_n}_{n-r \text{ zero eigenvalues}} = 0 \tag{2.20}$$

i.e. we re-order the columns of $\boldsymbol{V}$ in (2.19) so that λ_1 is the largest in absolute value, with the remaining non-zero eigenvalues arranged in descending order, followed by $n - r$ zero eigenvalues. Note that if $\boldsymbol{A}$ is full rank, then $r = n$ and there are no zero eigenvalues. The eigenvectors are re-ordered to correspond with the ordering of the eigenvalues. We refer to the eigenvector corresponding to the largest eigenvalue as the *principal eigenvector*.

2.1.4 The Eigendecomposition in Relation to the Fundamental Matrix Subspaces

In this section, we develop relationships between the eigendecomposition of a matrix and its range, nullspace, and rank.

Here we discuss only the case where $\boldsymbol{A}$ is square and symmetric so that $\boldsymbol{V}$ is orthonormal, and we entertain the possibility that the matrix $\boldsymbol{A}$ may be rank deficient, i.e. we are given that $\text{rank}(\boldsymbol{A}) = r \leq n$. We write the eigendecomposition of $\boldsymbol{A}$ as $\boldsymbol{A} = \boldsymbol{V}\boldsymbol{\Lambda}\boldsymbol{V}^T$. We partition $\boldsymbol{V}$ and $\boldsymbol{\Lambda}$ in the following block formats:

$$\boldsymbol{V} = \begin{bmatrix} \underset{r}{\boldsymbol{V}_1} & \underset{n-r}{\boldsymbol{V}_2} \end{bmatrix}$$

where

$$\begin{aligned} \boldsymbol{V}_1 &= [\boldsymbol{v}_1, \boldsymbol{v}_2, \ldots, \boldsymbol{v}_r] \in \mathbb{R}^{n \times r} \\ \boldsymbol{V}_2 &= [\boldsymbol{v}_{r+1}, \ldots, \boldsymbol{v}_n] \in \mathbb{R}^{n \times n-r}. \end{aligned}$$

The columns of $\boldsymbol{V}_1$ are eigenvectors corresponding to the first r larger (in magnitude) eigenvalues of $\boldsymbol{A}$, and the columns of $\boldsymbol{V}_2$ are eigenvectors corresponding to the $n - r$ smallest eigenvalues. We also have

$$\boldsymbol{\Lambda} = \begin{bmatrix} \boldsymbol{\Lambda}_1 & \boldsymbol{0} \\ \boldsymbol{0} & \boldsymbol{\Lambda}_2 \end{bmatrix}$$

where $\boldsymbol{\Lambda}_1$ is an $r \times r$ diagonal block containing r non-zero eigenvalues and $\boldsymbol{\Lambda}_2$ is an $n - r \times n - r$ diagonal block containing the remaining eigenvalues. The eigendecomposition of A may therefore be written in block form as

$$A = \underset{r \;\; n-r}{\begin{bmatrix} V_1 & V_2 \end{bmatrix}} \begin{bmatrix} \Lambda_1 & \mathbf{0} \\ \mathbf{0} & \Lambda_2 \end{bmatrix} \begin{bmatrix} V_1^T \\ V_2^T \end{bmatrix} \begin{matrix} r \\ n-r \end{matrix} . \tag{2.21}$$

In the notation used above, the explicit absence of a matrix element in an off-diagonal position implies that element is zero. In the next section, we show the partitioning of A in the form of (2.21) reveals a great deal about the structure of the matrix.

Nullspace and Range

Range: Under our stated conditions, i.e. that A is square and symmetric (so V is orthonormal) and A is of rank $r \leq n$, we show that V_1 from (2.21) above is an orthonormal basis for $R(A)$, Λ_1 is non-zero, and $\Lambda_2 = \mathbf{0}$. Recall from (1.2.3) that $R(A)$ is defined as

$$R(A) = \left\{ y \in \mathbb{R}^m \mid y = Ax, \text{ for } x \in \mathbb{R}^n \right\}.$$

From (2.21) above, we can write the vector $y = Ax$ as

$$\begin{aligned} y &= \begin{bmatrix} V_1 & V_2 \end{bmatrix} \begin{bmatrix} \Lambda_1 & \mathbf{0} \\ \mathbf{0} & \Lambda_2 \end{bmatrix} \begin{bmatrix} V_1^T \\ V_2^T \end{bmatrix} x \\ &= \begin{bmatrix} V_1 & V_2 \end{bmatrix} \begin{bmatrix} \Lambda_1 & \mathbf{0} \\ \mathbf{0} & \Lambda_2 \end{bmatrix} \begin{bmatrix} c_1 \\ c_2 \end{bmatrix} \\ &= \begin{bmatrix} V_1 & V_2 \end{bmatrix} \begin{bmatrix} \Lambda_1 c_1 \\ \Lambda_2 c_2 \end{bmatrix} \end{aligned} \tag{2.22}$$

where

$$c = V^T x = \begin{bmatrix} V_1^T \\ V_2^T \end{bmatrix} x = \begin{bmatrix} c_1 \\ c_2 \end{bmatrix}. \tag{2.23}$$

The vector $c_1 \in \mathbb{R}^r$ and $c_2 \in \mathbb{R}^{n-r}$. It is clear that as x varies through an n-dimensional space, c maps out an equivalent n-dimensional space (due to the orthonormal transformation), whereas c_1 and c_2 each map out r and $n - r$ dimensional subspaces, respectively. Since the dimension of $R(A)$ is by definition rank(A), which we have assigned to be r, y must be the set of all linear combinations of exactly r linearly independent vectors, which for our purposes are contained in V_1 with coefficients $\Lambda_1 c_1$. It follows directly from (2.22) that 1) $\Lambda_2 = \mathbf{0}$ and 2) that Λ_1 is non-zero and 3) that V_1 is an orthonormal basis for $R(A)$. Therefore when A is rank r, A must contain $n - r$ zero eigenvalues.

We can see that when A is rank deficient, it must contain zero eigenvalues in a different way as follows. If A is rank deficient, then its nullspace is non-empty and therefore we have, for any $\boldsymbol{v} \in \mathcal{N}(\boldsymbol{A})$,

$$\boldsymbol{A}\boldsymbol{v} = \boldsymbol{0}$$

which can be written as $\boldsymbol{A}\boldsymbol{v} = \boldsymbol{0}\boldsymbol{v}$, which shows that $\boldsymbol{0}$ is an eigenvalue and any vector in the nullspace is a corresponding eigenvector. Therefore any square, symmetric rank-deficient matrix must have at least one zero eigenvalue.

Nullspace Recall from (1.11) that the nullspace $\mathcal{N}(\boldsymbol{A})$ of $\boldsymbol{A}$ is defined as

$$\mathcal{N}(\boldsymbol{A}) = \left\{\boldsymbol{x} \in \mathbb{R}^n \mid \boldsymbol{A}\boldsymbol{x} = \boldsymbol{0}\right\}$$

for non-zero $\boldsymbol{x}$. From (2.22) and the fact that $\boldsymbol{\Lambda}_2 = \boldsymbol{0}$, we see that $\boldsymbol{A}\boldsymbol{x} = \boldsymbol{0}$ for non-zero $\boldsymbol{x}$ iff $\boldsymbol{c}_1 = \boldsymbol{0}$ and $\boldsymbol{c}_2$ is arbitrary but non-zero. Therefore from (2.23), if $\boldsymbol{x} \in \mathcal{N}(\boldsymbol{A})$ then we have

$$\boldsymbol{x} = \begin{bmatrix} \boldsymbol{V}_1 & \boldsymbol{V}_2 \end{bmatrix} \begin{bmatrix} \boldsymbol{0} \\ \boldsymbol{c}_2 \end{bmatrix} = \boldsymbol{V}_2 \boldsymbol{c}_2.$$

Therefore any $\boldsymbol{x} \in \mathcal{N}(\boldsymbol{A})$ is a linear combination of the columns of $\boldsymbol{V}_2$, and therefore $\boldsymbol{V}_2$ is an orthonormal basis for $\mathcal{N}(\boldsymbol{A})$. Since $\boldsymbol{V}_2$ has $n - r$ columns, the nullity of $\boldsymbol{A}$ is $n - r$, which is the number of zero eigenvalues of $\boldsymbol{A}$.

We have seen that $\boldsymbol{V}_1$ and $\boldsymbol{V}_2$ have r and $n - r$ columns, respectively. Since we have established that r is rank and $n - r$ is nullity, it follows (for the sake of square, symmetric matrices) that

$$\text{rank}(\boldsymbol{A}) + \text{nullity}(\boldsymbol{A}) = n,$$

which is the relation established from (1.12) in Chap. 1. This rationalisation extends easily to arbitrary matrices, but requires the singular value decomposition, which is an extension of the eigendecomposition and is treated in the following chapter.

Diagonalising a System of Equations

We now show that transforming the variables involved in a linear system of equations into the eigenvector basis diagonalises the system of equations. We are given a system of equations $\boldsymbol{A}\boldsymbol{x} = \boldsymbol{b}$, where $\boldsymbol{A}$ is assumed square and symmetric and hence can be decomposed using the eigendecomposition as $\boldsymbol{A} = \boldsymbol{V}\boldsymbol{\Lambda}\boldsymbol{V}^T$. Thus our system can be written in the form

$$\boldsymbol{\Lambda}\boldsymbol{c} = \boldsymbol{d}, \tag{2.24}$$

where $\boldsymbol{c} = \boldsymbol{V}^T\boldsymbol{x}$ and $\boldsymbol{d} = \boldsymbol{V}^T\boldsymbol{b}$. Since $\boldsymbol{\Lambda}$ is diagonal, (2.24) has a much simpler structure and hence is much easier to solve than the original form. We expand on this relation further in Chap. 3.

2.2 Covariance and Covariance Matrices

Here we "change gears" for a while and discuss the idea of covariance, which is a very important topic in any form of statistical analysis and signal processing. In Sect. 2.5, we combine the topics of eigen-analysis and covariance matrices. The definitions of covariances vary somewhat across various books and articles, but the fundamental idea remains unchanged. We start the discussion with a review of some fundamental definitions.

We are given two scalar random variables x_1 and x_2. Recall that the *mean* μ and *variance* σ^2 of a random variable are defined respectively as

$$\begin{aligned} \mu &= E[x] \quad \text{and} \\ \sigma^2 &= E[x - \mu]^2, \end{aligned}$$

where E is the expectation operator. The standard deviation σ is the positive square root of the variance. The covariance σ_{12} and correlation ρ_{12} between the variables x_1 and x_2 are defined respectively as

$$\begin{aligned} \sigma_{12} &= E\big[(x_1 - \mu_1)(x_2 - \mu_2)\big] \quad \text{and} \\ \rho_{12} &= \frac{E\big[(x_1 - \mu_1)(x_2 - \mu_2)}{\sigma_1 \sigma_2}\big], \end{aligned}$$

where subscripts 1 and 2 refer to the random processes x_1 and x_2 respectively. It is straightforward to show that $-1 \leq \rho_{12} \leq 1$ (*Hint:* use the Schwartz inequality). Expectations are only an abstract concept, since they require an infinite sample of data to evaluate. In the practical case where only a finite sample of length n is available, and the process is not strongly non-stationary, we replace the expectation operator with an average over the n available samples to obtain an estimate of the respective quantity. The fact it is an estimate is denoted by placing a hat; i.e. $\hat{}$ over the respective quantity. Thus, we have[6]

$$\begin{aligned} \hat{\mu} &= \frac{1}{n} \sum_{i=1}^{n} x(i) \\ \hat{\sigma}^2 &= \frac{1}{n} \sum_{i=1}^{n} (x(i) - \hat{\mu})^2 \\ \hat{\sigma}_{12} &= \frac{1}{n} \sum_{i=1}^{n} (x_1(i) - \hat{\mu}_1)(x_2(i) - \hat{\mu}_2). \end{aligned}$$

[6] It turns out that these estimates are biased downwards, so in some cases in the literature, the $\frac{1}{n}$ terms are replaced with $\frac{1}{n-1}$ to remove this bias.

We often dispense with the hat notation since in most cases the context is clear. The hats are used only if necessary to avoid ambiguity. The above equations can be written in a more compact manner if we assemble the available samples into a vector $\boldsymbol{x} \in \mathbb{R}^n$. In this case, we can write

$$\sigma_j^2 = \frac{1}{n}(\boldsymbol{x}_j - \boldsymbol{\mu}_j)^T(\boldsymbol{x}_j - \boldsymbol{\mu}_j), \quad j \in [1, 2] \tag{2.25}$$

$$\sigma_{12} = \frac{1}{n}(\boldsymbol{x}_1 - \boldsymbol{\mu}_1)^T(\boldsymbol{x}_2 - \boldsymbol{\mu}_2), \tag{2.26}$$

where in the above we define subtraction of a scalar from a vector as the scalar being subtracted from each element of the vector.

The presentation is easier in the case where the means are zero; in fact most of the real-life signals we deal with do have zero mean. Therefore from this point onwards, we assume either that the means of the variables are zero or that we have subtracted the means as a pre-processing step—i.e. $x_j \leftarrow x_j - \mu_j$. Variables of this sort are referred to as *mean-centred* data.

We offer an example to help in the interpretation of covariances and related topics. Let x_1 be a person's mean-centred height and x_2 be their corresponding mean-centred weight. Then with respect to a particular person, if x_1 is positive (above the mean value), then it is more likely that x_2 is also positive, and vice versa. Thus it is likely that x_1 and x_2 both have the same sign and therefore the product x_1x_2 is most often positive when evaluated over many distinct people. Occasionally we encounter a short, stout individual or a tall, slender person, and in this case the product x_1x_2 would be negative. By taking the expectation $\sigma_{12} = E[\boldsymbol{x}_1\boldsymbol{x}_2]$ over all possible people, *on average* we will obtain a positive value, i.e. in this case the covariance σ_{12} is positive. We express this idea by saying that these variables are "positively correlated".

Another example is where we let x_1 be the final mark a person receives in a specific course, and x_2 remains as their weight. Then, since the variables appear to be unrelated, we would expect that the product x_1x_2 is positive as often as it is negative, with equal average magnitudes. In this case, $\sigma_{12} \rightarrow 0$. Here we say the variables are *uncorrelated*.

For the final example we assume x_1 is the maximum speed at which a person can run and again x_2 remains as the person's corresponding weight. Then generally, the greater the person's weight for a fixed height, the slower they run. So in this case, the variables most often have opposite signs, so the covariance σ_{12} is negative. A non-zero covariance between two random variables implies that one variable affects the other, whereas a zero covariance implies there is no (second-order) relationship between them.

The situations corresponding to these three examples are depicted in what is referred to as a *scatterplot*. Three such scatterplots are shown in Fig. 2.2. Each point in the scatterplots correspond to a single measurement of the two variables $\boldsymbol{x} = [x_1, x_2]$. The scatterplot is characteristic of the underlying probability density function (pdf), which in this example is a multivariate Gaussian distribution to be discussed in Chap. 4. The variables have been normalised to zero mean and unit standard deviation. We see that when the covariance is positive as in Fig. 2.2a, (here the covariance has the value +0.8), the effect of the positive covariance is to cluster the samples within an ellipse which is oriented along the direction $[1, 1]^T$. When the covariance is zero (Fig. 2.2b), there is no directional discrimination when the variances are equal. When the covariance is negative, the samples are again clustered within an ellipse, but oriented along the direction $[1, -1]^T$.

As we have seen, in the example of Fig. 2.2a, the effect of a positive covariance between the variables is to cause the respective scatterplot to become elongated along the direction $[1, 1]^T$, where the elements have the same sign. In this case mean-centred observations where the height and weight have the same sign are relatively common, and therefore observations further from the mean along this direction have a higher probability than in other directions, and so the variance of the observations along this direction is relatively high. On the other hand, again due to the positive correlation, mean-centred observations along the direction $[1, -1]^T$] where height and weight have opposite signs that are far from the mean have a lower probability; with the result the variance in this direction is smaller (i.e. tall and skinny people occur more rarely than tall and correspondingly heavier people). This behaviour is what causes the scatterplot ellipse to elongate along the $[1, 1]^T$ direction. In cases where the governing distribution is not Gaussian, similar behaviour persists, although the scatterplots will not be elliptical in shape.

As a further example, take the limiting case in Fig. 2.2a where $\sigma_{12} \to 1$. Then, the knowledge of one variable completely specifies the other, and the scatterplot devolves into a straight line. In summary, we see that as the value of the covariance increases from zero, the scatterplot ellipse transitions from being circular (when the variance of the variables are equal) to becoming elliptical, with the eccentricity of the ellipse increasing with covariance until eventually the ellipse collapses into a line as $\sigma_{12} \to 1$.

2.2.1 Covariance Matrices

The covariance structure in the multivariate case of a vector $\boldsymbol{x}$ consisting of n random variables can be conveniently encapsulated into a *covariance matrix*, $\boldsymbol{R}_x \in \mathbb{R}^{n \times n}$. In the more general case where the data are not mean-centred, the covariance matrix corresponding to the (column) vector random variable $\boldsymbol{x}$ is defined as

$$\boldsymbol{R}_x \triangleq E\left[(\boldsymbol{x} - \boldsymbol{\mu})(\boldsymbol{x} - \boldsymbol{\mu})^T\right] \tag{2.27}$$

$$= E\begin{bmatrix} (x_1 - \mu_1)(x_1 - \mu_1) & \dots & \dots & (x_1 - \mu_1)(x_n - \mu_n) \\ (x_2 - \mu_2)(x_1 - \mu_1) & (x_2 - \mu_2)(x_2 - \mu_2) & \dots & (x_1 - \mu_1)(x_n - \mu_n) \\ \vdots & \vdots & \ddots & \vdots \\ (x_n - \mu_n)(x_1 - \mu_1) & \dots & \dots & (x_n - \mu_n)(x_n - \mu_n) \end{bmatrix}.$$

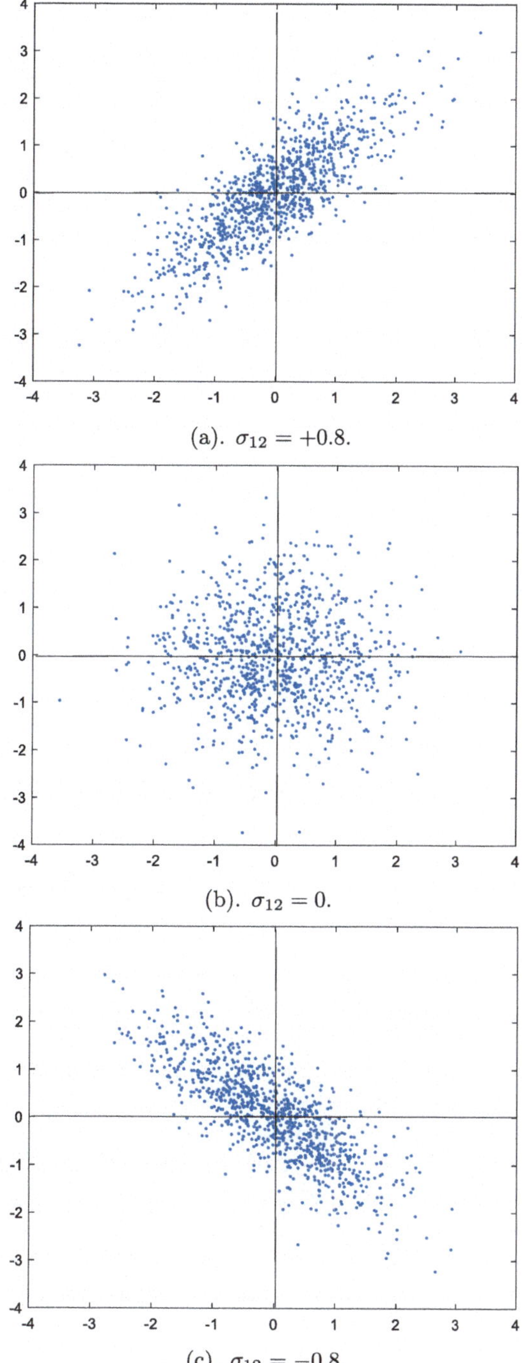

Fig. 2.2 Scatter plots for different cases of random vectors $[x_1, x_2]^T$ corresponding to different values of covariance, for mean–centered data. For each observation, the x_1 value is plotted against the corresponding x_2 value. The covariance values σ_{12} are +0.8, 0, −0.8 respectively for (**a**), (**b**), and (**c**). The axes are normalized to zero mean and standard deviation = 1. Each point in each figure represents one observation $[x_1, x_2]^T$ of the random vector $\boldsymbol{x}$. In each subfigure there are 1000 observations

where $\boldsymbol{\mu}$ is the vector mean. In the case where the data is complex, the covariance matrix is defined as $\boldsymbol{R}_x = E\left[(\boldsymbol{x}-\boldsymbol{\mu})(\boldsymbol{x}-\boldsymbol{\mu})^H\right]$, where superscript H indicates *Hermitian transpose*, i.e. the respective vector is transposed and complex conjugated.

We recognise the diagonal elements as the variances $\sigma_1^2 \ldots \sigma_n^2$ of the elements $x_1 \ldots x_n$, respectively, and the (i, j)th off-diagonal element as the covariance σ_{ij} between x_i and x_j. Since multiplication is commutative, $\sigma_{ij} = \sigma_{ji}$ and so $\boldsymbol{R}_x$ is symmetric. It is also apparent that covariance matrices are square. It therefore follows that its eigenvectors are orthogonal.

Let's say we have m available samples of the vector $\boldsymbol{x}_i,\ i = 1, \ldots, m$. To form a finite-sample estimate $\hat{\boldsymbol{R}}_x$ of $\boldsymbol{R}_x$, we formulate a matrix $\boldsymbol{X} \in \mathbb{R}^{m \times n}$. The ith row $\boldsymbol{x}_i^T$ of $\boldsymbol{X}$ contains all n variables (height and weight in our previous example) corresponding to the ith observation (person), whereas the jth column $\boldsymbol{x}_j$ of $\boldsymbol{X}$ contains all m values of the jth variable over all available observations. (In our previous examples, $m = 1000$ and $n = 2$). The matrix $\boldsymbol{X}$ formulated in this way is the standard format for organising data in machine learning and regression problems.

By replacing expectations by arithmetic averages, it follows from (2.27) that a finite sample estimate $\hat{\boldsymbol{R}}_x$ of $\boldsymbol{R}_x$ can be evaluated by forming

$$\hat{\boldsymbol{R}}_x = \frac{1}{m}\sum_{i=1}^{m}(\boldsymbol{x}_i-\boldsymbol{\mu})(\boldsymbol{x}_i-\boldsymbol{\mu})^T. \tag{2.28}$$

Equation (2.28) is a sum of outer products. Using the outer-product representation of matrix multiplication for the mean-centred case, we also have

$$\hat{\boldsymbol{R}}_x = \frac{1}{m}\sum_{i=1}^{m}\boldsymbol{x}_i\boldsymbol{x}_i^T = \frac{1}{m}\boldsymbol{X}^T\boldsymbol{X}. \tag{2.29}$$

Note that the rows $\boldsymbol{x}_i^T$ of $\boldsymbol{X}$ on the right side of (2.29) are the transpose of the $\boldsymbol{x}_i$ in the middle term. It would perhaps be more straightforward if we represented $\boldsymbol{X}$ in its transposed version where each $\boldsymbol{x}_i$ forms a column of $\boldsymbol{X}$. Then $\boldsymbol{X}$ would be an $n \times m$ matrix where each column is an observation and there would be a more direct correspondence between the summation term in the middle and the outer product term on the right. However the formulation in (2.29) is the only way to abide by the common conventions that vectors are columns and that $\boldsymbol{X}$ is formulated so that its rows consist of observations. From (2.29) it is apparent that every element of $\hat{\boldsymbol{R}}_x$ is an instance of either (2.25) or (2.26), depending on whether the respective element is on the main diagonal or not. Note that $m \geq n$ for $\hat{\boldsymbol{R}}$ to be full rank. The covariance matrices for each of our three examples discussed above are given as

$$\boldsymbol{R}_1 = \begin{bmatrix} 1 & +0.8 \\ +0.8 & 1 \end{bmatrix},\ \boldsymbol{R}_2 = \begin{bmatrix} 1 & 0 \\ 0 & 1 \end{bmatrix},\ \text{and } \boldsymbol{R}_3 = \begin{bmatrix} 1 & -0.8 \\ -0.8 & 1 \end{bmatrix}.$$

Note that 1's on each diagonal result because the variances of x_1 and x_2 have been normalised to unity.

2.3 Covariance Matrices of Stationary Time Series

Here, we extend the concept of covariance matrices to a discrete-time, one-dimensional, random stationary time series, *a.k.a.* a *random process*, represented by $x[k]$. The idea in this case is similar to that of the previous discussion, except the variables involved are sequential samples from a time series. In the following treatment, we assume mean-centred data. The objective is to characterise the covariance structure of the time series over an interval (window) of n samples. In this vein we form the row vectors $\boldsymbol{x}_i^T \in \mathbb{R}^n$, which consist of n sequential samples of the time series within a window of length n, as shown in Fig. 2.3. The value of n is to be determined by the application at hand. The windows generally overlap; in fact, they are typically displaced from one another by only one sample. Hence, the vector corresponding to each window is a *vector sample* from the random process $x[k]$. Examples of random processes are a speech signal or the signal received from an electroencephalogram (EEG) electrode, noise, or the signal transmitted from a cell phone, etc.

The word *stationary* as used above means the random process is one for which the corresponding joint n-dimensional probability density function describing the distribution of the vector sample (observation) $\boldsymbol{x}^T$ does not change with time. This means that all moments of the distribution (i.e. quantities such as the mean, the variance, and all covariances, as well as all other higher-order statistical characterisations) are invariant with time. Here however, we deal with a weaker form of stationarity referred to as *wide-sense stationary* (WSS). With these processes,

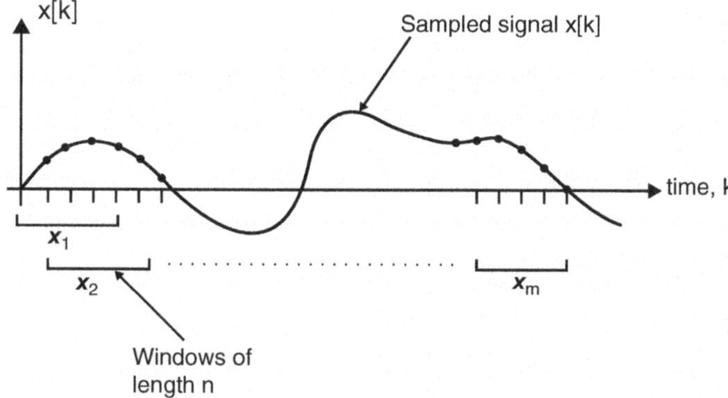

Fig. 2.3 The time series (i.e. observed signal) $x[k]$ is decomposed into windows of length n. The samples in the ith window comprise the row vector $\boldsymbol{x}_i^T$, $i = 1, 2, \ldots m$. Each window $\boldsymbol{x}_i^T$ forms a row in the $m \times n$ matrix $\boldsymbol{X}$

only the first two moments (mean, variances, and covariances) need be invariant with time. Strictly, the idea of a covariance matrix is only relevant for stationary or WSS processes, since expectations only have meaning if the underlying process is stationary. However, we see later that this condition can be relaxed in an approximate sense in the case of a slowly varying non-stationary signal, if the expectation is replaced with a time average over an interval over which the signal does not vary significantly.

Here, as in (2.27), $\boldsymbol{R}_x$ corresponding to a real, mean-centred, stationary, or WSS process $x[k]$ is defined as

$$\boldsymbol{R}_x \in \mathbb{R}^{n\times n} = E\left[\boldsymbol{x}_i\boldsymbol{x}_i^T\right] = E\begin{bmatrix} x_1x_1 & x_1x_2 & \cdots & x_1x_n \\ x_2x_1 & x_2x_2 & \cdots & x_2x_n \\ \vdots & \vdots & \ddots & \vdots \\ x_nx_1 & x_nx_2 & \cdots & x_nx_n \end{bmatrix}. \tag{2.30}$$

Taking the expectation over all windows, (2.30) tells us that the element $r(1, 1)$ of $\boldsymbol{R}_x$ is by definition $E[x_1^2]$, which is the *variance* σ_x^2 of the first element x_1 over all possible vector samples $\boldsymbol{x}_i$ of the process. But because of stationarity, $r(1, 1) = r(2, 2) = \ldots = r(n, n)$ which are all equal to σ_x^2. Thus all main diagonal elements of $\boldsymbol{R}_x$ are equal to the variance of the process. The element $r(1, 2) = E[x_1x_2]$ is the covariance between the first element of $\boldsymbol{x}_i$ and its second element taken over all possible windows. Thus this quantity is the covariance σ_{12} of the process and itself delayed by one sample. Because of stationarity, the elements $r(1, 2) = r(2, 3) = \ldots = r(n-1, n)$ and hence all elements on the first upper diagonal are equal to σ_{12}, for a time-lag of one sample. Since multiplication is commutative, $r(2, 1) = r(1, 2)$, and therefore all elements on the first lower diagonal are also all equal to this same covariance value. Using similar reasoning, all elements on the jth upper or lower diagonal are all equal to the covariance σ_{1j} value of the process for a time lag of j samples. This type of matrix, with equal elements along any diagonal, is referred to as a *Toeplitz matrix*.

If we compare the process shown in Fig. 2.3 with that shown in Fig. 2.4, we see that in the former case, the process is relatively slowly varying. Because we have assumed $x[k]$ to be mean-centred, adjacent samples of the process in Fig. 2.3 will have the same sign most of the time, and hence all elements on the first upper and lower diagonal of $\boldsymbol{R}_x$ will be equal to σ_{12}, which will be a positive number in this case, coming close to the value σ_x^2. The same can be said for all elements on the second upper and lower diagonal, except they will not be quite so close to σ_x^2. Thus, we see that for the slowly varying process of Fig. 2.3, the diagonals decay fairly slowly away from the main diagonal value.

However, for the process shown in Fig. 2.4, adjacent samples are uncorrelated with each other. This means that adjacent samples are just as likely to have opposite signs as they are to have the same signs. On average, the terms with positive values have the same magnitude as those with negative values. Thus, when the expectations $E[x_ix_{i+1}, E[x_ix_{i+2}] \ldots$ are taken, the resulting averages approach zero. In this case,

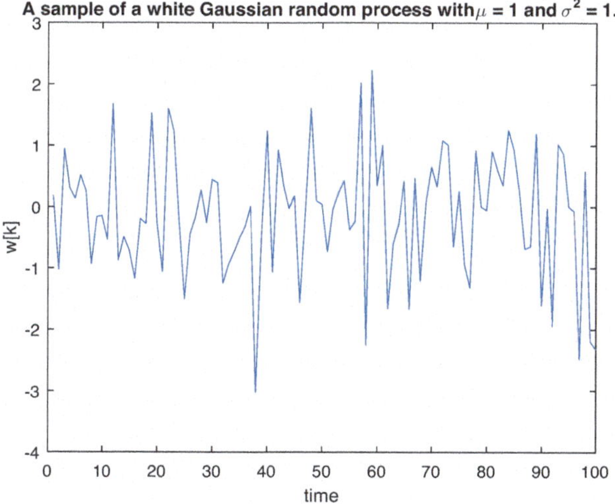

Fig. 2.4 A sample of a white Gaussian discrete-time process, with mean $\mu = 0$ and variance $\sigma^2 = 1$

we see the elements of $\boldsymbol{R}_x$ concentrate around the main diagonal, and therefore $\boldsymbol{R}_x = \sigma_x^2 \boldsymbol{I}$. We note that *all* the eigenvalues of $\boldsymbol{R}_x$ are equal to the value σ_x^2. Because of this property, such processes are referred to as *white*, in analogy to white light, whose spectral components are all of equal magnitude.

When $x[k]$ is stationary, the sequence $\{r(1, 1), r(1, 2), \ldots, r(1, n)\}$ is the autocorrelation function of the process, for lags 0 to $n - 1$. In the Gaussian case, the process is completely characterised[7] by the autocorrelation function. In fact, it may be shown [7] that the Fourier transform of the autocorrelation function is the *power spectral density* of the process. Further discussion on this aspect of random processes is beyond the scope of this treatment; the interested reader is referred to the reference.

In practice where only a finite number of observations are available, we form the matrix $\boldsymbol{X} \in \mathbb{R}^{m \times n}$ as in (2.29), whose rows are the $\boldsymbol{x}_i^T$ formed from the windows of data as in Fig. 2.3. Then, an estimate $\hat{\boldsymbol{R}}_x$ of $\boldsymbol{R}_x$ which represents the stationary random process is formed in an exactly analogous manner to (2.29)

$$\hat{\boldsymbol{R}}_x = \frac{1}{m} \boldsymbol{X}^T \boldsymbol{X}. \tag{2.31}$$

A method [4] for modifying $\hat{\boldsymbol{R}}$ to track mildly non-stationary environments is outlined in Problem 11.

[7] By "characterising" a random process, we mean that its joint probability density function is completely specified.

Some Properties of R_x

1. $\boldsymbol{R}_x$ is square and symmetric. In the complex case, it is Hermitian symmetric; i.e. $r_{ij} = r_{ji}^*$, where * denotes complex conjugation.
2. If the process $x[k]$ is stationary or wide-sense stationary, then $\boldsymbol{R}_x$ is Toeplitz. This means that all the elements on a given diagonal of the matrix are equal. If you understand this property, then you have a good understanding of the nature of covariance matrices.
3. If $\boldsymbol{R}_x$ is diagonal with equal elements, then the elements of $\boldsymbol{x}$ are uncorrelated and x is said to be a *white process*. If the magnitudes of the off-diagonal elements of $\boldsymbol{R}_x$ are significant with respect to those on the main diagonal, the process is said to be *correlated* or *coloured*.
4. $\boldsymbol{R}_x$ is *positive semi-definite*. This implies that all the eigenvalues are greater than or equal to zero. We will discuss positive definiteness and positive semi-definiteness in Chap. 4.

So far in this chapter, we have discussed eigenvalues, eigenvectors, and covariance matrices in some detail. In the following two sub-sections we present two practical examples from signal processing where eigen-analysis is performed on covariance matrices to considerable advantage. These examples are *(i)* array signal processing and *(ii)* principal component analysis (PCA).

2.4 Array Processing

One of the major subfields of signal processing is *array processing*, which amongst other things has to do with estimating *directions of arrival* of plane waves onto arrays of sensors. Here we present the "MUSIC" (MUltiple SIgnal Classification) algorithm [10] for this purpose. A broader treatment of the array signal processing field is given in [5].

The following paragraph develops the electromagnetic model of the signal vector $\boldsymbol{x}(t)$ received from the array. For those readers who are not familiar with fundamental electromagnetics, this paragraph may be skipped over. The main point of relevance for the following paragraph is the formulation of (2.32), which is the basis of the following mathematical development.

Consider a linear array of M sensors (e.g. antennas, microphones, etc.) as shown in Fig. 2.5, with a uniform inter-element spacing of d. Let there be $K < M$ plane waves incident onto the array as shown. Assume for the moment that only one wave is incident onto the array with a positive angle and whose amplitude at the first sensor is normalised to one. Then, from the physics shown in Fig. 2.5, each element of the signal vector $\boldsymbol{x}(n)$, obtained by sampling each element of the array simultaneously, is a time-delayed version of the signal received at the first element. If we assume the incident waves are narrow band, then the signal envelope does not change appreciably during the time the wavefront traverses the array. Then in this narrow band case, from fundamental electromagnetics, the signals received from

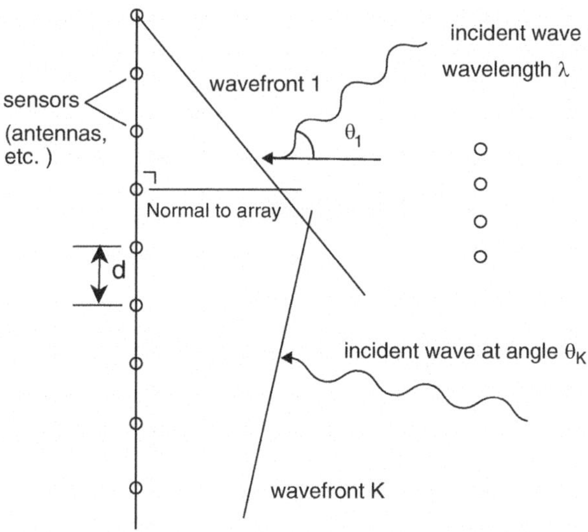

Fig. 2.5 Physical description of incident signals onto an array of sensors

adjacent elements are phase-shifted versions of each other, where the phase shift ϕ is given as $\phi = \frac{2\pi d}{\lambda} \sin(\theta)$, where d is the inter-element spacing, λ is the wavelength of the propagating wave, and θ is the physical angle of the wave relative to the line of the array. If $d \leq \lambda/2$, then there is a one-to-one correspondence between the quantities ϕ and θ, so in this condition it suffices to estimate only ϕ. The received vector $\boldsymbol{x}(n)$ may then be described as $\boldsymbol{x}(n) = [1, e^{j\phi}, e^{j2\phi}, \ldots, e^{j(M-1)\phi}]^T$, where $j = \sqrt{-1}$. When there are K incident signals, with corresponding amplitudes $a_k, k = 1, \ldots, K$, and distinct electrical angles $\phi_1, \ldots, \phi_K$ (where the ϕ may be either positive or negative), the effects of the K incident signals each add linearly together, each weighted by the corresponding amplitude a_k, to form the received signal vector $\boldsymbol{x}(n)$. The resulting received signal vector at time n, including the noise can then be written in the form

$$\underset{(M\times 1)}{\boldsymbol{x}_n} = \underset{(M\times K)}{\boldsymbol{S}} \underset{(K\times 1)}{\boldsymbol{a}_n} + \underset{(M\times 1)}{\boldsymbol{w}_n}, \qquad n = 1, \ldots, N, \tag{2.32}$$

where

$\boldsymbol{w}_n$ = M-length noise vector at time n whose elements are uncorrelated random variables with zero mean and variance σ^2, i.e. $\text{cov}(\boldsymbol{w}) = \sigma^2 \boldsymbol{I}$. The vector $\boldsymbol{w}$ is assumed uncorrelated with the signal.

$\boldsymbol{S} = [\boldsymbol{s}_1 \ldots \boldsymbol{s}_K]$

$\boldsymbol{s}_k = [1, e^{j\phi_k}, e^{j2\phi_k}, \ldots, e^{j(M-1)\phi_k}]^T, k = 1, \ldots, K$, are referred to as *steering vectors*.

$\phi_k, k = 1, \ldots, K$ are the electrical phase-shift angles corresponding to the incident signals. The ϕ_k are assumed to be distinct.

$\boldsymbol{a}_n = [a_1 \ldots a_K]^T$ is a vector of uncorrelated random variables, describing the amplitudes of each of the incident signals at time n.

The MUSIC algorithm requires that $K < M$. Before we discuss the implementation of the MUSIC algorithm *per se*, we analyse the covariance matrix $\boldsymbol{R}$ of the received signal $\boldsymbol{x}$

$$\begin{aligned}\boldsymbol{R} = E[\boldsymbol{x}\boldsymbol{x}^H] &= E\left[(\mathbf{Sa}+\mathbf{w})(\boldsymbol{a}^H\mathbf{S}^H+\mathbf{w}^H)\right] \\ &= \boldsymbol{S}E[\boldsymbol{a}\boldsymbol{a}^H]\mathbf{S}^H + \sigma^2\boldsymbol{I} \\ &= \boldsymbol{S}\boldsymbol{A}\boldsymbol{S}^H + \sigma^2\boldsymbol{I}\end{aligned} \tag{2.33}$$

where $\boldsymbol{A} = E[\boldsymbol{a}\boldsymbol{a}^H]$. The second line follows because the noise is uncorrelated with the signal, thus forcing the cross-terms to be zero. In the last line of (2.33), we have also used that fact that the covariance matrix of the noise is $\sigma^2\boldsymbol{I}$. We refer to the first term of (2.33) as $\boldsymbol{R}_o$, which is the contribution to the covariance matrix due only to the *signal* component. The matrix $\boldsymbol{A} \in \mathbb{C}^{K\times K}$ is full rank if the incident signals are not fully correlated. In this case, $\boldsymbol{R}_o \in \mathbb{C}^{M\times M}$ is rank $K < M$. Therefore $\boldsymbol{R}_o$ has K non-zero eigenvalues and $M-K$ zero eigenvalues.

From the definition of one of the zero eigenvectors, we have

$$\begin{aligned}\boldsymbol{R}_o\boldsymbol{v}_i &= \mathbf{0} \\ \text{or } \mathbf{SAS}^H\boldsymbol{v}_i &= \mathbf{0}, \qquad i = K+1,\ldots,M.\end{aligned}$$

Since $\boldsymbol{A}$ and $\boldsymbol{S}$ are assumed full rank, we must have

$$\boldsymbol{S}^H\boldsymbol{V}_N = \mathbf{0}, \tag{2.34}$$

where $\boldsymbol{V}_N = [\boldsymbol{v}_{K+1},\ldots,\boldsymbol{v}_M]$. (These eigenvectors are referred to as the *noise* eigenvectors. More on this later.) From the definition of the steering vector $\boldsymbol{s}_k$, it follows that if $\phi_o \in [\phi_1,\ldots,\phi_K]$, then

$$\left[1, e^{-j\phi_o}, e^{-j2\phi_o},\ldots, e^{j(-M-1)\phi_o}\right]\boldsymbol{V}_N = \mathbf{0}. \tag{2.35}$$

Up to now, we have considered only the noise-free case. What happens when the noise component $\sigma^2\boldsymbol{I}$ is added to $\boldsymbol{R}_o$ to give $\boldsymbol{R}$ in (2.33)? From **Property 3** of this chapter, we see that if the eigenvalues of $\boldsymbol{R}_o$ are λ_i, then because here we dealing with expectations and because the noise is white, those of $\boldsymbol{R}$ are $\lambda_i + \sigma^2$. The eigenvectors remain unchanged with the noise contribution, and therefore (2.35) still holds when noise is present, under the current assumptions.

With this background in place, we can now discuss the implementation of the MUSIC algorithm for estimating directions of arrival $\phi_1\ldots\phi_K$ of plane waves incident onto arrays of sensors, given a finite number N of snapshots. We simultaneously sample all array elements at N distinct points in time to obtain vector samples $\boldsymbol{x}_n \in \mathbb{C}^{M\times 1}, n = 1,\ldots,N$. Our objective is to estimate the directions of arrival ϕ_k of the plane waves relative to the array, by observing only the received signal. The basic idea follows from (2.35), where $\boldsymbol{s}(\phi_o) \perp \boldsymbol{V}_N$ iff $\phi_o \in [\phi_1,\ldots,\phi_k]$.

We form an estimate $\hat{\boldsymbol{R}}$ of $\boldsymbol{R}$ based on the available snapshots as follows:

$$\hat{\boldsymbol{R}} = \frac{1}{N} \sum_{n=1}^{N} \boldsymbol{x}_n \boldsymbol{x}_n^H,$$

and then extract the $M - K$ noise eigenvectors $\hat{\boldsymbol{V}}_N$, which are those associated with the smallest $M - K$ eigenvalues of $\hat{\boldsymbol{R}}$. Because of the finite N and the presence of noise, (2.35) only holds approximately for the true ϕ_o. Thus, a reasonable estimate of the desired directions of arrival may be obtained by finding values of the variable ϕ for which the expression on the left of (2.35) is minimum instead of exactly zero. Thus, we determine K estimates $\hat{\boldsymbol{\phi}}$ which locally satisfy

$$\hat{\phi} = \arg\min_{\phi} \left\| \boldsymbol{s}^H(\phi) \hat{\boldsymbol{V}}_N \right\|_2^2. \tag{2.36}$$

By convention, it is desirable to express (2.36) as a spectrum-like function, where a peak instead of a null represents a desired signal. Thus, the MUSIC "spectrum" $P(\phi)$ is defined as

$$P(\phi) = \frac{1}{\boldsymbol{s}(\phi)^H \hat{\boldsymbol{V}}_N \hat{\boldsymbol{V}}_N^H \boldsymbol{s}(\phi)}.$$

It will look something like what is shown in Fig. 2.6, when $K = 2$ incident signals. Estimates of the directions of arrival $\boldsymbol{\phi}_k$ are then taken as the peaks of the MUSIC spectrum.

Signal and Noise Subspaces The MUSIC algorithm opens up some insight into the use of the eigendecomposition that will be of use later on. Let us define the *signal subspace* S_S as

$$S_S = \text{span}\,[\boldsymbol{v}_1, \ldots, \boldsymbol{v}_K] \tag{2.37}$$

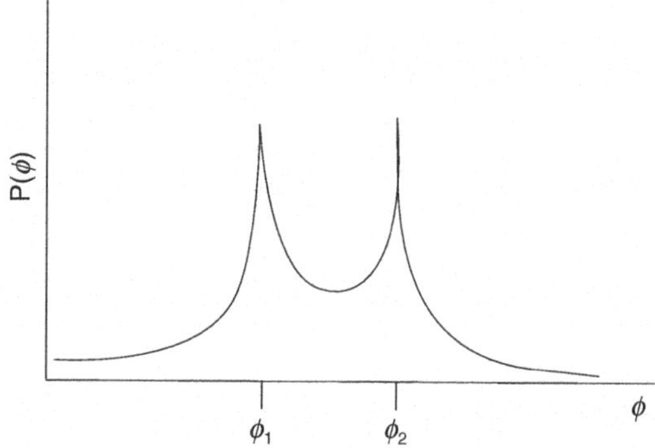

Fig. 2.6 MUSIC spectrum $P(\phi)$ for the case $K = 2$ signals

and the *noise subspace* S_N as

$$S_N = \text{span}\,[\boldsymbol{v}_{K+1}, \ldots, \boldsymbol{v}_M]\,. \tag{2.38}$$

From (2.33), all columns of $\boldsymbol{R}_o = \boldsymbol{R} - \sigma^2 \boldsymbol{I}$ are linear combinations of $\boldsymbol{S}$. Therefore

$$R(\boldsymbol{R}_o) = R(\boldsymbol{S}). \tag{2.39}$$

We have seen earlier in this chapter that the eigenvectors associated with the non-zero eigenvectors form a basis for $R(\boldsymbol{R}_o)$. Therefore

$$R(\boldsymbol{R}_o) \in \mathcal{S}_s. \tag{2.40}$$

Comparing (2.39) and (2.40), we see that $\boldsymbol{S} \in S_S$. From (2.32) we see that any received signal vector $\boldsymbol{x}(n)$, in the absence of noise, is a linear combination of the columns of $\boldsymbol{S}$. Thus, any noise-free signal resides completely in S_S. This is the origin of the term "signal subspace". Further, in the presence of noise, provided $N \to \infty$, (2.37) and (2.38) still hold and then any component of the received signal residing in S_N must be entirely due to the noise, although noise can also reside in the signal subspace. This is the origin of the term "noise subspace". In the case where N is finite, the eigenvectors are only approximations to their ideal values, and (2.37) and (2.38) hold only approximately. This results in some leakage between the two subspaces and (2.34) holding only approximately, resulting in some error in the estimates of ϕ, as one would expect under these circumstances. We note that the signal and noise subspaces are orthogonal complement subspaces of each other. The ideas surrounding this example lead to the ability to de-noise a signal in some situations, as we see in a subsequent example.

2.5 Principal Component Analysis (PCA)

PCA is the second example, which shows how eigen-analysis can be applied to covariance matrices to solve real-life signal processing problems. PCA is nothing more than a straightforward change of basis that has very useful properties as we see later. Changes of basis are in common use in signal processing; different bases are used to reveal unique properties of a discrete-time signal $x[k]$. For example, the discrete Fourier transform (DFT) is an orthonormal transformation of a segment $\boldsymbol{x} \in \mathbb{R}^N$ of $x[k]$ in the $N \times N$ orthogonal basis $\boldsymbol{W}$ shown below. The columns are rotating complex exponentials whose normalised frequencies are $p/N,\ p = 0, \ldots, N-1$ Hz. The resulting coefficients $\boldsymbol{X}$ represent the contribution of these various frequency components to the composition of $x[k]$. The vector $\boldsymbol{x}$ can be represented as a linear combination of complex exponential basis functions in the following way:

$$
\begin{aligned}
x &= \frac{1}{N}\begin{bmatrix} 1 & 1 & 1 & \dots & 1 \\ 1 & e^{\frac{j2\pi}{N}} & e^{\frac{j4\pi}{N}} & \dots & e^{\frac{j(N-1)2\pi}{N}} \\ 1 & e^{j\frac{4\pi}{N}} & e^{\frac{j8\pi}{N}} & \dots & e^{j\frac{2(N-1)2\pi}{N}} \\ \vdots & \vdots & \vdots & & \vdots \\ 1 & e^{\frac{j(N-1)2\pi}{N}} & e^{\frac{j2(N-1)2\pi}{N}} & \dots & e^{\frac{j(N-1)^2 2\pi}{N}} \end{bmatrix} \begin{bmatrix} X_0 \\ X_1 \\ X_2 \\ \vdots \\ X_{N-1} \end{bmatrix} \\
&= \frac{1}{N} \boldsymbol{W}\boldsymbol{X},
\end{aligned}
\tag{2.41}
$$

where $\boldsymbol{W}$ is as shown and $j = \sqrt{-1}$ and the X_i are the coefficients of $\boldsymbol{x}$ in the DFT basis. It is easily verified that $\boldsymbol{W}$ is close to orthonormal, in that $\boldsymbol{W}^H\boldsymbol{W} = N\boldsymbol{I}$. With this, and using the familiar rules for change of bases from Sect. 1.2.5, the coefficients $\boldsymbol{X}$ are given as

$$
\boldsymbol{X} = \boldsymbol{W}^H\boldsymbol{x} \text{ and } \boldsymbol{x} = \frac{1}{N}\boldsymbol{W}\boldsymbol{X}.
$$

From the second relation above, we can see that the coefficient X_i is the weighting of the ith complex exponential in the representation of $\boldsymbol{x}$. A detailed description of discrete-time signal processing and the DFT is given in Appendix 1 of this volume. A further reference on this topic is [6].

Another basis in common use is the *discrete wavelet transform*. This is another form of orthonormal transform which is used primarily for representing the frequency decomposition of a non-stationary signal $x(t)$ as a function of time; that is, $x(t)$ is represented as a time-varying spectrum in the so-called *time-frequency* plane. The wavelet transform is an orthonormal transformation with a multi-resolution property, i.e. specific coefficients of the transformation exhibit high resolution in the time domain but lower resolution in the frequency domain, whereas other coefficients exhibit the opposite behaviour. For more detail on the wavelet transform, the reader is referred to the tutorial presentation in [1].

In the case of **principal component analysis**, which is sometimes referred to as the *Karhunen-Loeve transform*, we are given a set of m vector observations $\boldsymbol{x}_i \in \mathbb{R}^n, i = 1, \dots m$, where $m \geq n$. We assume the data is mean-centred, as before. The available data are arranged into a matrix $\boldsymbol{X} \in \mathbb{R}^{m\times n}$ as in Sect. 2.2. The ith row of $\boldsymbol{X}$ is $\boldsymbol{x}_i^T$. With PCA, we define the PCA coefficients $\boldsymbol{\theta}_i \in \mathbb{R}^r$ as $\boldsymbol{\theta}_i = \boldsymbol{Q}_1^T\boldsymbol{x}_i$, where $\boldsymbol{Q}_1$ is an incomplete $n \times r$ orthonormal matrix, where $r \leq n$ (ideally, $r \ll n$). We choose $\boldsymbol{Q}_1$ so that the quantity $E_i\ ||\boldsymbol{\theta}_i||_2^2$ is maximised, where the expectation is over the observations i. This objective is equivalent to maximising the sum of variances of the r elements of $\boldsymbol{\theta}_i$.

The rationale for this choice of basis is that it enables compression of the data in $\boldsymbol{X}$. The data sample $\boldsymbol{x}_i \in \mathbb{R}^n$ can be represented using only the r coefficients of $\boldsymbol{\theta}_i$, instead of the n coefficients of $\boldsymbol{x}_i$, with small error. The choice of the parameter r is also discussed later, but for now it suffices to say that it has to do with the level of compression offered by the PCA approach.

With the PCA method, we find the incomplete orthonormal r-dimensional basis $\boldsymbol{Q}_1 \in \mathbb{R}^{n \times r}$, which maximises the quantity $E_i||\boldsymbol{\theta}_i||_2^2$, where $\boldsymbol{\theta}_i$ are the PCA coefficients and are given as

$$\boldsymbol{\theta}_i \in \mathbb{R}^r = \boldsymbol{Q}_1^T \boldsymbol{x}_i. \tag{2.42}$$

In this vein, the optimal PCA matrix $\boldsymbol{Q}_1$ may be determined as the solution $\boldsymbol{Q}^\star$ to the following optimisation problem:

$$\boldsymbol{Q}^\star = \arg\max_{\boldsymbol{Q}_1} E_i||\boldsymbol{\theta}_i||_2^2 = \arg\max_{\boldsymbol{Q}_1} E_i||\boldsymbol{Q}_1^T \boldsymbol{x}_i||_2^2, \text{ subject to } \boldsymbol{Q}_1^T \boldsymbol{Q}_1 = \boldsymbol{I}. \tag{2.43}$$

The constraint $\boldsymbol{Q}_1^T \boldsymbol{Q}_1 = \boldsymbol{I}$ is necessary to prevent $||\boldsymbol{Q}_1||_2$ from growing to infinity and to force orthonormality amongst the $\boldsymbol{q}$. The simplest way of solving this problem is to find the columns $\boldsymbol{q}_j$, $j = 1, \ldots, r$ of $\boldsymbol{Q}_1$ one at a time. For each column, we address the question "What is the $\boldsymbol{q}$ so that the variance $E_i(\theta_i)^2 = E_i(\boldsymbol{q}^T \boldsymbol{x}_i)^2$ is maximum, such that the $\boldsymbol{q}$ form an orthonormal set?" After the r basis vectors are determined, we can combine them into the matrix $\boldsymbol{Q}_1$. We now present this concept in a formal mathematical framework.

Lemma 2.1 *The vector $\boldsymbol{q}^\star$, defined as*

$$\boldsymbol{q}^\star = \arg\max_{\boldsymbol{q}} E(\theta_i)^2 = \arg\max_{\boldsymbol{q}} E[\boldsymbol{q}^T \boldsymbol{x}_i]^2, \textit{ subject to } \boldsymbol{q}^T \boldsymbol{q} = 1. \tag{2.44}$$

is given as the largest eigenvector $\boldsymbol{v}_1$ of the covariance matrix $E_i[\boldsymbol{x}_i \boldsymbol{x}_i^T] \triangleq \boldsymbol{R}_x$.

Proof For each $\boldsymbol{q}$, we wish to solve

$$\begin{aligned}
\boldsymbol{q}^* &= \arg\max_{\boldsymbol{q}} E_i(\boldsymbol{x}_i^T \boldsymbol{q})^2 \\
&= \arg\max_{\boldsymbol{q}} E_i[\boldsymbol{q}^T \boldsymbol{x}_i \boldsymbol{x}_i^T \boldsymbol{q}] \\
&= \arg\max_{\boldsymbol{q}} \boldsymbol{q}^T E_i[\boldsymbol{x}_i \boldsymbol{x}_i^T] \boldsymbol{q} \\
&= \arg\max_{\boldsymbol{q}} \boldsymbol{q}^T \boldsymbol{R}_x \boldsymbol{q} \text{ s.t. } ||\boldsymbol{q}||_2^2 = 1.
\end{aligned} \tag{2.45}$$

Equation (2.45) is a constrained optimisation problem which may be solved by the method of Lagrange multipliers.[8] Briefly, this method may be stated as follows. Given some objective function $G(\boldsymbol{x})$ which we wish to minimise or maximise with respect to $\boldsymbol{x}$, subject to an equality constraint function $f(\boldsymbol{x}) = 0$, we formulate the Lagrangian function $L(\boldsymbol{x})$ which is defined as

$$L(\boldsymbol{x}) = G(\boldsymbol{x}) + \lambda f(\boldsymbol{x}),$$

[8] There is a good description of the method of Lagrange multipliers in Wikipedia.

where λ is referred to as a Lagrange multiplier, whose value is to be determined. The constrained solution $\boldsymbol{x}^*$ then satisfies

$$\frac{dL(\boldsymbol{x}^*)}{d\boldsymbol{x}} = \boldsymbol{0}. \tag{2.46}$$

Applying the Lagrange multiplier method to (2.45), the objective function $G(f) = \boldsymbol{q}^T \boldsymbol{R}_x \boldsymbol{q}$, and the constraint function $f(x) = 1 - \boldsymbol{q}^T \boldsymbol{q}$. Thus the Lagrangian is given by

$$L(\boldsymbol{q}) = \boldsymbol{q}^T \boldsymbol{R}_x \boldsymbol{q} + \lambda(1 - \boldsymbol{q}^T \boldsymbol{q}). \tag{2.47}$$

As shown in the Appendix of this chapter, and from [9], the derivative of the first term is given by

$$\frac{d}{d\boldsymbol{q}} \boldsymbol{q}^T \boldsymbol{R}_x \boldsymbol{q} = 2\boldsymbol{R}_x \boldsymbol{q}.$$

It is straightforward to show that with respect to the second term

$$\frac{d}{d\boldsymbol{q}}(1 - \boldsymbol{q}^T \boldsymbol{q}) = -2\boldsymbol{q}.$$

Substituting the above two derivative terms into (2.46) and rearranging, the desired solution $\boldsymbol{q}^*$ satisfies

$$\boldsymbol{R}_x \boldsymbol{q} = \lambda \boldsymbol{q}. \tag{2.48}$$

We therefore have the significant result that the stationary points of the constrained optimisation problem (2.45) are given by the eigenvectors $\boldsymbol{v}_i$ of $\boldsymbol{R}_x$. From Property 9 to follow, we show that the variances $E[\theta(j)]^2$ are equal to their corresponding eigenvalue λ_j, $j = 1, \ldots, n$. Therefore, to maximise the total variance of $\boldsymbol{\theta}$, which is $E[\theta(1)]^2 + E[\theta(2)]^2 + \ldots + E[\theta(r)]^2$, it is seen that the solution $\boldsymbol{Q}^\star$ to (2.44) satisfies

$$\boldsymbol{Q}^\star = \boldsymbol{V}_1 \triangleq [\boldsymbol{v}_1, \boldsymbol{v}_2, \ldots, \boldsymbol{v}_r], \tag{2.49}$$

where $\boldsymbol{V}_1$ consists of the eigenvectors corresponding to the largest $r < n$ eigenvalues. It is worthy of note that use of the PCA method requires that the underlying process from which $\boldsymbol{X}$ is formed is stationary. Thus, the covariance matrix $\boldsymbol{R}_x$ and consequently the eigenvectors $\boldsymbol{V}$ and eigenvalues λ are constant over all observations of the data. □

We now discuss why the above analysis is useful in the signal processing context. To address this question, we consider an approximation $\hat{\boldsymbol{x}}_i$ to $\boldsymbol{x}_i$ as

$$\hat{\boldsymbol{x}}_i = \boldsymbol{V}_1 \boldsymbol{\theta}_i. \tag{2.50}$$

where

$$\boldsymbol{\theta}_i = \boldsymbol{V}_1^T \boldsymbol{x}_i. \tag{2.51}$$

Since $\boldsymbol{V}_1$ is fixed, note that $\hat{\boldsymbol{x}}_i \in \mathbb{R}^n$ is represented in terms of the $r < n$ coefficients $\boldsymbol{\theta}_i \in \mathbb{R}^r$, and so in this way $\hat{\boldsymbol{x}}_i$ is a compressed version of the original $\boldsymbol{x}_i$. But the most interesting property of the PCA representation is that the 2-norm error $\epsilon_r = E\left|\left|\boldsymbol{x}_i - \hat{\boldsymbol{x}}_i\right|\right|_2^2$ in representing $\boldsymbol{x}_i$ is minimum with respect to basis, i.e. there is no basis other than $\boldsymbol{V}_1$ for which ϵ_r is smaller. The fact that ϵ_r is minimised by the PCA representation is the key to data compression, and also as we see later, it can be highly effective at denoising signals. In addition, because the PCA coefficients $\boldsymbol{\theta}$ are a parsimonious representation of the original data, they serve as effective features for machine learning applications.

To show that $\boldsymbol{\epsilon}_r$ is indeed minimum, we proceed as follows. Let $\boldsymbol{V}$ be the full $n \times n$ eigenvector matrix, which we partition as $\boldsymbol{V} = [\boldsymbol{V}_1 \ \boldsymbol{V}_2]$ where $\boldsymbol{V}_1$ and $\boldsymbol{V}_2$ have r and $n - r$ columns, respectively. We can then define the full vector $\bar{\boldsymbol{\theta}}$ of n PCA coefficients as

$$\bar{\boldsymbol{\theta}}_i = \boldsymbol{V}^T \boldsymbol{x}_i = \begin{bmatrix} \boldsymbol{V}_1^T \\ \boldsymbol{V}_2^T \end{bmatrix} \boldsymbol{x}_i \stackrel{\Delta}{=} \begin{bmatrix} \boldsymbol{\theta}_i \\ \boldsymbol{\theta}_2 \end{bmatrix} \begin{matrix} r \\ n-r \end{matrix}, \tag{2.52}$$

where we have omitted the subscript i on $\boldsymbol{\theta}_2$ for notational clarity. From our earlier discussions, there is no other basis for which $E||\boldsymbol{\theta}_i||_2^2$ is greater for a fixed value of r. Because $\boldsymbol{V}$ is an orthonormal transformation, for each observation i we have $||\boldsymbol{\theta}_i||_2^2 + ||\boldsymbol{\theta}_2||_2^2 = ||\boldsymbol{x}_i||_2^2$. Since $||\boldsymbol{x}||_2^2$ is invariant to the choice of basis and $E||\boldsymbol{\theta}_i||_2^2$ is maximum, $E||\boldsymbol{\theta}_2||_2^2 = \epsilon_r$ must be minimum with respect to basis.

We now offer additional insight and interpretation of the PCA representation. We use the scatterplot of Fig. 2.2a as a typical example. We see that as a direct consequence of the (positive) correlation between x_1 and x_2, the variation of samples along the principal eigenvector direction $\boldsymbol{v}_1 = \frac{1}{\sqrt{2}}[1, 1]^T$ is greater than that along the second eigenvector direction $\boldsymbol{v}_2 = \frac{1}{\sqrt{2}}[1, -1]^T$. Therefore for any observation i, the value $\theta_i(1)$[9] tends to be larger in magnitude than $\theta_i(2)$ most of the time, and so the corresponding variances have the relation $E[\theta(1)^2] \geq E[\theta(2)^2]$. Now consider the $x_2 - x_3$ plane where $n \geq 3$ and assume that a non-zero correlation also exists between x_2 and x_3. We can apply the same argument as we did in the $x_1 - x_2$ plane with the result that $E[\theta(2)^2] \geq E[\theta(3)^2]$. Continuing this process to n dimensions, provided correlations between the respective variables are non-zero, we have

$$E[\theta(1)^2] > E[\theta(2)^2] > \ldots > E[\theta(n)^2]. \tag{2.53}$$

[9] The notation $\theta_i(j)$ refers to the jth element of $\boldsymbol{\theta}_i$, whereas the notation $\boldsymbol{\theta}_i$ or $\boldsymbol{\theta}_2$ refers to the partitioning of (2.52).

Based on the above, if the correlations between the elements of $\boldsymbol{x}$ are large enough, there exists an integer $r \leq n$ such that the variances $E[\theta(r+1)^2], \ldots, E[\theta(n)^2]$ become negligible in value compared to the variances $E[\theta(1)^2], \ldots, E[\theta(r)^2]$. On this basis, we assign $\boldsymbol{\theta}_i$ and $\boldsymbol{\theta}_2$ from (2.52), respectively, to be

$$\boldsymbol{\theta}_i = [\theta_i(1), \ldots, \theta_i(r)] \tag{2.54}$$

and

$$\boldsymbol{\theta}_2 = [\theta_i(r+1), \ldots, \theta_i(n)]. \tag{2.55}$$

Then from (2.52) we can write

$$\boldsymbol{x}_i = \boldsymbol{V}\bar{\boldsymbol{\theta}}_i = \boldsymbol{V}_1\boldsymbol{\theta}_i + \boldsymbol{V}_2\boldsymbol{\theta}_2. \tag{2.56}$$

But because $\boldsymbol{\theta}_2$ has been declared negligible in comparison to $\boldsymbol{\theta}_i$, we can write the approximation $\hat{\boldsymbol{x}}_i$ to $\boldsymbol{x}_i$ from (2.50) as

$$\hat{\boldsymbol{x}}_i = \boldsymbol{V}_1\boldsymbol{\theta}_i. \tag{2.57}$$

In this manner we can see why the error ϵ_r in the representation $\hat{\boldsymbol{x}}_i$ is indeed minimum. The quantity $\hat{\boldsymbol{x}}_i$ is the *principal component* representation of $\boldsymbol{x}_i$. Because $r \leq n$, and sometimes we have $r \ll n$ if $\boldsymbol{x}_i$ is sufficiently correlated, and because we assume $\boldsymbol{V}_1$ is constant, we can represent the n-length vector $\boldsymbol{x}_i$ by $\hat{\boldsymbol{x}_i}$ which requires only the r coefficients of $\boldsymbol{\theta}_i$ instead of the n parameters needed to represent $\boldsymbol{x}$ directly, with minimum mean-squared error. Thus significant degrees of data compression with very small mean-squared error can be achieved in favourable circumstances.

Note that it is the correlation between the variables that causes the major axes of the scatterplot ellipses to be disparate in length, as shown in Fig. 2.2a. This disparity in turn is what gives rise to the phenomenon where the variances $E[\theta(j)^2]$ become small with increasing j, as in (2.53). The larger the correlation, the faster the coefficients decay, which in turn permits a greater level of acceptable compression.

Following the convention of the MUSIC algorithm, we refer to the subspace $R(\boldsymbol{V}_1)$ as the *signal subspace* $\mathcal{S}$, since all useful components of the observed signal reside in this subspace. Likewise, the subspace span$[\boldsymbol{v}_{r+1}, \ldots, \boldsymbol{v}_n]$ is referred to as the *noise subspace*, since this subspace contains primarily the noise components of the observed signal. Using (2.51) for $\boldsymbol{\theta}_i$, and substituting this into (2.50), we obtain the result

$$\hat{\boldsymbol{x}} = \boldsymbol{V}_1\boldsymbol{V}_1^T\boldsymbol{x}. \tag{2.58}$$

Since $\hat{\boldsymbol{x}} \in \mathcal{S}$, $\hat{\boldsymbol{x}}$ may be viewed as a projection of $\boldsymbol{x}$ into $\mathcal{S}$. This projection operation is accomplished by pre-multiplication of $\boldsymbol{x}$ by the matrix $\boldsymbol{V}_1\boldsymbol{V}_1^T$. This matrix is referred to as a *projector*. Projectors are explored in more detail in Chap. 3. (Note that since $\boldsymbol{V}_1$ is tall, $\boldsymbol{V}_1\boldsymbol{V}_1^T \neq \boldsymbol{I}$.)

The choice of r is application dependent and involves a trade-off between accuracy of the reconstructed $\hat{\boldsymbol{x}}_i$ vs. the level of data compression—the larger the value of r, the more accurate is the reconstruction, but the lower is the level of data compression and vice versa. Since the eigenvalues are available, the expected error in $\hat{\boldsymbol{x}}_i$ can determined from Property 10 and (2.60) and therefore r can be adjusted to best suit the needs of the application from knowledge of the eigenvalues. It is interesting to note that as the elements of $\boldsymbol{x}$ become more correlated, the corresponding scatterplot ellipse becomes more elongated (in the Gaussian case), with the result that the PCA coefficients become more concentrated in the first few values. This has the effect either of being able to reduce r for the same reconstruction error or reduce the error for the same r. This makes intuitive sense, since as the correlations between the elements increase, the process becomes more predictable and thus easier to compress.

Consider a plot of the eigenvalue λ_i vs. its index i.[10] In the ideal case, the first r λs would have significant, positive, non-zero values, followed by $n - r$ negligible values. Thus there would be a sudden drop in value between the rth and the $(r + 1)$th eigenvalue. Therefore one way of determining r is to locate that index where the drop occurs. However, in the practical case, due to noise, finite data, etc., this behaviour is seldom evident. Instead, information theoretic criteria such as the Akaike information criterion (AIC), the minimum description length algorithm (MDL), or the Bayesian information criterion (BIC) have been shown to work well in model order selection problems, such as the choice of r in this case. See [3] for further details. All these methods work on the principle of trading off model complexity vs. model fit. In the present case, model complexity is a function of the number of parameters used to describe the model and is a direct function of r. Model fit is a function of the mean-squared error in the representation of $\boldsymbol{x}$ by $\hat{\boldsymbol{x}}$. Complexity is a monotonically increasing function of r, whereas error is a monotone decreasing function. Each method involves evaluating some carefully constructed linear combination of these functions. The minimum value of this linear combination as a function of r gives us the most suitable value of r.

2.5.1 Implementation of the PCA Method

To implement the PCA method, given a set of m observations $\boldsymbol{x}_i \in \mathbb{R}^n, i = 1, \ldots, m$, or equivalently the data matrix $\boldsymbol{X} \in \mathbb{R}^{m \times n}$ as previously discussed in Sect. 2.2, we proceed as follows:

- Form the estimated covariance matrix $\hat{\boldsymbol{R}}_x$ from (2.29) or (2.31) and then compute the eigendecomposition to yield the eigenvector basis $\boldsymbol{V}$ and the correspond-

[10] As discussed in Chap. 4, the covariance matrix $\boldsymbol{R}$ is positive or positive semi-definite, meaning that its eigenvalues are all greater than or equal to zero.

ing eigenvalues λ_i. Sort the eigenvalues (and corresponding eigenvectors) in descending order.

- Determine a suitable value of r as discussed earlier.
- Once r is determined, form $\boldsymbol{V}_1 = \boldsymbol{V}(:, 1 : r)$, where Matlab® notation has been used.
- Form $\boldsymbol{\theta}_i \in \mathbb{R}^r$ as $\boldsymbol{\theta}_i = \boldsymbol{V}_1^T \boldsymbol{x}_i$ for each i. It is important to note that the data are stored or represented as the r variables in $\boldsymbol{\theta}_i$ rather than as the n variables in $\boldsymbol{x}_i$; otherwise the data compression property PCA would be lost.
- To recover an approximation $\hat{\boldsymbol{x}}_i$ to the original data, we calculate $\hat{\boldsymbol{x}}_i = \boldsymbol{V}_1 \boldsymbol{\theta}_i$.

2.5.2 Properties of the PCA Representation

Here we introduce several properties of PCA analysis to lend a more rigorous framework for the understanding of this topic.

Property 8 *The coefficients $\boldsymbol{\theta}$ are uncorrelated.*

To prove this, we evaluate the covariance matrix $\boldsymbol{R}_{\theta\theta}$ of $\boldsymbol{\theta}$, using the definition of covariance from (2.27), where here $\boldsymbol{\mu} = \boldsymbol{0}$ as follows:

$$\begin{aligned} \boldsymbol{R}_\theta &= E\left(\boldsymbol{\theta}\boldsymbol{\theta}^T\right) \\ &= E\left(\boldsymbol{V}^T \boldsymbol{x}\boldsymbol{x}^T \boldsymbol{V}\right) \\ &= \boldsymbol{V}^T \boldsymbol{R}_x \boldsymbol{V} \\ &= \boldsymbol{\Lambda}. \end{aligned} \tag{2.59}$$

Since $\boldsymbol{R}_\theta$ is equal to the *diagonal* eigenvalue matrix $\boldsymbol{\Lambda}$ of $\boldsymbol{R}_x$, the PCA coefficients are uncorrelated.

Property 9 *The variance of the jth PCA coefficient θ_j is equal to the jth eigenvalue λ_j of $\boldsymbol{R}_x$.*

The proof follows directly from Property 8 and the fact that the jth diagonal element of $\boldsymbol{R}_\theta$ is the variance of θ_j.

Remarks From this property, we can infer that the length of the jth semi-axis of the scatterplot ellipse is directly proportional to the positive square root of the ith eigenvalue.

Property 10 *The mean-squared error $\epsilon_r = E||\boldsymbol{x}_i - \hat{\boldsymbol{x}}_i||_2^2$ in the PCA representation $\hat{\boldsymbol{x}}_i$ of $\boldsymbol{x}_i$ using r components is given by*

$$\epsilon_r = E||\boldsymbol{x}_i - \hat{\boldsymbol{x}}_i||_2^2 = \sum_{i=r+1}^{n} \lambda_i, \tag{2.60}$$

which corresponds to the sum of the truncated eigenvalues.[11]

Proof

$$\begin{aligned}\epsilon_r^* = E||\boldsymbol{x}_i - \hat{\boldsymbol{x}}_i||_2^2 &= E||\boldsymbol{V\theta} - \boldsymbol{V\theta}_1||_2^2 \\ &= E||(\boldsymbol{\theta} - \boldsymbol{\theta}_1)||_2^2 \\ &= \sum_{i=r+1}^{n} E(\theta_i)^2 \\ &= \sum_{i=r+1}^{n} \lambda_i, \end{aligned} \tag{2.61}$$

where $\boldsymbol{\theta}_1 = [\theta(1), \ldots, \theta(r), 0, \ldots, 0]$ where $n - r$ zeros have been padded, and in the last line we have used Property 9. The second line follows due to the fact that the 2-norm is invariant to multiplication by an orthonormal matrix. This property indicates that increasing r decreases the error ϵ_r at the expense of a loss of compression. □

Property 11 *The eigenvector basis provides the minimum ϵ_r for a given value of r (repeated from Sect. 2.5).*

Proof Recall

$$\boldsymbol{\theta} = \boldsymbol{V}^T\boldsymbol{x} = \begin{bmatrix} \boldsymbol{V}_1 \\ \boldsymbol{V}_2 \end{bmatrix} \boldsymbol{x} = \begin{bmatrix} \boldsymbol{\theta}_1 \\ \boldsymbol{\theta}_2 \end{bmatrix}$$

where $\boldsymbol{\theta}_2$ are the coefficients which become truncated. Then from our earlier discussions, there is no other basis for which $E||\boldsymbol{\theta}_1||_2^2$ is greater. Because $\boldsymbol{V}$ is an orthonormal transformation, $||\boldsymbol{\theta}_1||_2^2 + ||\boldsymbol{\theta}_2||_2^2 = ||\boldsymbol{x}||_2^2$. Since $||\boldsymbol{x}||_2^2$ is invariant to the choice of basis and $E||\boldsymbol{\theta}_1||_2^2$ is maximum, $E||\boldsymbol{\theta}_2||_2^2 = \epsilon_r^*$ must be minimum which respect to basis. □

We can offer an additional example illustrating the compression phenomenon. Consider the extreme case where the process becomes so correlated that all elements of its covariance matrix approach the same value. (This will happen if the process $x[k]$ is non-zero and does not vary with time. In this case, we must consider the fact that the mean is non-zero). Then all columns of the covariance matrix are equal, and the rank of $\boldsymbol{R}_x$ is one, and therefore only one eigenvalue is non-zero. Then *all* the power of the process is concentrated into only the first PCA coefficient and therefore all the later coefficients are small (zero in this specific case) in comparison to the first coefficient, and the process is highly compressible.

[11] We see later in Chap. 4 that $\boldsymbol{R}_x$ is positive definite and so all the eigenvalues are positive. Thus each truncated term can only increase the error.

In contrast, in the opposite extreme when $x[k]$ is white, the corresponding $\boldsymbol{R}_x = \sigma_x^2 \boldsymbol{I}$. Thus all the eigenvalues are equal to σ_x^2, and none of them are small compared to the others. In this case there can be no compression (truncation) without inducing significant distortion in (2.61). Also, since we have n repeating eigenvectors, the eigenvectors are not unique. In fact, any orthonormal set of m vectors is an eigenvector basis for a white signal.

2.6 Examples of PCA Analysis

2.6.1 Compression of a Low-Pass Random Process

Here we present a simulation example using PCA analysis, to illustrate the effectiveness of this technique. A process $x[k]$ was simulated by passing a unit-variance zero-mean white noise sequence $w[k]$ through a third-order digital low-pass Butterworth filter with a relatively low normalised cutoff frequency (0.1 Hz). A white noise sample and its corresponding filtered version are shown in Fig. 2.7. Appendix 1 of this volume gives a treatment on the fundamentals of digital signal processing and digital filtering for those who may be unfamiliar with this subject matter.

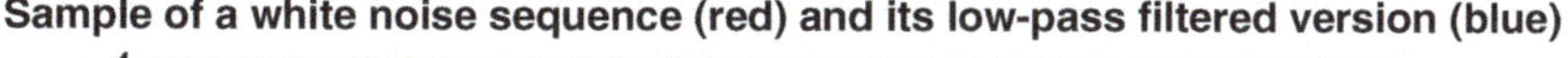

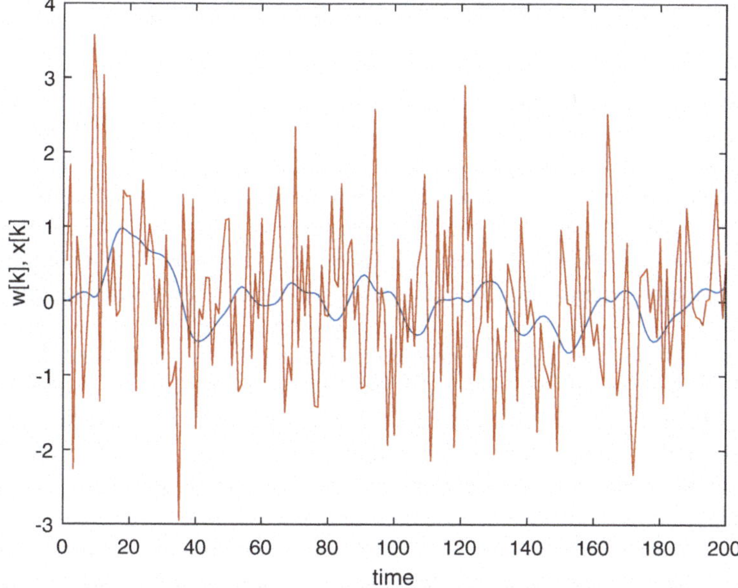

Fig. 2.7 A sample of a white noise sequence (red) and its corresponding filtered version $x[k]$ (blue). The white noise sequence is a Gaussian random process with $\mu = 0$ and $\sigma^2 = 1$, generated using the matlab® command "randn"

It is clear from Fig. 2.7 that since successive samples of the white noise sequence are uncorrelated, they bear no resemblance to one another. On the other hand, the filter removes the high-frequency components from the input, and so the resulting output process $x[k]$ varies more smoothly in time and therefore exhibits a significant correlation structure between neighbouring samples. As a result, we expect to be able to accurately represent $x[k]$ using only a few principal eigenvector components and thus be able to achieve significant compression gains. The methodology for this example follows that of Sect. 2.3, i.e. vector samples $\boldsymbol{x}_i^T \in \mathbb{R}^n$ are extracted from the sequence $x[k]$ in the manner shown in Fig. 2.3 and assembled into rows of the data matrix $\boldsymbol{X}$. The sample covariance matrix $\hat{\boldsymbol{R}}_x$ was then computed from $\boldsymbol{X}$ as in (2.31) for the value $n = 10$ and an eigendecomposition performed. Listed below are the 10 eigenvalues of $\hat{\boldsymbol{R}}_x$:

Eigenvalues

$$
\begin{array}{l}
0.5468 \\
0.1975 \\
0.1243 \times 10^{-1} \\
0.5112 \times 10^{-3} \\
0.2617 \times 10^{-4} \\
0.1077 \times 10^{-5} \\
0.6437 \times 10^{-7} \\
0.3895 \times 10^{-8} \\
0.2069 \times 10^{-9} \\
0.5761 \times 10^{-11}
\end{array}
$$

Inspection of the eigenvalues above indicates that a large part of the total variance is contained in the first two eigenvalues. We therefore choose $r = 2$. From Property 10 the error ϵ_r for $r = 2$ is thus evaluated from the above data as $\sum_{i=3}^{10} \lambda_i = 0.0130$, which may be compared to the value 0.7573, which is the total eigenvalue sum and the total variance of the signal. The normalised error is $\frac{0.0130}{0.7573} = 0.0171$. Because this error may be considered a low enough value, only the first $r = 2$ components may be considered significant. In this case, we have a compression gain of $10/2 = 5$.

Since $\hat{\boldsymbol{x}}_i = \boldsymbol{V}_1 \boldsymbol{\theta}_i$, i.e. $\hat{\boldsymbol{x}}_i$ is a linear combination of the columns of $\boldsymbol{V}_1$, and since $\hat{\boldsymbol{x}}_i$ is a function of time over the interval spanning the ith window, the eigenvectors themselves must also represent waveforms in time. As such, Fig. 2.8 shows plots of the eigenvectors as discrete-time waveforms, with the elements of the respective $\boldsymbol{v}$ acting as its samples. In this case, we would expect that any observation $\boldsymbol{x}_i$ can be expressed accurately as a linear combination of only the first two eigenvector waveforms shown in Fig. 2.8, whose coefficients $\boldsymbol{\theta}_i$ are given by (2.51). In Fig. 2.9 we show samples of the true observation $\boldsymbol{x}_i$ shown as a waveform in time, compared with the reconstruction $\hat{\boldsymbol{x}}_i$ formed from (2.50) using only the first $r = 2$ eigenvectors. It is seen that the difference between the true and reconstructed vector samples is small, as expected.

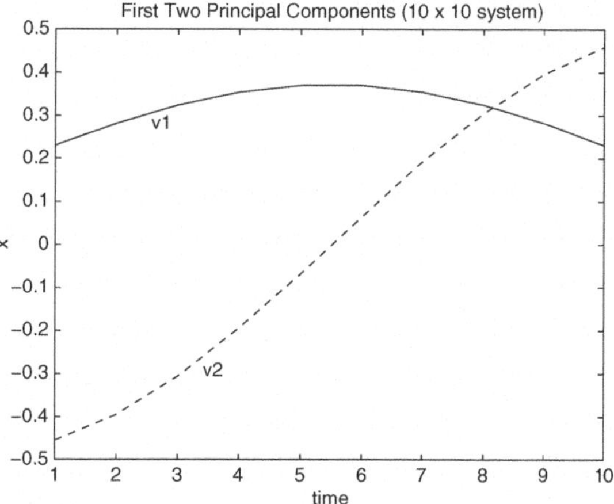

Fig. 2.8 First two eigenvector components as functions of time, for Butterworth low-pass filtered noise example

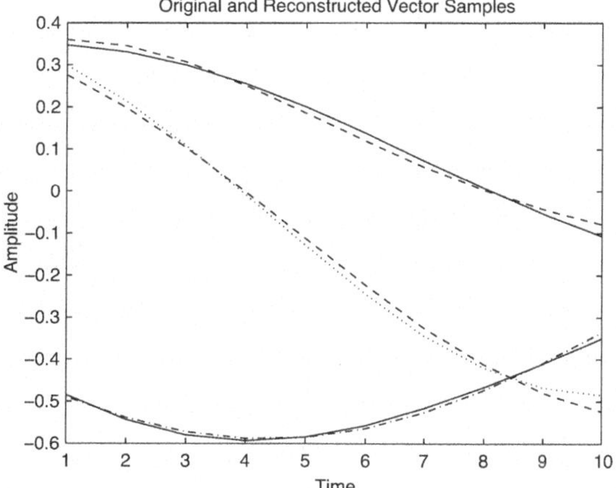

Fig. 2.9 Original vector samples of $\boldsymbol{x}$ as functions of time (solid), compared with their reconstruction using only the first two eigenvector components (dotted). Three vector samples are shown

2.6.2 Example: Denoising Signals

We now present an example showing how the PCA process can be used to denoise a signal. If the signal of interest has significant correlation structure, we can take advantage of the fact that the signal resides in a low-dimensional signal subspace defined by $\boldsymbol{V}_1$, just as it is in the previous case where we were interested in compression. In cases where the noise subspace is sufficiently distinct from the signal subspace, which tends to be the case when the signal has significant correlation structure, reconstructing the signal using only the signal subspace components has the effect of suppressing a large part of the noise. This example is similar to the previous compression example, except here the effect we are interested in is denoising rather than compression.

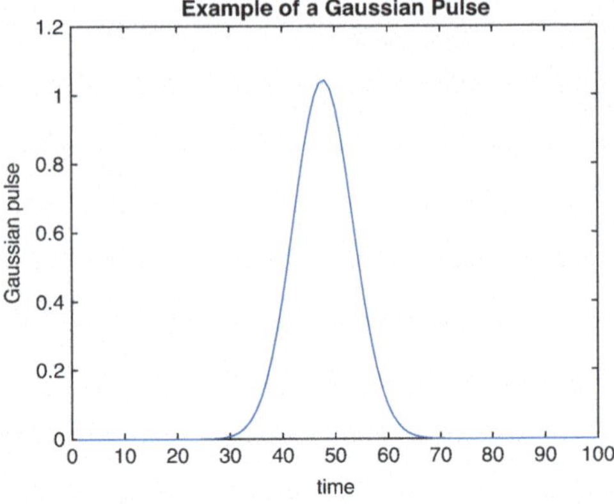

Fig. 2.10 The prototype Gaussian pulse

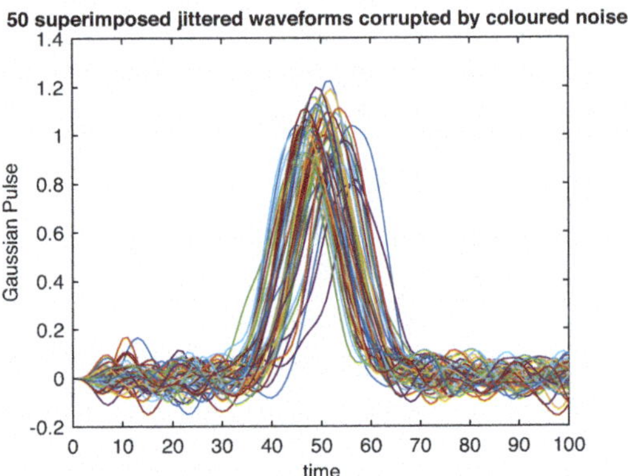

Fig. 2.11 50 superimposed, simulated pulses corrupted by timing jitter, amplitude variation, and additive coloured noise

We illustrate the concept through a simulation experiment. We follow the methodology of Sect. 2.5.1. We consider a Gaussian pulse whose samples are encapsulated into a row vector $\boldsymbol{x}_i^T \in \mathbb{R}^n$, as shown in Fig. 2.10. We observe $m = 1000$ such pulses, where each pulse is subject to timing jitter, amplitude variation, and additive coloured noise. The duration n of the pulse was chosen to be $n = 100$ samples. The observations are collected into a 1000×100 matrix $\boldsymbol{X}$, where the ith row $\boldsymbol{x}_i^T$ represents the 100 samples from the ith distinct pulse. 50 of the 1000 pulses, which are subject to timing jitter, amplitude variation, and additive coloured noise, are shown superimposed in Fig. 2.11. The covariance matrix estimate $\hat{\boldsymbol{R}} \in \mathbb{R}^{n \times n} = 1/m\ \boldsymbol{X}^T\boldsymbol{X}$ was then calculated, from which the eigenvector matrix $\boldsymbol{V}_1 \in \mathbb{R}^{n \times r}$ was extracted according to the conventional PCA process as discussed. It was empirically determined by trial and error that the best value for r in this case is 3.

We denoise the waveforms using the same procedure as in the compression example. The PCA coefficients $\boldsymbol{\theta}_i \in \mathbb{R}^r$ corresponding to the ith sample $\boldsymbol{x}_i$ can be determined using $\boldsymbol{\theta}_i = \boldsymbol{V}_1^T \boldsymbol{x}_i$. The denoised waveforms $\hat{\boldsymbol{x}}_i$ can then be reconstructed using (2.50) as $\hat{\boldsymbol{x}}_i = \boldsymbol{V}_1 \boldsymbol{\theta}_i$.

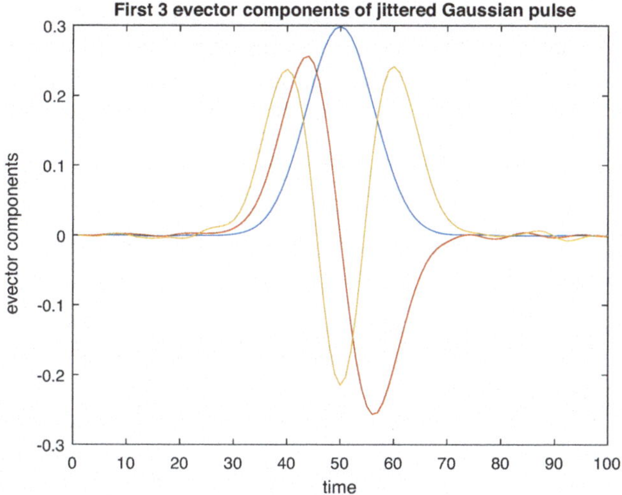

Fig. 2.12 The three principal eigenvectors for the denoising example, shown as functions of time

Fig. 2.13 A comparison between the original noise-free waveform (dotted, red), the waveform corrupted by coloured noise and jitter, (blue, dash-dot), and the denoised version, shown in (black, dashed)

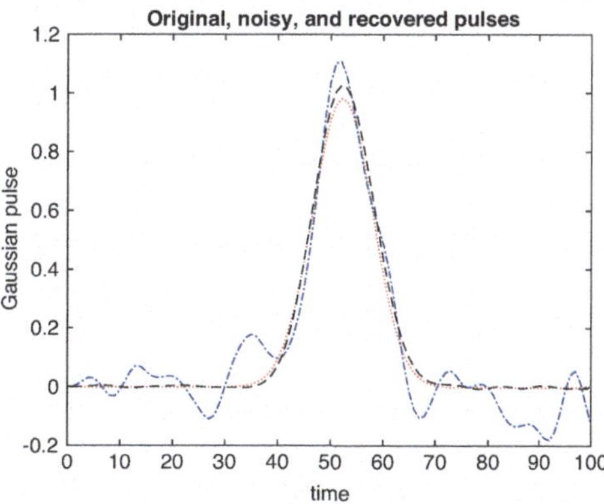

The first $r = 3$ principal eigenvectors are shown superimposed in Fig. 2.12, where the eigenvector may again be interpreted as a function of time (i.e. a waveform) over the 100 sample interval. A typical denoised waveform corresponding to an arbitrarily chosen row is shown in Fig. 2.13, showing the original waveform (dotted, red), the same waveform corrupted by additive coloured noise (blue, dash–dot), and the corresponding restored, denoised version shown in black (dashed). It may be seen that the quality of the recovered signal is quite remarkable, in the presence of substantial timing jitter and noise, using only $r = 3$ eigenvector components. The denoising effect works most effectively when the signal is highly correlated (thus increasing the concentration of the eigenvalues in the signal subspace), and the noise components are concentrated elsewhere in the underlying n-dimensional space.

In the finite data case, the principal eigenvectors of $\hat{\boldsymbol{R}}$ are only an approximation to the true signal subspace basis, with the result that there will be some degree of noise leaking into the estimated signal subspace. Thus the denoising process we have described in this section is not exact; however, in most cases in practice the level of noise is suppressed considerably.

2.6.3 Example: Classification Using PCA Coefficients

Here we introduce a simulated machine learning example where we wish to classify between two classes of observations using PCA coefficients as features. PCA coefficients are potentially useful as features for classification, since, as we have seen from previous examples, PCA coefficients have the capability of characterising the behaviour of a correlated random process with only a few parameters. In this simulation experiment, the data samples (observations) for both classes are Gaussian-distributed random vectors $\boldsymbol{x} \in \mathbb{R}^n$, where in this case $n = 4$. The true $n \times n$ covariance matrices $\boldsymbol{R}_1$ and $\boldsymbol{R}_2$ for each class are given, respectively, as

$$\boldsymbol{R}_1 = \begin{bmatrix} 1 & 0.95 & 0 & 0 \\ 0.95 & 1 & 0 & 0 \\ 0 & 0 & 1 & 0 \\ 0 & 0 & 0 & 1 \end{bmatrix}, \quad \boldsymbol{R}_2 = \begin{bmatrix} 1 & -0.95 & 0 & 0 \\ -0.95 & 1 & 0 & 0 \\ 0 & 0 & 1 & 0 \\ 0 & 0 & 0 & 1 \end{bmatrix}.$$

It is apparent from the entries [1, 2] and [2, 1] of $\boldsymbol{R}_1$ that variables 1 and 2 of class 1 have a strong positive correlation with each other. Likewise, the negative entries in corresponding positions of $\boldsymbol{R}_2$ indicate that variables 1 and 2 of class 2 have a strong negative correlation. The variables 3 and 4 of each class are uncorrelated with each other and with the variables 1 and 2. The means of all variables are zero. Consistent with the discussion in Sect. 2.2, the scatterplot of the class 1 variables in the plane consisting of the first two variables (the 1–2 plane) is an ellipse whose major axis is oriented along the direction $[1, 1]^T$. Likewise, due to the negative values of the entries of $\boldsymbol{R}_2$, the class 2 scatterplot in the 1–2 plane is oriented along the axis $[1, -1]$. It is clear that in the 1–2 plane, the two classes are not completely separable due to the overlap of the samples around the origin. However, in this simulation the overlap region contains a relatively small portion of the total number samples, and so a high level of discrimination between the classes can still be made. The complete four-dimensional scatterplots of the classes contain two additional dimensions of uncorrelated noise. Thus, without any a priori knowledge, it may be difficult to visualise the overall behaviour of the data from the four-dimensional observations, especially as the number of noisy dimensions becomes large. However, in this example we demonstrate that using PCA coefficients as features we can achieve high levels of accuracy in a straightforward manner.

To perform the simulation, we first must generate simulated class 1 and class 2 random data samples $\boldsymbol{x}$ whose covariance matrices are $\boldsymbol{R}_1$ and $\boldsymbol{R}_2$, respectively. In this simulation, the classes are discriminated solely on the basis of their correlation behaviour. The following schema generates this data in the form of $m \times n$ matrices $\boldsymbol{X}_1$ and $\boldsymbol{X}_2$. While the schema is described for the generation of $\boldsymbol{X}_1$ data, $\boldsymbol{X}_2$ is generated in an analogous fashion. Note that Matlab® notation has been adopted in some places.

- for $i = 1 : m$
  - draw $\boldsymbol{w}_i = \text{randn}(n, 1)$, where $\text{randn}(j, k)$ is a matlab® command which generates a $j \times k$ matrix of *iid* Gaussian random variables with mean 0 and variance 1.
  - assign $\boldsymbol{x}_i = \boldsymbol{G}_1^T \boldsymbol{w}_i$, where $\boldsymbol{G}_1$ is the Cholesky factor of $\boldsymbol{R}_1$, such that $\boldsymbol{G}_1^T \boldsymbol{G}_1 = \boldsymbol{R}_1$. The result of this operation is that $\text{cov}(\boldsymbol{x}) = \boldsymbol{R}_1$. More on Cholesky factors in Chap. 5.
  - assign the ith row of $\boldsymbol{X}_1$ as $\boldsymbol{X}_1(i, :) = \boldsymbol{x}_i^T$.
- end
- form $\hat{\boldsymbol{R}}_1 = \frac{1}{m} \boldsymbol{X}_1^T \boldsymbol{X}_1$.
- perform an eigendecomposition on $\hat{\boldsymbol{R}}_1$ to yield $\boldsymbol{V}_1$, whose columns are ordered to correspond to the eigenvalues sorted in descending order. The choice of r is discussed in the following.
- repeat the above process to form class 2 data, yielding $\boldsymbol{X}_2$, $\hat{\boldsymbol{R}}_2$, and $\boldsymbol{V}_2$, using the respective class 2 quantities instead of their class 1 equivalents.

In this simulation, $m = 1000$ and $n = 4$.

We are now left with the task of selecting which columns of $\boldsymbol{V}_1$ and $\boldsymbol{V}_2$ yield PCA coefficients (features) which are most discriminative between the two classes. This task is referred to as *feature selection* [3]. There are two types of feature selection algorithms, supervised and unsupervised, where the distinction is that in the supervised case only, labels are made available indicating to which class each data sample belongs. In our case, we execute a supervised feature selection process, explained as follows. We first form a matrix $\boldsymbol{X} \in \mathbb{R}^{2m \times n}$ as

$$\boldsymbol{X} = \begin{bmatrix} \boldsymbol{X}_1 \\ \boldsymbol{X}_2 \end{bmatrix}.$$

We then form the matrix $\boldsymbol{\Theta} \in \mathbb{R}^{2m \times 2n}$ of PCA coefficients as

$$\boldsymbol{\Theta} = \boldsymbol{X}\boldsymbol{V},$$

where $\boldsymbol{V} = [\boldsymbol{V}_1 \quad \boldsymbol{V}_2]$, which are the eigenvector matrices of $\boldsymbol{R}_1$ and $\boldsymbol{R}_2$, respectively. The column $\boldsymbol{\theta}_i = \boldsymbol{X}\boldsymbol{v}_i$, $i = 1, \ldots, 2n$ represents the ith candidate feature vector containing the $2m$ PCA coefficients over all the available data from both classes. The top half of $\boldsymbol{\Theta}$ contains the PCA coefficients for class 1 and the bottom half for class 2. Good features are those which change most significantly between the top and bottom halves of $\boldsymbol{\Theta}$. To determine the r most discriminative columns of $\mathbf{\Theta}$ for our supervised learning case, we use the minimum redundancy maximum relevance (mrMR) criterion [8]. We provide the necessary labels by appending a column to $\mathbf{\Theta}$, as follows:

$$\mathbf{\Theta} \leftarrow [\mathbf{\Theta}\ \boldsymbol{\ell}]$$

where the first m entries of $\boldsymbol{\ell}$ contain the label "1" indicating class 1, whereas the remaining m entries contain the value 2 for class 2. The objective of supervised feature selection is to find those columns $\mathbf{\Theta}$ which result in strong clustering behaviour between the classes, i.e. that the scatterplots of the data in the respective feature space separate into the two classes as cleanly as possible. Briefly, the mrMR algorithm accomplishes this task in the following manner. First, the matrix $\mathbf{\Theta}$ in the above form is input to the mrMR algorithm, which works in an iterative fashion. In the first iteration, the column $\boldsymbol{\theta}_i$ of $\mathbf{\Theta}$ that exhibits the most statistical dependence (max relevance) with the label vector $\boldsymbol{\ell}$ is chosen as the most significant first feature. Then in subsequent iterations, the column with the highest statistical dependency with $\boldsymbol{\ell}$, amongst those columns not previously chosen, in combination with the minimum statistical dependency amongst those columns already chosen in previous iterations, is chosen as the next feature. The process iterates in this fashion until r features are chosen.

In our present example, $r = 2$, and the mrMR algorithm identified the first (principal) column $\boldsymbol{v}_1$ of $\boldsymbol{V}_1$ and the first principal column $\boldsymbol{v}_a$ of $\boldsymbol{V}_2$ as the most distinguishing features. These columns are combined into a matrix $\boldsymbol{V}_r = [\boldsymbol{v}_1\ \boldsymbol{v}_a]$, and the corresponding $m \times r$ matrix $\mathbf{\Theta}_r$ of most discriminative PCA coefficients is then given as $\mathbf{\Theta}_r = \boldsymbol{X}\boldsymbol{V}_r$.

Figure 2.14 shows the scatterplot of the data samples, where each data point corresponds to a row of $\mathbf{\Theta}_r$. The axis labels θ_1 and θ_2, respectively, indicate the entries from the first and second column of $\mathbf{\Theta}_r$. Class 1 points are shown with

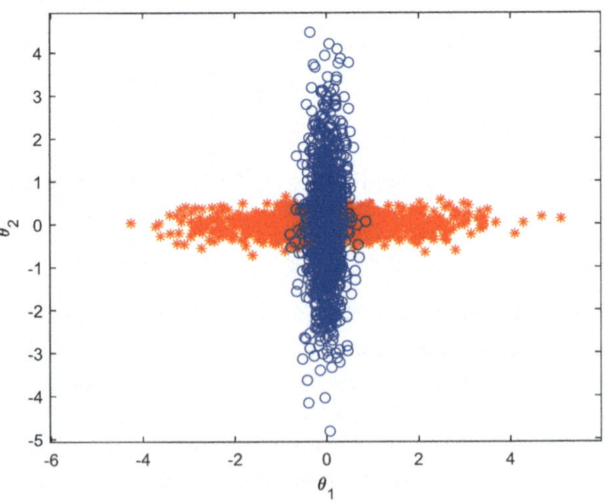

Fig. 2.14 A scatterplot of the PCA coefficients for our classification simulation. Class 1 samples are indicated with a red $*$ and class 2 samples with a blue o. The classes can be separated with 87.2% accuracy using a k nearest neighbour classifier

a red $*$ symbol, whereas class 2 points are indicated with a blue o. The overall classification accuracy between the two classes provided by the k-nearest neighbour (KNN) algorithm [3] is 87.2%. Numerical values of the two most discriminative eigenvectors $\boldsymbol{v}_1$ and $\boldsymbol{v}_a$ of $\boldsymbol{V}_r$ corresponding to the true covariance matrices $\boldsymbol{R}_1$ and $\boldsymbol{R}_2$ are given as follows:

$$\boldsymbol{v}_1 = \left[\frac{1}{\sqrt{2}}, \frac{1}{\sqrt{2}}, 0, 0\right]^T,$$

$$\boldsymbol{v}_2 = \left[\frac{1}{\sqrt{2}}, -\frac{1}{\sqrt{2}}, 0, 0\right]^T.$$

The eigenvector values corresponding to the estimated covariances $\hat{\boldsymbol{R}}_1$ and $\hat{\boldsymbol{R}}_2$ obtained from our simulation run are very close to those given above. These eigenvector values may be interpreted in the following way. Due to their strong, positive correlation, variables 1 and 2 of data samples from class 1 are very likely to have the same sign; likewise due to their strong, negative correlation, variables 1 and 2 of samples from class 2 are very likely to have opposite signs. Variables 3 and 4 of both classes are white noise random variables and as such are irrelevant to the classification between the classes. As a consequence, these latter variables are given zero weighting by both selected eigenvectors. From these observations, it follows that the value of the first PCA coefficient $\theta_1 = \boldsymbol{v}_1^T\boldsymbol{x}$, where $\boldsymbol{x}$ is a sample from class 1, will have a large value most of the time, whereas $\theta_2 = \boldsymbol{v}_a^T\boldsymbol{x}$ will have a small value. The situation is reversed when $\boldsymbol{x}$ is a sample from class 2. The result is that the scatterplot of class 1 data is stretched out along the θ_1-axis, whereas the class 2 scatterplot is aligned along the θ_2-axis, as is evident from Fig. 2.14.

2.6.4 A Critique of PCA vs. Wavelet Analysis

One of the practical difficulties with PCA compression is that the eigenvector basis set $\boldsymbol{V}$ is *data driven*, i.e. the basis $\boldsymbol{V}$ is derived from the data itself. As a consequence, it is usually not available at the reconstruction stage in practical cases when the observed signal is moderately to severely non-stationary, as is often the case with speech or video signals. In this case, the covariance matrix estimate $\hat{\boldsymbol{R}}$ changes with time, hence so do the eigenvectors. Provision of the eigenvector set for signal reconstruction is expensive in terms of information storage and so is undesirable. Wavelet transforms, which can be regarded as another form of orthonormal transformation, can replace the eigenvector basis in many cases. Wavelet transforms generate time-frequency decompositions of signals and so are particularly useful in the non-stationary case. For further details on the wavelet decomposition, see, e.g. [2, 12]. Wavelet transforms have the ability to concentrate signal energy into localised regions of the time-frequency plane, and while not optimal in comparison to PCA coefficients in the stationary case, the wavelet transform can still be effective for data compression and denoising, etc. The advantage is that the wavelet basis, unlike the eigenvector basis, is constant and invariant with time, implying that it is unnecessary to store the eigenvector basis functions for signal reconstruction. The current MPEG and JPEG standards for compression of audio and video signals use the wavelet transform for compression.

On the other hand, in many instances where denoising is the objective, the PCA basis may be more effective than wavelets. In these cases where real-time performance is not required, the signal is not severely non-stationary, and a large sample of data is available from which a stable covariance matrix may be calculated, the covariance and eigenvector matrices are readily computed off-line, and so denoising with the PCA basis is straightforward to implement. Also, because of the optimality of the eigenvector basis, denoising with PCA coefficients is more efficient. Further, as we observe in Chaps. 8 and 9, PCA analysis is invaluable in analysing least squares problems when the underlying data is poorly conditioned. A further and very useful advantage of the PCA representation is that it can be used to diagonalise a quadratic form, thus rendering the corresponding problem into a much simpler form. This advantage is explained further in Chap. 4.

In conclusion, in non-stationary environments where compression is the motivation, PCA analysis may not be the preferred choice. However, in non-real-time situations where adequate quantities of data are available for stable evaluation of the required covariance matrices, and in poorly conditioned model-building situations (to be explained later in Chaps. 8 and 9), PCA analysis is an excellent choice.

2.7 Matrix Norms

Now that we have some understanding of eigenvectors and eigenvalues, we can present the *matrix norm*. The matrix norm is related to the vector norm: it is a function which maps $\mathbb{R}^{m \times n}$ into $\mathbb{R}$. A matrix norm must obey the same properties as a vector norm. Since a norm is only strictly defined with a *vector* quantity as an argument, a matrix norm is defined by mapping a matrix into a vector and evaluating the resulting vector norm. This is accomplished by evaluating the quantity $||\boldsymbol{Ax}||_p$, where $\boldsymbol{x}$ is some suitable vector and p denotes the type of norm. However, the quantity $||\boldsymbol{Ax}||_p$ varies in norm as $\boldsymbol{x}$ changes direction and norm. We can resolve this ambiguity by choosing the direction of $\boldsymbol{x}$ so that $||\boldsymbol{Ax}||_p$ is maximum. This approach is justified since the norm expresses how large a quantity can be. We expand on this idea as follows:

Matrix p-Norms A matrix p-norm is defined in terms of a vector p-norm. The matrix p-norm of an arbitrary matrix $\boldsymbol{A}$, denoted $||\boldsymbol{A}||_p$, is defined as

$$||\boldsymbol{A}||_p = \sup_{\boldsymbol{x} \neq 0} \frac{||\boldsymbol{Ax}||_p}{||\boldsymbol{x}||_p} \tag{2.62}$$

where "sup" means *supremum*; i.e. the largest value of the argument over all values of $\boldsymbol{x} \neq \boldsymbol{0}$. Since a property of a vector norm is $||c\boldsymbol{x}||_p = |c| \, ||\boldsymbol{x}||_p$ for any scalar

c, we can choose c in (2.62) so that $||\boldsymbol{x}||_p = 1$. Then, an equivalent statement to (2.62) is

$$||\boldsymbol{A}||_p = \max_{||\boldsymbol{x}||_p=1} ||\boldsymbol{A}\boldsymbol{x}||_p \,. \tag{2.63}$$

For the specific case where $p = 2$ for $\boldsymbol{A}$ square and symmetric, it follows from (2.63) and (2.45) that $||\boldsymbol{A}\boldsymbol{x}||_2 = \lambda_1$. More generally, it is shown in the next lecture for an *arbitrary* matrix $\boldsymbol{A}$ that

$$||\boldsymbol{A}||_2 = \sigma_1 \tag{2.64}$$

where σ_1 is the largest *singular value* of $\boldsymbol{A}$. This quantity results from the *singular value decomposition*, to be discussed in the next chapter.

Matrix norms for other values of p, for arbitrary $\boldsymbol{A}$, are given as

$$||\boldsymbol{A}||_1 = \max_{1\le j\le n} \sum_{i=1}^{m} |a_{ij}| \qquad \text{(maximum column sum)} \tag{2.65}$$

and

$$||\boldsymbol{A}||_\infty = \max_{1\le i\le m} \sum_{j=1}^{n} |a_{ij}| \qquad \text{(maximum row sum).} \tag{2.66}$$

Frobenius Norm The Frobenius norm is the 2-norm of the vector obtained by concatenating all the rows (or columns) of the matrix $\boldsymbol{A}$:

$$||\boldsymbol{A}||_F = \left[\sum_{i=1}^{m}\sum_{j=1}^{n} |a_{ij}|^2\right]^{1/2}$$

Properties of Matrix Norms

1. Consider the matrix $\boldsymbol{A} \in \mathbb{R}^{m\times n}$ and the vector $\boldsymbol{x} \in \mathbb{R}^n$. Then,

$$||\mathbf{Ax}||_p \le ||\boldsymbol{A}||_p \, ||\boldsymbol{x}||_p$$

   This property follows by dividing both sides of the above by $||\boldsymbol{x}||_p$, and applying (2.62).

2. If $\boldsymbol{Q}$ and $\boldsymbol{Z}$ are orthonormal matrices of appropriate size, then

$$||\mathbf{QAZ}||_2 = ||\boldsymbol{A}||_2$$

   and

$$||\mathbf{QAZ}||_F = ||\boldsymbol{A}||_F$$

   Thus, we see that the matrix 2-norm and Frobenius norm are invariant to pre- and post- multiplication by an orthonormal matrix.

3. Further,

$$||\boldsymbol{A}||_F^2 = \text{tr}\left(\boldsymbol{A}^T\boldsymbol{A}\right)$$

where tr(·) denotes the *trace* of a matrix, which is the sum of its diagonal elements. While we are considering *trace*, an important property of the trace operator is

$$\text{tr}\,(\boldsymbol{A}\boldsymbol{B}) = \text{tr}\,(\boldsymbol{B}\boldsymbol{A}) \tag{2.67}$$

for any pair of matrices $\boldsymbol{A}$, $\boldsymbol{B}$ whose dimensions are $n \times k$ and $k \times n$, respectively, $k, n \in \mathcal{Z}$ [12].

Appendix

Differentiation of a Quadratic Form

Consider a square matrix $\boldsymbol{A} \in \mathbb{R}^{n\times n}$ (not necessarily symmetric). Then the quantity $\boldsymbol{x}^T\boldsymbol{A}\boldsymbol{x}$ is referred to as a *quadratic form*, which is a scalar quantity which we discuss further in Chap. 4. To differentiate a scalar function $f(\boldsymbol{x})$ by a vector $\boldsymbol{x}$, we differentiate $f(\boldsymbol{x})$ by each element of $\boldsymbol{x}$ in turn and then assemble the results back into a vector. Thus the derivative in this case has the same dimensions as $\boldsymbol{x}$.

To differentiate the quadratic form $f(\boldsymbol{x}) = \boldsymbol{x}^T\boldsymbol{A}\boldsymbol{x}$ with respect to x_i, we use the vector version of the product rule[13]. If $f(\boldsymbol{x}) = a(\boldsymbol{x})b(\boldsymbol{x})$, then

$$\frac{df(\boldsymbol{x})}{dx_i} = a(\boldsymbol{x})\frac{db(\boldsymbol{x})}{dx_i} + \frac{da(\boldsymbol{x})}{dx_i}b(\boldsymbol{x}). \tag{2.68}$$

For the problem at hand, we assign $a(\boldsymbol{x}) = \boldsymbol{x}^T$, and $b(\boldsymbol{x}) = \boldsymbol{A}\boldsymbol{x}$. Then it is readily verified that the first term of (2.68) is given by

$$a(\boldsymbol{x})\frac{db(\boldsymbol{x})}{dx_i} = \boldsymbol{x}^T\bar{\boldsymbol{a}}_i = \bar{\boldsymbol{a}}_i^T\boldsymbol{x},$$

where $\bar{\boldsymbol{a}}_i$ is the ith column of $\boldsymbol{A}$. This follows since the respective quantity is a scalar. The second is

$$\frac{da(\boldsymbol{x})}{dx_i}b(\boldsymbol{x}) = \underline{\boldsymbol{a}}_i\boldsymbol{x},$$

[12] $\mathcal{Z}$ is the set of positive integers, excluding zero.

[13] This may be proved in an analogous manner to the scalar case.

where $\underline{a}_i$ is the ith row of A. Combining the results from both terms for $i = 1, \ldots, n$ into a vector, we have

$$\begin{aligned} \frac{df(x)}{dx} &= A^T x + Ax \\ &= (A^T + A)x. \end{aligned} \tag{2.69}$$

In the case when A is symmetric, then

$$\frac{df(x)}{dx} = 2Ax.$$

This result is loosely analogous to the scalar case where the derivative $\frac{d}{dx}ax^2 = 2ax$, where $a \in \mathbb{R}$.

Problems

1. Consider a *real* skew-symmetric matrix (i.e. one for which $A^T = -A$). Prove its eigenvalues are pure imaginary, and the eigenvectors are mutually orthogonal.
2. If the eigendecomposition of a square, symmetric matrix A is given by $A = V\Lambda V^T$, what is the eigendecomposition of A^{-1}?
3. Consider two matrices A and B. Under what conditions is $AB = BA$?
4. We are given a matrix A whose eigendecomposition is $A = V\Lambda V^T$. Find the eigenvalues and eigenvectors of the matrix $C = BAB^{-1}$ in terms of those of A, where B is any invertible matrix of appropriate size. C is referred to as a similarity transform of A.
5. Consider a non-white random process $x[k]$ of duration m. The sequence $f[k]$ of duration $n <= m$ operates on x[k] to give an output sequence y[k] according to

$$y[k] = \sum_i x[k-i]f[i]$$

where $x[k] = 0$ for $k > m$ or $k < 0$. Using the x-data in file Ch2Q5.mat, which is available in the supplementary material described in the opening pages of this chapter, find $f[k]$ of length $n = 10$ so that $||y||_2^2$ is minimised, subject to $||f||_2^2 = 1$.
6. The supplementary material mentioned in Problem 5 above also includes a file Ch2Q6.mat, which contains a matrix $X \in \mathbb{R}^{m \times n}$ of data corresponding to the example of Sect. 2.6.2. Each row is a time-jittered Gaussian pulse corrupted by coloured noise. Here, $m = 1000$ and $n = 100$, as per the example. Using your preferred programming language, produce a denoised version of the signal represented by the first row of X.

7. The supplementary material described in the opening pages of this chapter includes a file Ch2Q7.mat. It contains a matrix $\boldsymbol{X}$ whose columns contain two superimposed Gaussian pulses with additive noise. Using methods discussed in the course, estimate the position of the peaks of the Gaussian pulses.
8. Construct a 6×6 symmetric matrix $\boldsymbol{A}$ of rank 3.
   (a) Express a basis for $\mathcal{N}(\boldsymbol{A})$ using the eigendecomposition and verify using matlab® that a vector in the respective subspace is indeed in the nullspace.
   (b) Find a vector $\boldsymbol{y}$ for which $\boldsymbol{A}\boldsymbol{x} = \boldsymbol{y}$ exactly. Fully characterise the solution $\boldsymbol{x}$ in this case in terms of the eigendecomposition of $\boldsymbol{A}$.
   (c) Under what conditions does an exact solution not exist?
9. We are given a sequence of vector samples $\boldsymbol{x}_i = a_i \boldsymbol{g}, i = 1 \ldots, N$ of length K samples, where $a_i \in \mathbb{R}$ is a zero mean random variable with variance σ_a^2 and $\boldsymbol{g} \in \mathbb{R}^K$ is a Gaussian pulse waveform in time with unit norm, similar to that shown in Fig. 2.10. The width (σ) and position (μ) of the pulse are invariant with i. We form the sample covariance matrix $\hat{\boldsymbol{R}}$ over the N observations as $\hat{\boldsymbol{R}} = \frac{1}{N} \sum_{i=1}^{N} \boldsymbol{x}_i \boldsymbol{x}_i^T$.
   (a) What is the rank of $\hat{\boldsymbol{R}}$?
   (b) What is the first eigenvector of $\hat{\boldsymbol{R}}$, as $N \to \infty$? Describe the remaining eigenvectors.
   (c) White noise is added to $\boldsymbol{x}_i$ so that $\boldsymbol{x}_i = a_i \boldsymbol{g} + \sigma \boldsymbol{w}$, where $\boldsymbol{w} \in \mathbb{R}^K$ is a zero-mean noise vector with uncorrelated elements; i.e. $E[\boldsymbol{w}\boldsymbol{w}^T] = \sigma^2 \boldsymbol{I}$. The vector $\boldsymbol{w}$ is uncorrelated with the signal component $\boldsymbol{g}$. Address the two questions above for this present case.
10. Prove that m must be greater than or equal to n for $\hat{\boldsymbol{R}}$ in (2.29) to be full rank.
11. Consider the following alternative formulation for $\hat{\boldsymbol{R}}$ that enables it to adapt to a non-stationary environment:

$$\hat{\boldsymbol{R}}(k) = (1 - \lambda)\boldsymbol{x}(k)\boldsymbol{x}^T(k) + \lambda \hat{\boldsymbol{R}}(k-1),$$

   where k is the time index and $0 < \lambda < 1$ is a parameter that controls the adaptation rate. Explain how the method operates, and what is the effect of varying λ? What happens to the observation $\boldsymbol{x}(k_o)$, where k_o is constant, as time increases past the value k_o?
12. Prove (2.67).
13. The random process $x_1[k]$ is narrow-band, mean-centred, whose spectrum is centred at 0 Hz. The process $x_2[k]$ is narrow band, mean-centred, whose spectrum is centred at a normalised frequency of 1/2 Hz. What is the sign of the covariances between adjacent samples of each process?
14. Explain why the matrix $\boldsymbol{X}$ approaches rank deficiency as the columns become more correlated.
15. An $n \times n$ *circulant* matrix $\boldsymbol{X}$ is one defined by a single (first) column $\boldsymbol{x} = [x_0, x_1, \ldots, x_{n-1}]$, which can contain arbitrary real or complex values. Each successive column is a cyclic permutation of the previous column, i.e. all

elements of the first column (except the last) fall down one position to form the second column, and the last element assumes the first position. The remaining columns are formed in a corresponding manner. Thus $\boldsymbol{X}$ has the form

$$\boldsymbol{X} = \begin{bmatrix} x_0 & x_{n-1} & \cdots & x_2 & x_1 \\ x_1 & x_0 & x_{n-1} & & x_2 \\ \vdots & x_1 & x_0 & & \vdots \\ x_{n-2} & & \ddots & \ddots & x_{n-1} \\ x_{n-1} & x_{n-2} & \cdots & x_1 & x_0 \end{bmatrix}$$

Show that the eigenvectors of $\boldsymbol{X}$ are discrete Fourier transform (DFT) basis vectors given as columns of $\boldsymbol{W}$ in (2.41) or in (A.20) of the Appendix of this volume. Show also that the eigenvalues are the DFT coefficients corresponding to the sequence $\boldsymbol{x}$. *Hint:* Assume the sequence formed by $\boldsymbol{x}^T$ is one period of an infinite periodic sequence. Then apply Property 16 from the Appendix.

References

1. P.M. Bentley, J.T.E. McDonnell, Wavelet transforms: an introduction. Electron. Commun. Eng. J. **6**(4), 175–186 (1994)
2. H. Guo, C.S. Burrus, R.A. Gopinath, *Introduction To Wavelets And Wavelet Transforms—A Primer* (Prentice Hall, Upper Saddle River, NJ, 1998)
3. T. Hastie, R. Tibshirani, J. Friedman, *The Elements of Statistical Learning: Data Mining, Inference, and Prediction* (Springer Science & Business Media, New York, 2009)
4. S. Haykin, *Adaptive Filter Theory*, 4th edn. (Prentice Hall, Englewood Cliffs, NJ, USA, 2001)
5. S. Haykin et al., Some aspects of array signal processing. IEE Proc. F **139**(1), 1–26 (1992)
6. A.V. Oppenheim, R.W. Shafer, *Discrete-Time Signal Processing*, 3rd edn. (Pearson Education Ltd., Upper Saddle River, NJ, 2010)
7. A. Papoulis, *Random Variables and Stochastic Processes* (McGraw Hill, New York, 1994)
8. H. Peng, F. Long, C. Ding, Feature selection based on mutual in- formation criteria of max-dependency, max-relevance, and min-redundancy. IEEE Trans. Pattern Anal. Mach. Intell. **27**(8), 1226–1238 (2005)
9. K.B. Petersen, M.S. Pedersen, et al., *The Matrix Cookbook*, vol. 7 (Technical University of Denmark, 2008)
10. R. Schmidt, Multiple emitter location and signal parameter estimation. IEEE Trans. Antennas Propagat. **34**(3), 276–280 (1986)
11. N.K. Sinha, G.J. Lastman, *Microcomputer-Based Numerical Methods for Science and Engineering* (Oxford University Press, Oxford, 1988)
12. G. Strang, T. Nguyen, *Wavelets and Filter Banks* (SIAM, 1996)

Chapter 3
The Singular Value Decomposition (SVD)

3.1 Development of the SVD

We have found so far that the eigendecomposition is a useful analytic tool. However, it is only applicable on *square symmetric* matrices. We now consider the SVD, which may be considered a generalisation of the ED to arbitrary matrices. Thus, with the SVD, all the analytical uses of the ED which before were restricted to symmetric matrices may now be applied to any form of matrix, regardless of size, whether it is symmetric or nonsymmetric, rank deficient, etc.

Theorem 3.1 *Let* $\mathbf{A} \in \mathbb{R}^{m \times n}$ *be a rank r matrix* ($r \leq p = \min(m, n)$)*. Then* $\mathbf{A}$ *can be decomposed according to the singular value decomposition as*

$$\mathbf{A} = \mathbf{U}\boldsymbol{\Sigma}\mathbf{V}^T \tag{3.1}$$

where $\boldsymbol{U} \in \mathbb{R}^{m \times m}$ *and* $\boldsymbol{V} \in \mathbb{R}^{n \times n}$ *are orthonormal, and*

$$\boldsymbol{\Sigma} = \begin{array}{c} r \\ m-r \end{array} \begin{bmatrix} \tilde{\boldsymbol{\Sigma}} & \mathbf{0} \\ \mathbf{0} & \mathbf{0} \end{bmatrix}_{\;r \quad n-r}$$

where $\tilde{\boldsymbol{\Sigma}} = diag(\sigma_1, \sigma_2, \ldots, \sigma_r)$ *and*

$$\sigma_1 \geq \sigma_2 \geq \sigma_3 \ldots \geq \sigma_p \geq 0.$$

The matrix $\boldsymbol{\Sigma}$ must be of dimension $\mathbb{R}^{m \times n}$ (i.e. the same size as $\boldsymbol{A}$), to ensure the product in (3.1) exists. It is therefore padded with appropriately sized zero blocks to augment it to the required size.

J. Reilly, *Fundamentals of Linear Algebra for Signal Processing*,
https://doi.org/10.1007/978-3-031-68915-4_3

Since U and V are orthonormal, we may also write (3.1) in the form

$$\underset{m\times m}{\mathbf{U}^T}\ \underset{m\times n}{\mathbf{A}}\ \underset{n\times n}{\mathbf{V}} = \underset{m\times n}{\mathbf{\Sigma}} \tag{3.2}$$

where $\mathbf{\Sigma}$ is a diagonal matrix. The values σ_i, which are defined to be positive, are referred to as the *singular values* of $\boldsymbol{A}$. The columns $\mathbf{u}_i$ and $\mathbf{v}_i$ of $\boldsymbol{U}$ and $\boldsymbol{V}$, respectively, are called the left and right *singular vectors* of $\boldsymbol{A}$.

Proof Consider the square symmetric positive semi-definite matrix $\boldsymbol{A}^T\boldsymbol{A}$[1]. Let the eigenvalues greater than zero be $\sigma_1^2, \sigma_2^2, \ldots, \sigma_r^2$, $r \le p = \min(m, n)$. Then, applying the eigendecomposition on $\boldsymbol{A}^T\boldsymbol{A}$, there exists an orthonormal matrix $\boldsymbol{V} \in \mathbb{R}^{n\times n}$ such that

$$\boldsymbol{V}^T\boldsymbol{A}^T\boldsymbol{A}\boldsymbol{V} = \begin{bmatrix} \tilde{\mathbf{\Sigma}}^2 & \mathbf{0} \\ \mathbf{0} & \mathbf{0} \end{bmatrix}. \tag{3.3}$$

where $\tilde{\mathbf{\Sigma}}^2 = \mathrm{diag}[\sigma_1^2, \ldots, \sigma_r^2]$. We now partition $\boldsymbol{V}$ as $[\boldsymbol{V}_1 \ \ \boldsymbol{V}_2]$, where $\boldsymbol{V}_1 \in \mathbb{R}^{n\times r}$. Then (3.3) has the form

$$\underset{n}{\begin{bmatrix} \boldsymbol{V}_1^T \\ \boldsymbol{V}_2^T \end{bmatrix}} \boldsymbol{A}^T\boldsymbol{A} \begin{bmatrix} \underset{r}{\boldsymbol{V}_1} & \underset{n-r}{\boldsymbol{V}_2} \end{bmatrix} = \begin{bmatrix} \tilde{\mathbf{\Sigma}}^2 & \mathbf{0} \\ \mathbf{0} & \mathbf{0} \end{bmatrix}. \tag{3.4}$$

Then by equating corresponding blocks in (3.4), we have

$$\boldsymbol{V}_1^T\boldsymbol{A}^T\boldsymbol{A}\boldsymbol{V}_1 = \tilde{\mathbf{\Sigma}}^2 \quad (r\times r) \tag{3.5}$$

$$\boldsymbol{V}_2^T\boldsymbol{A}^T\boldsymbol{A}\boldsymbol{V}_2 = \mathbf{0}. \quad (n-r)\times(n-r). \tag{3.6}$$

From (3.5), we can write

$$\tilde{\mathbf{\Sigma}}^{-1}\boldsymbol{V}_1^T\boldsymbol{A}^T\boldsymbol{A}\boldsymbol{V}_1\tilde{\mathbf{\Sigma}}^{-1} = \boldsymbol{I}. \tag{3.7}$$

Then, we define the matrix $\boldsymbol{U}_1 \in \mathbb{R}^{m\times r}$ from (3.7) as

$$\boldsymbol{U}_1 = \boldsymbol{A}\boldsymbol{V}_1\tilde{\mathbf{\Sigma}}^{-1}. \tag{3.8}$$

Then, noting that the product of the first three terms in (3.7) is the transpose of the product of the latter three terms, we have $\boldsymbol{U}_1^T\boldsymbol{U}_1 = \mathbf{I}$ and it follows that

$$\boldsymbol{U}_1^T\boldsymbol{A}\boldsymbol{V}_1 = \tilde{\mathbf{\Sigma}}. \tag{3.9}$$

[1] The concept of *positive semi-definiteness* is discussed in the next chapter. It means all the eigenvalues are greater than or equal to zero.

From (3.6) we also have

$$AV_2 = \mathbf{0}. \tag{3.10}$$

We now choose a matrix U_2 so that $U \in \mathbb{R}^{m\times m} = [U_1 \;\; U_2]$ is orthonormal. Then from (3.8) and because $U_1 \perp U_2$, we have

$$U_2^T U_1 = U_2^T A V_1 \tilde{\Sigma}^{-1} = \mathbf{0}. \tag{3.11}$$

Since $\tilde{\Sigma}$ is full rank by design, we have

$$U_2^T A V_1 = \mathbf{0}. \tag{3.12}$$

Combining (3.9), (3.10), and (3.12), we have

$$U^T A V = \begin{bmatrix} U_1^T A V_1 & U_1^T A V_2 \\ U_2^T A V_1 & U_2^T A V_2 \end{bmatrix} = \begin{bmatrix} \tilde{\Sigma} & \mathbf{0} \\ \mathbf{0} & \mathbf{0} \end{bmatrix} \tag{3.13}$$

which was to be shown. □

The proof can be repeated using an eigendecomposition on the matrix $AA^T \in \mathbb{R}^{m\times m}$ instead of on $A^T A$. In this case, the roles of the orthonormal matrices V and U are interchanged, and the non-zero eigenvalues remain unchanged.

3.1.1 Partitioning the SVD

Here we formalise the partitioning of the SVD that was implicit in the proof of the previous section. Following the convention used for ordering eigenvalues, for convenience of notation, we arrange the singular values as

$$\underbrace{\underset{\text{max}}{\sigma_1} \geq \cdots \geq \underset{\text{min}}{\sigma_r}}_{r \text{ non-zero s.v.'s}} > \underbrace{\sigma_{r+1} = \cdots = \sigma_p}_{p-r \text{ zero s.v.'s}} = 0$$

We also partition the U and V as before in the previous section. We can then write the SVD of A in the form

$$A = \begin{bmatrix} U_1 & U_2 \end{bmatrix} \begin{bmatrix} \Sigma_r & \mathbf{0} \\ \mathbf{0} & \mathbf{0} \end{bmatrix} \begin{bmatrix} V_1^T \\ V_2^T \end{bmatrix} \tag{3.14}$$

where where $\boldsymbol{\Sigma}_r \in \mathbb{R}^{r \times r} = \text{diag}(\sigma_1, \ldots, \sigma_r)$, and $\boldsymbol{U}$ is partitioned as

$$\boldsymbol{U} = \underset{}{\left[\underset{r}{\boldsymbol{U}_1}\ \underset{m-r}{\boldsymbol{U}_2}\right]} m \tag{3.15}$$

$\boldsymbol{V}$ is partitioned in an analogous manner

$$\boldsymbol{V} = \left[\underset{r}{\boldsymbol{V}_1}\ \underset{n-r}{\boldsymbol{V}_2}\right] n. \tag{3.16}$$

3.1.2 Relationship Between SVD and ED

It is clear that the eigendecomposition and the singular value decomposition share many properties in common. The price we pay for being able to perform a diagonal decomposition on an *arbitray* matrix is that we need two orthonormal matrices instead of just one, as is the case for square symmetric matrices. In this section, we explore further relationships between the ED and the SVD.

Using the partitioning from (3.16), we can write

$$\begin{aligned} \boldsymbol{A}^T\boldsymbol{A} &= \left[\boldsymbol{V}_1\ \boldsymbol{V}_2\right]\begin{bmatrix} \tilde{\boldsymbol{\Sigma}} & \mathbf{0} \\ \mathbf{0} & \mathbf{0} \end{bmatrix}\boldsymbol{U}^T\boldsymbol{U}\begin{bmatrix} \tilde{\boldsymbol{\Sigma}} & \mathbf{0} \\ \mathbf{0} & \mathbf{0} \end{bmatrix}\begin{bmatrix} \boldsymbol{V}_1^T \\ \boldsymbol{V}_2^T \end{bmatrix} \\ &= \boldsymbol{V}_1\tilde{\boldsymbol{\Sigma}}^2\boldsymbol{V}_1^T \\ &= \boldsymbol{V}\begin{bmatrix} \tilde{\boldsymbol{\Sigma}}^2 & \mathbf{0} \\ \mathbf{0} & \mathbf{0} \end{bmatrix}\boldsymbol{V}^T \end{aligned}$$

Comparing the line above to the eigendecomposition of the square, symmetric matrix $\boldsymbol{A}^T\boldsymbol{A}$, it is apparent that the eigenvectors $\boldsymbol{V}$ of the matrix $\boldsymbol{A}^T\boldsymbol{A}$ are the right singular vectors of $\boldsymbol{A}$ and that the square of the singular values of $\boldsymbol{A}$ form the corresponding eigenvalues. If rank($\boldsymbol{A}$) = r, regardless of its shape (tall, square or short) and from the partitioning of $\boldsymbol{V}$ in (3.16), if $n - r \geq 2$, there are repeated zero eigenvalues, and $\boldsymbol{V}_2$ is not unique in this case.

Further, using the form $\boldsymbol{A}\boldsymbol{A}^T$ instead of $\boldsymbol{A}^T\boldsymbol{A}$ and using the paritioning of (3.15), it is straightforward to show that

$$\boldsymbol{A}\boldsymbol{A}^T = \left[\boldsymbol{U}_1\ \boldsymbol{U}_2\right]\begin{bmatrix} \tilde{\boldsymbol{\Sigma}}^2 & \mathbf{0} \\ \mathbf{0} & \mathbf{0} \end{bmatrix}\begin{bmatrix} \boldsymbol{U}_1^T \\ \boldsymbol{U}_2^T \end{bmatrix} = \boldsymbol{U}_1\tilde{\boldsymbol{\Sigma}}^2\boldsymbol{U}_1^T,$$

which indicates that the eigenvectors of $\boldsymbol{A}\boldsymbol{A}^T$ are the left singular vectors $\boldsymbol{U}$ of $\boldsymbol{A}$ and the squared singular values of $\boldsymbol{A}$ form the corresponding eigenvalues of $\boldsymbol{A}\boldsymbol{A}^T$. If rank($\boldsymbol{A}$) $= r$ and from the partitioning of $\boldsymbol{U}$ in (3.15), if $m - r \geq 2$, then there are repeated zero eigenvalues, and $\boldsymbol{U}_2$ is not unique in this case.

3.1.3 Properties and Interpretations of the SVD

The above partition reveals many interesting properties of the SVD:

rank$(A) = r$

From (3.14), by multiplying matrices from the right-hand side, we have $\boldsymbol{A} = \boldsymbol{U}_1\boldsymbol{B}$, where $\boldsymbol{B} = \tilde{\boldsymbol{\Sigma}}\boldsymbol{V}_1^T$ is $r \times n$ and full rank. It is then clear that all columns of $\boldsymbol{A}$ are linear combinations of $\boldsymbol{U}_1 \in \mathbb{R}^{m\times r}$. Therefore $R(\boldsymbol{A})$ spans r dimensions and so $\text{rank}(\boldsymbol{A}) = r$. If $r < p = \min(m, n)$, then there are $p - r$ zero singular values.

Determination of rank when $\sigma_1, \ldots, \sigma_r$ are distinctly greater than zero and when $\sigma_{r+1}, \ldots, \sigma_p$ are exactly zero is easy. But often in practice, due to finite precision arithmetic and fuzzy data, σ_r may be very small, and σ_{r+1} may be not quite zero. Hence, in practice, determination of rank is not so easy. A common method is to declare rank $\boldsymbol{A} = r$ if $\sigma_{r+1} \le \epsilon$, where ϵ is a small number specific to the problem considered.

$\mathcal{N}(\boldsymbol{A}) = R(\boldsymbol{V}_2)$

The proof is analogous to the eigenvector case discussed in Chap. 2. Recall the nullspace $\mathcal{N}(\mathbf{A}) = \{\boldsymbol{x} \mid \boldsymbol{A}\boldsymbol{x} = \mathbf{0}\}$. From (3.14) it is clear that $\boldsymbol{A}\boldsymbol{x} = \mathbf{0}$ for non-zero $\boldsymbol{x}$ iff $\boldsymbol{x} \in R(\boldsymbol{V}_2)$.

$R(\mathbf{A}) = R(\mathbf{U}_1)$

From the paragraph on rank above, we have $\boldsymbol{A} = \boldsymbol{U}_1\boldsymbol{B}$ and $\boldsymbol{B}$ is full rank. Hence, the ranges of $\boldsymbol{A}$ and $\boldsymbol{U}_1$ are equivalent, and $\boldsymbol{U}_1$ is a basis for the column space (range) of $\boldsymbol{A}$.

$R(\mathbf{A}^T) = R(\mathbf{V}_1)$

Recall that $R(\mathbf{A}^T)$ is the set of all linear combinations of rows of $\boldsymbol{A}$. If we transpose the expression for $\boldsymbol{A}$ in (3.14) and apply the same argument that justifies $R(\boldsymbol{A}) = R(\boldsymbol{U}_1)$ above, we get the desired result. Hence $\boldsymbol{V}_1$ is a basis for the row space of $\boldsymbol{A}$.

$R(\boldsymbol{A})_\perp = R(\boldsymbol{U}_2)$

We have seen that $R(\mathbf{A}) = R(\mathbf{U}_1)$. Since from (3.14), $\boldsymbol{U}_1 \perp \boldsymbol{U}_2$, then $\boldsymbol{U}_2$ is a basis for the orthogonal complement of $R(\boldsymbol{A})$, hence the result.

$||\mathbf{A}||_2 = \boldsymbol{\sigma}_1 = \boldsymbol{\sigma}_{\max}$

This is straightforward to see from the definition of the 2-norm and the ellipsoid example to follow in Sect. 3.1.3.

Inverse of A

If the SVD of a square invertible matrix $\boldsymbol{A}$ is given, it is easy to find the inverse. In this case, we have $\sigma_1 \ldots \sigma_n > 0$. The inverse of $\boldsymbol{A}$ is given from the SVD, using the familiar rules, as

$$\boldsymbol{A}^{-1} = \boldsymbol{V}\boldsymbol{\Sigma}^{-1}\boldsymbol{U}^T.$$

The evaluation of $\boldsymbol{\Sigma}^{-1}$ is simple because it is square and diagonal. Note that this treatment indicates that the singular values of $\boldsymbol{A}^{-1}$ are $[\sigma_n^{-1}, \sigma_{n-1}^{-1}, \ldots, \sigma_1^{-1}]$ in that order. The only difficulty with this approach is that in general, finding the SVD is more costly in computational terms than finding the inverse by more conventional means.

The SVD Diagonalises Any System of Equations

Consider the system of equations $\boldsymbol{A}\boldsymbol{x} = \boldsymbol{b}$, for an arbitrary matrix $\boldsymbol{A}$. Using the SVD of $\boldsymbol{A}$, we have

$$\boldsymbol{U}\boldsymbol{\Sigma}\boldsymbol{V}^T\boldsymbol{x} = \boldsymbol{b}. \tag{3.17}$$

We represent $\boldsymbol{b}$ in the basis $\boldsymbol{U}$, and $\boldsymbol{x}$ in the basis $\boldsymbol{V}$. We therefore have

$$\boldsymbol{c} = \begin{matrix} r \\ m-r \end{matrix} \begin{bmatrix} \boldsymbol{c}_1 \\ \boldsymbol{c}_2 \end{bmatrix} = \begin{bmatrix} \boldsymbol{U}_1^T \\ \boldsymbol{U}_2^T \end{bmatrix} \boldsymbol{b} \tag{3.18}$$

and

$$\boldsymbol{d} = \begin{matrix} r \\ n-r \end{matrix} \begin{bmatrix} \boldsymbol{d}_1 \\ \boldsymbol{d}_2 \end{bmatrix} = \begin{bmatrix} \boldsymbol{V}_1^T \\ \boldsymbol{V}_2^T \end{bmatrix} \boldsymbol{x}. \tag{3.19}$$

Substituting the above into (3.17), the system of equations becomes

$$\boldsymbol{\Sigma}\boldsymbol{d} = \boldsymbol{c}. \tag{3.20}$$

This shows that as long as we choose the correct bases, *any* system of equations can become diagonal. This property represents the power of the SVD; it allows us to transform arbitrary algebraic structures into their simplest forms.

Equation (3.20) can be expanded as

$$\begin{bmatrix} \tilde{\boldsymbol{\Sigma}} & \mathbf{0} \\ \mathbf{0} & \mathbf{0} \end{bmatrix} \begin{bmatrix} \boldsymbol{d}_1 \\ \boldsymbol{d}_2 \end{bmatrix} = \begin{bmatrix} \boldsymbol{c}_1 \\ \boldsymbol{c}_2 \end{bmatrix}. \tag{3.21}$$

The above equation reveals several interesting facts about the solution of the system of equations. First, if $m > n$ ($\boldsymbol{A}$ is tall) and $\boldsymbol{A}$ is full rank, then the right blocks of zeros in $\boldsymbol{\Sigma}$, as well as the quantity $\boldsymbol{d}_2$, are both empty. In this case, the system of equations can be satisfied exactly only if $\boldsymbol{c}_2 = \mathbf{0}$. This implies that $\boldsymbol{U}_2^T \boldsymbol{b} = \mathbf{0}$ or that $\boldsymbol{b} \in R(\boldsymbol{U}_1) = R(\boldsymbol{A})$ for an exact solution to exist. This result makes sense, since in this case the quantity $\boldsymbol{Ax}$ is a linear combination of the columns of $\boldsymbol{A}$ and therefore the resulting $\boldsymbol{b}$ must be in $R(\boldsymbol{A})$.

If $m < n$ ($\boldsymbol{A}$ is short) and full rank, then the bottom blocks of zeros in $\boldsymbol{\Sigma}$, as well as $\boldsymbol{c}_2$ in (3.21) are both empty. In this case we have $\boldsymbol{d}_1 = \tilde{\boldsymbol{\Sigma}}^{-1} \boldsymbol{c}_1$. Because $\boldsymbol{d}_2$ interacts with a zero-block in (3.21), its value is arbitrary. We can write these relationships in the form

$$\begin{bmatrix} \boldsymbol{d}_1 \\ \boldsymbol{d}_2 \end{bmatrix} = \begin{bmatrix} \boldsymbol{V}_1^T \boldsymbol{x} \\ \boldsymbol{V}_2^T \boldsymbol{x} \end{bmatrix} = \begin{bmatrix} \tilde{\boldsymbol{\Sigma}}^{-1} \boldsymbol{c}_1 \\ \boldsymbol{d}_2 \end{bmatrix},$$

where we have used (3.21) for $\boldsymbol{d}_1$. Multiplying through by $\boldsymbol{V} = [\boldsymbol{V}_1 \ \boldsymbol{V}_2]$, we have

$$\begin{aligned} \boldsymbol{Vd} = \boldsymbol{x} &= \begin{bmatrix} \boldsymbol{V}_1 & \boldsymbol{V}_2 \end{bmatrix} \begin{bmatrix} \boldsymbol{\Sigma}^{-1} \boldsymbol{U}_1^T \boldsymbol{b} \\ \boldsymbol{d}_2 \end{bmatrix} \\ &= \boldsymbol{V}_1 \boldsymbol{\Sigma}^{-1} \boldsymbol{U}_1^T \boldsymbol{b} + \boldsymbol{V}_2 \boldsymbol{d}_2 \end{aligned} \tag{3.22}$$

where in the top line we have substituted (3.18) for $\boldsymbol{c}_1$. Thus we see that the solution consists of a "basic" component, which is the first term above. This term is closely related to the pseudo-inverse, which we discuss in some detail in Chap. 8. Since $\boldsymbol{V}_2$ is a basis for $\mathcal{N}(\boldsymbol{A})$ and $\boldsymbol{d}_2$ is an arbitrary $n - r$ vector, the second term above contributes an arbitrary component in the nullspace of $\boldsymbol{A}$ to the solution $\boldsymbol{x}$. Thus $\boldsymbol{x}$ is not unique in this case. It is straightforward to verify that the quantity $\boldsymbol{AV}_2\boldsymbol{d}_2 = \mathbf{0}$, so the addition of the second term does not affect the fact we have an exact solution.

If $\boldsymbol{A}$ is not full rank, then none of the zero blocks in (3.21) are empty. This implies that both scenarios above apply in this case.

Defining Relationships for the ED and the SVD

For the ED, if $\boldsymbol{A}$ is symmetric, we have

$$\boldsymbol{A} = \boldsymbol{Q}\boldsymbol{\Lambda}\boldsymbol{Q}^T \rightarrow \boldsymbol{AQ} = \boldsymbol{Q}\boldsymbol{\Lambda},$$

where $\boldsymbol{Q}$ is the matrix of eigenvectors and $\boldsymbol{\Lambda}$ is the diagonal matrix of eigenvalues. Writing this relation column-by-column, we have the familiar eigenvector/eigenvalue relationship:

$$\boldsymbol{A}\boldsymbol{q}_i = \lambda_i \boldsymbol{q}_i \quad i = 1, \ldots, n. \tag{3.23}$$

For the SVD, we have

$$\mathbf{A} = \mathbf{U}\boldsymbol{\Sigma}\mathbf{V}^T \rightarrow \mathbf{A}\mathbf{V} = \mathbf{U}\boldsymbol{\Sigma}$$

or

$$\boldsymbol{A}\boldsymbol{v}_i = \sigma_i \boldsymbol{u}_i \quad i = 1, \ldots, p, \tag{3.24}$$

where $p = \min(m, n)$. Also, since $\boldsymbol{A}^T = \boldsymbol{V}\boldsymbol{\Sigma}\boldsymbol{U}^T \rightarrow \boldsymbol{A}^T\boldsymbol{U} = \boldsymbol{V}\boldsymbol{\Sigma}$, we have

$$\boldsymbol{A}^T\boldsymbol{u}_i = \sigma_i \boldsymbol{v}_i \quad i = 1, \ldots, p. \tag{3.25}$$

Thus, by comparing (3.23), (3.24), and (3.25), we see the singular vectors and singular values obey a relation which is similar to that which defines the eigenvectors and eigenvalues. However, we note that in the SVD case, the fundamental relationship expresses left singular values in terms of right singular values and vice versa, whereas the eigenvectors are expressed in terms of themselves. These SVD relations are used in Chap. 8 to develop the *partial least squares* regression method.

Exercise Compare the ED and the SVD on a square symmetric matrix, when (i) $\boldsymbol{A}$ is positive definite and (ii) when $\boldsymbol{A}$ has some positive and some negative eigenvalues.

Ellipsoidal Interpretation of the SVD

The singular values of $\boldsymbol{A}$, where $\mathbf{A} \in \mathbb{R}^{m \times n}$ are the lengths of the semi-axes of the hyperellipsoid E given by

$$E = \{\mathbf{y} \mid \mathbf{y} = \mathbf{A}\mathbf{x}, ||\mathbf{x}||_2 = 1\}.$$

That is, E is the set of points mapped out as $\boldsymbol{x}$ takes on all possible values such that $||\mathbf{x}||_2 = 1$, as shown in Fig. 3.1. To appreciate this point, we look at the set of $\boldsymbol{y}$ corresponding to $\{\boldsymbol{x} \mid ||\boldsymbol{x}||_2 = 1\}$. We take

$$\begin{aligned} \boldsymbol{y} &= \boldsymbol{A}\boldsymbol{x} \\ &= \boldsymbol{U}\boldsymbol{\Sigma}\boldsymbol{V}^T\boldsymbol{x}. \end{aligned} \tag{3.26}$$

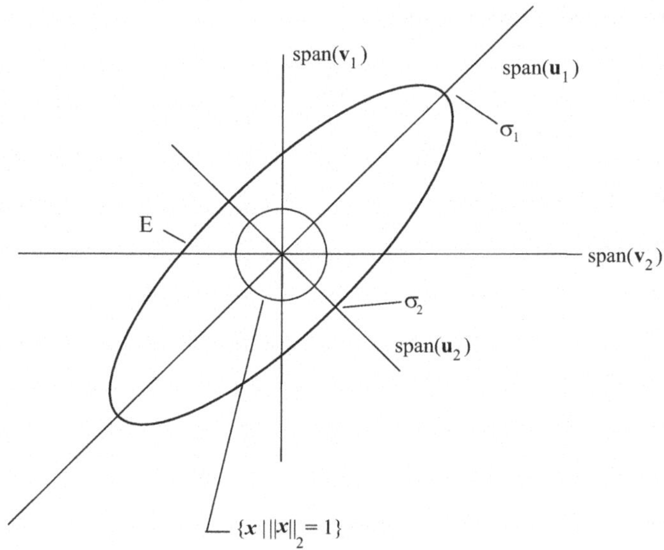

Fig. 3.1 The ellipsoidal interpretation of the SVD. The locus of points $E = \{y \mid y = Ax, ||x||_2 = 1\}$ defines an ellipse. The principal axes of the ellipse are aligned along the left singular vectors u_i, with lengths equal to the corresponding singular value

We change bases for both x and y. Define

$$c = U^T y$$
$$d = V^T x.$$

Then (3.26) becomes

$$c = \Sigma d, \text{ or } \Sigma^{-1} c = d, \tag{3.27}$$

where A is assumed full rank.

Due to the orthonormal transformation, we note that $||d||_2 = 1$ if $||x||_2 = 1$. Thus, our problem is transformed into observing the set $\{c\}$ corresponding to the set $\{d \mid ||d||_2 = 1\}$. The set $\{c\}$ can be determined by evaluating the squared 2-norms on each side of (3.27):

$$\sum_{i=1}^{p} \left(\frac{c_i}{\sigma_i}\right)^2 = \sum_{i=1}^{p} (d_i)^2 = 1.$$

We see that the set $\{c\}$ is indeed the canonical form of an ellipse in the basis U. Thus, the principal axes of the ellipse are aligned along the columns u_i of U, with lengths equal to the corresponding singular value σ_i. This interpretation of the SVD is useful later in our study of *condition numbers*.

3.2 Orthogonal Projections

Consider an orthonormal $\boldsymbol{Q} \in \mathbb{R}^{m \times m}$ which we partition as $[\boldsymbol{Q}_1 \ \boldsymbol{Q}_2]$, where $\boldsymbol{Q}_1 \in \mathbb{R}^{m \times n}$, $m > n$, so $\boldsymbol{Q}_1$ is tall. Let a subspace $\mathcal{S}$ be defined as $\mathcal{S} = R(\boldsymbol{Q}_1)$. We wish to formulate a matrix $\boldsymbol{P}$ so that the vector $\boldsymbol{P}\boldsymbol{y} \in \mathcal{S}$, where $\boldsymbol{y} \in \mathbb{R}^m$ is arbitrary, i.e. $\boldsymbol{P}$ *projects* $\boldsymbol{y}$ into $\mathcal{S}$.

We can express any vector $\boldsymbol{y} \in \mathbb{R}^m$ in the complete $m \times m$ orthonormal basis $\boldsymbol{Q}$ according to the usual rules for changing bases, as $\boldsymbol{y} = \boldsymbol{Q}\boldsymbol{c}$, where $\boldsymbol{c} = \boldsymbol{Q}^T\boldsymbol{y}$. Therefore

$$\begin{aligned} \boldsymbol{y} = \boldsymbol{Q}\boldsymbol{c} &= [\boldsymbol{Q}_1 \quad \boldsymbol{Q}_2] \begin{bmatrix} \boldsymbol{c}_1 \\ \boldsymbol{c}_2 \end{bmatrix} \begin{matrix} n \\ m-n \end{matrix} \\ &\triangleq \boldsymbol{y}_S + \boldsymbol{y}_C \end{aligned} \tag{3.28}$$

where

$$\begin{aligned} \boldsymbol{y}_S &= \boldsymbol{Q}_1\boldsymbol{c}_1 \in \mathcal{S} \text{ and} \\ \boldsymbol{y}_C &= \boldsymbol{Q}_2\boldsymbol{c}_2 \in \mathcal{S}_\perp. \end{aligned} \tag{3.29}$$

If we multiply both sides of (3.28) by $\boldsymbol{Q}_1^T$, we obtain

$$\begin{aligned} \boldsymbol{Q}_1^T\boldsymbol{y} &= \boldsymbol{Q}_1^T [\boldsymbol{Q}_1 \ \boldsymbol{Q}_2] \begin{bmatrix} \boldsymbol{c}_1 \\ \boldsymbol{c}_2 \end{bmatrix} \\ &= \boldsymbol{c}_1. \end{aligned}$$

Substituting this value for $\boldsymbol{c}_1$ into (3.29), we obtain

$$\boldsymbol{y}_S = \boldsymbol{Q}_1\boldsymbol{Q}_1^T\boldsymbol{y}.$$

It is therefore apparent that the matrix $\boldsymbol{P}$ which maps $\boldsymbol{y}$ into $\boldsymbol{y}_S$ is given by

$$\boldsymbol{P} = \boldsymbol{Q}_1\boldsymbol{Q}_1^T. \tag{3.30}$$

We can represent the vectors $\boldsymbol{y}$, $\boldsymbol{y}_S$, and $\boldsymbol{y}_C$ in a right-angled triangle configuration as shown in Fig. 3.2, where $\boldsymbol{y}$ is the hypotenuse. From this diagram, we can interpret the vector $\boldsymbol{y}_S$ as the result of dropping a perpendicular from $\boldsymbol{y}$ into $\mathcal{S}$. The principle of orthogonality [2] states that the shortest distance in the 2-norm sense from a point $\boldsymbol{y}$ into a subspace is obtained by dropping a perpendicular from $\boldsymbol{y}$ into the subspace. If follows therefore that $\boldsymbol{y}_C$ is the vector with the shortest distance from $\boldsymbol{y}$ into $\mathcal{S}$ in the 2-norm sense. This is the origin of the term *orthogonal* projection.

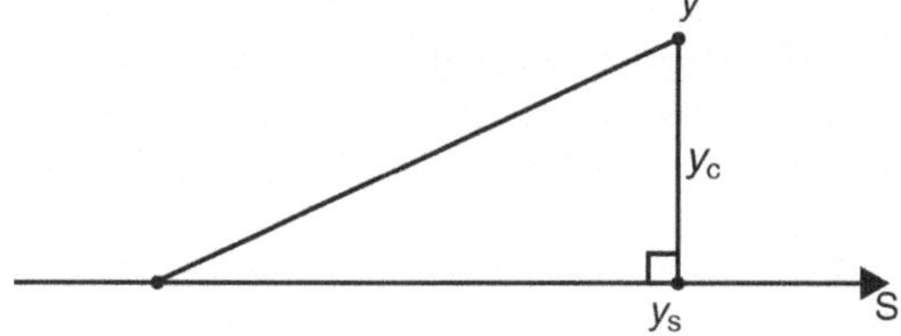

Fig. 3.2 The geometry of the orthogonal projection operation

We can extend the definition of a projector in the following way. Let $\boldsymbol{X}$ be such that $R(\boldsymbol{X}) = \mathcal{S}$, where $\boldsymbol{X} \in \mathbb{R}^{m \times n}$ is full rank, $m > n$. Now take the $n \times n$ matrix $\boldsymbol{R} = \boldsymbol{X}^T \boldsymbol{X}$. It is possible[2] to compute a full-rank $n \times n$ *square-root* matrix $\boldsymbol{C}$ such that $\boldsymbol{C}^T \boldsymbol{C} = \boldsymbol{R}$.

Theorem 3.2 *The matrix* $\boldsymbol{W} \in \mathbb{R}^{m \times n} = \boldsymbol{X}\boldsymbol{C}^{-1}$ *has orthonormal columns.*

Proof The matrix $\boldsymbol{W}$ has orthonormal columns if $\boldsymbol{W}^T \boldsymbol{W} = \boldsymbol{I}$. To show this, we take

$$\begin{aligned} \boldsymbol{W}^T \boldsymbol{W} &= \boldsymbol{C}^{-T} \boldsymbol{X}^T \boldsymbol{X} \boldsymbol{C}^{-1} \\ &= \boldsymbol{C}^{-T} \boldsymbol{R} \boldsymbol{C}^{-1} = \boldsymbol{C}^{-T} \boldsymbol{C}^T \boldsymbol{C} \boldsymbol{C}^{-1} \\ &= \boldsymbol{I}. \end{aligned}$$

□

We note that $R(\boldsymbol{W}) = \mathcal{S}$ because the columns of $\boldsymbol{W}$ are linear combinations of the columns of $\boldsymbol{X}$, which is full rank and in $\mathcal{S}$ by hypothesis. Since $\boldsymbol{W}$ has orthonormal columns, we can replace $\boldsymbol{Q}_1$ with $\boldsymbol{W}$ in (3.30), to obtain the alternate definition for a projector:

$$\begin{aligned} \boldsymbol{P} &= \boldsymbol{W}\boldsymbol{W}^T \\ &= \boldsymbol{X}\boldsymbol{C}^{-1}\boldsymbol{C}^{-T}\boldsymbol{X}^T \\ &= \boldsymbol{X}\boldsymbol{R}^{-1}\boldsymbol{X}^T \\ &= \boldsymbol{X}(\boldsymbol{X}^T\boldsymbol{X})^{-1}\boldsymbol{X}^T. \end{aligned} \tag{3.31}$$

Equation (3.31) is the usual definition of a projector defined using a tall, full-rank matrix in $\mathcal{S}$.

It is straightforward to show that $\boldsymbol{P}$ is the unique projector onto $\mathcal{S}$, regardless of the basis used to form it, provided it spans $\mathcal{S}$. To show this, we substitute $\boldsymbol{Y} = \boldsymbol{X}\boldsymbol{B}$ into (3.31), where $\boldsymbol{B}$ is an $n \times n$ invertible matrix, to obtain an alternative full-rank matrix to $\boldsymbol{X}$. It is apparent that if $\boldsymbol{X} \in \mathcal{S}$, then so is $\boldsymbol{Y}$. Straightforward manipulation

[2] The Cholesky decomposition, which we will study in Chap. 5, is one such example.

reveals that using $\boldsymbol{Y}$ instead of $\boldsymbol{X}$ in (3.31) yields exactly the same result for $\boldsymbol{P}$. Therefore $\boldsymbol{P}$ is unique, regardless of the basis used to form it.

It is shown [1] that if a matrix $\boldsymbol{P}$ satisfies these conditions:

1. $R(\mathbf{P}) = \mathcal{S}$
2. $\mathbf{P}^2 = \mathbf{P}$
3. $\mathbf{P}^T = \mathbf{P}$

then $\boldsymbol{P}$ is a projection matrix onto $\mathcal{S}$. The above are *sufficient* conditions for a projector. This means that while these conditions are enough to specify a projector, there may be other conditions which also specify a projector. But since we have now proved the projector is unique, these conditions are also *necessary*.

It is readily verified that both definitions for the projector, i.e. $\boldsymbol{P} = \boldsymbol{Q}_1 \boldsymbol{Q}_1^T$ and $\boldsymbol{P} = \boldsymbol{X}(\boldsymbol{X}^T \boldsymbol{X})^{-1} \boldsymbol{X}^T]$, satisfy the above properties.

Property 1 is necessary if the projector is to project into $\mathcal{S}$. Property 2 implies that if we take the projection operation $\boldsymbol{y}_S = \boldsymbol{P}\boldsymbol{y}$, and then project $\boldsymbol{y}_S$ into $\mathcal{S}$ again by evaluating $\boldsymbol{P}\boldsymbol{y}_S$, the result of the second operation should still be $\boldsymbol{y}_S$. That means $\boldsymbol{y}_S = \boldsymbol{P}\boldsymbol{y} = \boldsymbol{P}\boldsymbol{P}\boldsymbol{y}$. Hence we require $\boldsymbol{P}^2 = \boldsymbol{P}$. A matrix satisfying this property is called an *idempotent* matrix and is the fundamental property of a projector.

Property 3 above is included since we want the projector to apply to row vectors as well as column vectors, both of length m, i.e. a subspace can be defined using row vectors as well as column vectors. In this case, we want both quantities $(\boldsymbol{y}^T \boldsymbol{P})^T$ and $\boldsymbol{P}\boldsymbol{y}$ to be vectors in $\mathcal{S}$, and so therefore $\boldsymbol{P}$ should be symmetric.

3.2.1 The Orthogonal Complement Projector

Consider the vectors $\boldsymbol{y}$, $\boldsymbol{y}_S$, and $\boldsymbol{y}_C$ as defined previously. Then

$$\begin{aligned} \boldsymbol{y} &= \boldsymbol{y}_s + \boldsymbol{y}_c \\ &= \boldsymbol{P}\boldsymbol{y} + \boldsymbol{y}_c. \end{aligned}$$

Therefore we have

$$\begin{aligned} \boldsymbol{y} - \boldsymbol{P}\boldsymbol{y} &= \boldsymbol{y}_c \\ (\boldsymbol{I} - \boldsymbol{P})\,\boldsymbol{y} &= \boldsymbol{y}_c. \end{aligned}$$

It follows that if $\boldsymbol{P}$ is a projector onto $\mathcal{S}$, then the matrix $(\boldsymbol{I} - \boldsymbol{P})$ is a projector onto $S_\perp$. It is easily verified that this matrix satisfies the all required properties for this projector.

3.2.2 Orthogonal Projections and the SVD

Suppose we have a matrix $\mathbf{A} \in \mathbb{R}^{m \times n}$ of rank r. Then, using the SVD partitions of (3.14), we have these useful relations:

1. $\boldsymbol{V}_1 \boldsymbol{V}_1^T$ is the orthogonal projector onto $[\mathcal{N}(\boldsymbol{A})]^{\perp} = R(\boldsymbol{A}^T)$.
2. $\boldsymbol{V}_2 \boldsymbol{V}_2^T$ is the orthogonal projector onto $\mathcal{N}(\boldsymbol{A})$.
3. $\boldsymbol{U}_1 \boldsymbol{U}_1^T$ is the orthogonal projector onto $R(\boldsymbol{A})$.
4. $\boldsymbol{U}_2 \boldsymbol{U}_2^T$ is the orthogonal projector onto $[R(\boldsymbol{A})]_{\perp} = \mathcal{N}(\boldsymbol{A}^T)$.

The above may be justified since it is straightforward to show each of these matrices are indeed projectors, and each of the respective SVD partitions are an orthonormal basis for the subspace onto which they are projecting.

3.3 An Approximation Theorem [1]

This approximation theorem is the basis for our discussions on latent variable methods, as presented in Chap. 8. First, we realise that, because $\mathbf{\Sigma}$ is diagonal, the SVD of $\boldsymbol{A}$ provides a "sum of outer-products" representation as in (1.16)

$$\boldsymbol{A} = \boldsymbol{U}\boldsymbol{\Sigma}\boldsymbol{V}^T = \sum_{i=1}^{k} \sigma_i \boldsymbol{u}_i {\boldsymbol{v}_i}^T, \tag{3.32}$$

where $\text{rank}(\boldsymbol{A}) = k \leq p = \min(m, n)$. Given $\mathbf{A} \in \mathbb{R}^{m \times n}$, then what is the reduced-rank matrix $\boldsymbol{B} \in \mathbb{R}^{m \times n}$ with rank $r < k$ that is closest to $\boldsymbol{A}$ in Frobenius norm? What is the corresponding 2-norm distance? This question is answered in the following theorem.

Theorem 3.3 *Define a truncated (rank-reduced) version $\mathbf{A}_r$ of $\mathbf{A}$ in the following way:*

$$\boldsymbol{A}_r = \boldsymbol{U}_1 \boldsymbol{\Sigma}_r \boldsymbol{V}_1^T = \sum_{i=1}^{r} \sigma_i \boldsymbol{u}_i \boldsymbol{v}_i^T, \qquad r < k, \tag{3.33}$$

where $\boldsymbol{U}_1$ and $\boldsymbol{V}_1$ contain the first r columns of $\boldsymbol{U}$ and $\boldsymbol{V}$, respectively, and $\boldsymbol{\Sigma}_r$ is diagonal containing the first r singular values of $\boldsymbol{A}$. Then

$$\arg \min_{rank(B)=r} ||\boldsymbol{A} - \boldsymbol{B}||_F = \boldsymbol{A}_r. \tag{3.34}$$

Further, the respective minimum 2-norm distance $||\boldsymbol{A} - \boldsymbol{A}_r||_2 = \sigma_{r+1}$.

In words, this says the closest rank $r < k$ matrix $\boldsymbol{B}$ matrix to $\boldsymbol{A}$ in the Frobenius norm sense is given by $\boldsymbol{A}_r$. $\boldsymbol{A}_r$ is formed from $\boldsymbol{A}$ by truncating contributions in (3.32) associated with the smallest singular values. This idea may be seen as a generalisation of PCA, where we construct low-rank approximations to matrices instead of compressing vectors.

Proof The matrix $\boldsymbol{B}$ from (3.34) is formulated by orthogonally projecting $\boldsymbol{A}$ onto a subspace $\mathcal{S}$ of dimension r; i.e., we choose $\boldsymbol{B} = \boldsymbol{P}\boldsymbol{A}$, where $\boldsymbol{P}$ is a projector onto $\mathcal{S}$. Thus, any matrix formed from vectors residing in $\mathcal{S}$ can be at most rank r. For any column $\boldsymbol{a}_i$ of $\boldsymbol{A}$, from the Pythagorean theorem as depicted in Fig. 3.2 we have $||\boldsymbol{a}_i||_2^2 \geq ||\boldsymbol{P}\boldsymbol{a}_i||_2^2$. Because $||\boldsymbol{A}||_F^2 = \sum_i ||\boldsymbol{a}_i||_2^2$, we have $||\boldsymbol{A}||_F^2 = \sum_i ||\boldsymbol{a}_i||_2^2 \geq \sum_i ||\boldsymbol{P}\boldsymbol{a}||_2^2 = ||\boldsymbol{P}\boldsymbol{A}||_F^2$. Therefore $||\boldsymbol{A}||_F^2 \geq ||\boldsymbol{P}\boldsymbol{A}||_F^2$. Thus the norm $||\boldsymbol{A} - \boldsymbol{P}\boldsymbol{A}||_F^2$ can be minimized by choosing $\boldsymbol{P}$ so that $||\boldsymbol{P}\boldsymbol{A}||_F^2$ is maximum over all projectors of rank r.

This problem has already been addressed by Lemma 2.1 of Sect. 2.5 in the PCA context. By extending the results of this lemma, it is straightforward to show that an optimal orthonormal basis $\boldsymbol{Q}_1 \in \mathbb{R}^{m\times r}$ for $\mathcal{S}$ is given as $\boldsymbol{Q}_1 = \boldsymbol{U}_1$, as defined by (3.33). It follows that

$$\boldsymbol{P} = \boldsymbol{U}_1\boldsymbol{U}_1^T \tag{3.35}$$

and that the desired matrix $\boldsymbol{A}_r$ can be expressed as

$$\boldsymbol{A}_r = \boldsymbol{U}_1\boldsymbol{U}_1^T\boldsymbol{A} = \boldsymbol{P}\boldsymbol{A}, \tag{3.36}$$

which in the present context, has maximum Frobenius norm.

We now simplify (3.36). We partition $\boldsymbol{U}$ and $\boldsymbol{V}$, respectively, as $[\boldsymbol{U}_1 \;\; \boldsymbol{U}_2]$ and $[\boldsymbol{V}_1 \;\; \boldsymbol{V}_2]$, where the first component consists of the first r columns of the respective matrix. Then from (3.36), we may write

$$\begin{aligned}
\boldsymbol{A}_r = \boldsymbol{U}_1\boldsymbol{U}_1^T\boldsymbol{A} &= \boldsymbol{U}_1\boldsymbol{U}_1^T\boldsymbol{U}\boldsymbol{\Sigma}\boldsymbol{V}^T \\
&= \boldsymbol{U}_1\boldsymbol{U}_1^T[\boldsymbol{U}_1 \; \boldsymbol{U}_2]\boldsymbol{\Sigma}\boldsymbol{V}^T \\
&= \boldsymbol{U}_1\left[\boldsymbol{I} \; \boldsymbol{0}\right]\begin{bmatrix} \boldsymbol{\Sigma}_r & \boldsymbol{0} \\ \boldsymbol{0} & \boldsymbol{\Sigma}_2 \end{bmatrix}\boldsymbol{V}^T \\
&= \boldsymbol{U}_1\left[\boldsymbol{\Sigma}_r \; \boldsymbol{0}\right]\begin{bmatrix} \boldsymbol{V}_1^T \\ \boldsymbol{V}_2^T \end{bmatrix} \\
&= \boldsymbol{U}_1\boldsymbol{\Sigma}_r\boldsymbol{V}_1^T \\
&= \sum_{i=1}^{r} \sigma_i \boldsymbol{u}_i \boldsymbol{v}_i^T
\end{aligned} \tag{3.37}$$

where $\boldsymbol{\Sigma}_2$ in the third line contains all the remaining singular values $\sigma_{r+1}, \ldots, \sigma_n$ (some of which may be zero). Thus by comparing (3.37) to (3.33), it is apparent that $\boldsymbol{A}_r$ is rank r with maximum Frobenius norm and therefore $\boldsymbol{A}_r$ minimises the norm $||\boldsymbol{A} - \boldsymbol{B}||_F^2$ with respect to $\boldsymbol{B}$, which was to be shown. □

To determine the minimum value of this norm, we note that $\boldsymbol{U}^T \boldsymbol{A}_r \boldsymbol{V} = \text{diag}(\sigma_1 \ldots \sigma_r, 0 \ldots 0)$ from which it follows that rank$(\boldsymbol{A}_r)$ is indeed r and that

$$
\begin{aligned}
||\boldsymbol{A} - \boldsymbol{A}_k||_2 &= \left|\left|\boldsymbol{U}^T(\boldsymbol{A} - \boldsymbol{A}_r)\boldsymbol{V}\right|\right|_2 \\
&= ||\,\text{diag}(0 \ldots 0, \sigma_{k+1} \ldots \sigma_r, 0 \ldots 0)||_2 \\
&= \sigma_{k+1}
\end{aligned}
\tag{3.38}
$$

where the first line follows from the fact the 2-norm of a matrix is invariant to pre- and post-multiplication by an orthonormal matrix (properties of matrix p-norms, Chap. 2). □

It is interesting to note that we can obtain an identical result to (3.37) by operating on the *rows* of $\boldsymbol{A}$ instead of columns as we did above. By using a transposed version of the treatment of Lemma 2.1 of Sect. 2.5, we can show that the projector which maximizes $||\boldsymbol{A}^T \boldsymbol{P}||_F^2$ is given as

$$
\boldsymbol{P} = \boldsymbol{V}_1 \boldsymbol{V}_1^T, \tag{3.39}
$$

where $\boldsymbol{V}_1$ is defined in (3.33). This leaves us with an alternate definition for $\boldsymbol{A}_r$ given as

$$
\boldsymbol{A}_r = \boldsymbol{A}^T \boldsymbol{V}_1 \boldsymbol{V}_1^T. \tag{3.40}
$$

It is readily verified that by performing an analogous procedure as discussed previously to simplify the version of $\boldsymbol{A}_r$ given above, we end up with an identical result to that given by (3.37). Thus we may either project rows onto $R(\boldsymbol{V}_1)$ or columns onto $R(\boldsymbol{U}_1)$ and obtain exactly the same result.

Example As an example of this approximation theorem, we apply the above results to approximate the Lena image with another image of lower rank, which is closest to the true version. Here we treat the 512×512 image as a matrix and perform a truncated SVD, retaining only the first k terms. The SVD of this truncated image matrix is then

$$
\boldsymbol{I} = \sum_{i=1}^{k} \sigma_i \boldsymbol{u}_i \boldsymbol{v}_i^T, \tag{3.41}
$$

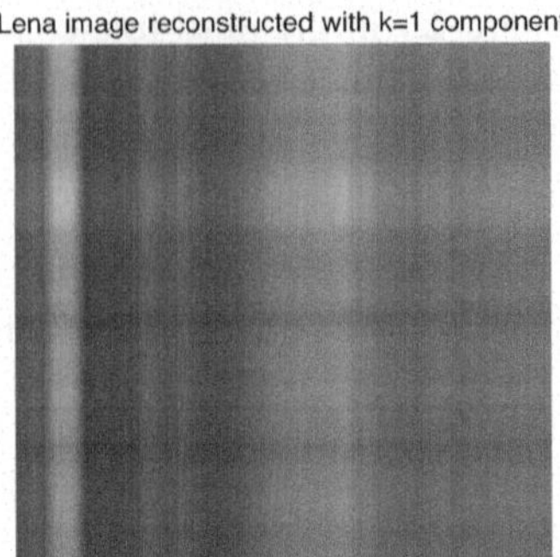

Fig. 3.3 Lena image reconstructed with only $k = 1$ component

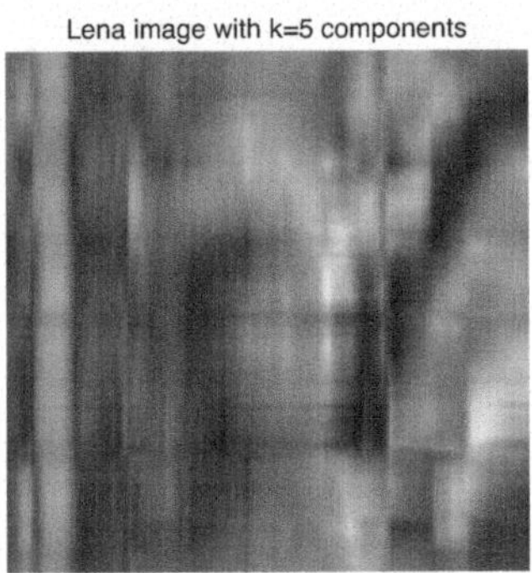

Fig. 3.4 Lena image reconstructed with $k = 5$ components

Fig. 3.5 Lena image reconstructed with $k = 10$ components

where $\boldsymbol{u}_i$ and $\boldsymbol{v}_i$ are the columns of $\boldsymbol{U}$ and $\boldsymbol{V}$ of the true image, respectively, and the σ_i are the singular values. The true image is represented by $k = 512$; however, here we reconstruct the image for a sequence of lower values of k, in order to obtain the closest rank(k) matrix to the original matrix. Figures 3.3, 3.4, 3.5, 3.6, 3.7 show the reconstructed images for a sequence of k-values ranging from 1 to 50.

Fig. 3.6 Lena image reconstructed with $k = 20$ components

Fig. 3.7 Lena image reconstructed with $k = 50$ components

Appendices

Alternative Proof of the SVD (from [1])

Consider two vectors $\boldsymbol{x}$ and $\boldsymbol{y}$ where $||\mathbf{x}||_2 = ||\mathbf{y}||_2 = 1$, such that $\mathbf{Ax} = \boldsymbol{\sigma}\mathbf{y}$, where $\sigma = ||\mathbf{A}||_2$. The fact that such vectors $\boldsymbol{x}$ and $\boldsymbol{y}$ can exist follows from the definition of the matrix 2-norm. We define orthonormal matrices $\boldsymbol{U}$ and $\boldsymbol{V}$ so that $\boldsymbol{x}$ and $\boldsymbol{y}$ form their first columns, as follows:

$$\mathbf{U} = [\mathbf{y}, \mathbf{U}_1]$$
$$\mathbf{V} = [\mathbf{x}, \mathbf{V}_1]$$

That is, $\mathbf{U}_1$ consists of a set of non-unique orthonormal columns which are mutually orthogonal to themselves and to $\boldsymbol{y}$; similarly for $\boldsymbol{V}_1$.

We then define a matrix $\boldsymbol{A}_1$ as

$$\begin{aligned} \mathbf{U}^T\mathbf{A}\mathbf{V} &= \mathbf{A}_1 \\ &= \begin{bmatrix} \mathbf{y}^T \\ \mathbf{U}_1{}^T \end{bmatrix} \mathbf{A}[\mathbf{x}, \mathbf{V}_1] \end{aligned} \tag{3.42}$$

The matrix $\boldsymbol{A}_1$ has the following structure:

$$\underbrace{\begin{bmatrix} \mathbf{y}^T \\ \mathbf{U}_1^T \end{bmatrix}}_{\text{orthonormal}} \mathbf{A} \underbrace{\begin{bmatrix} \mathbf{x} \ \mathbf{V}_1 \end{bmatrix}}_{\text{orthonormal}} = \begin{bmatrix} \mathbf{y}^T \\ \mathbf{U}_1{}^T \end{bmatrix} \begin{bmatrix} \sigma \mathbf{y} \ \mathbf{AV}_1 \end{bmatrix}$$

$$\begin{array}{c} \sigma y^T y \ \ y^T A V_1 \\ \downarrow \quad \downarrow \end{array}$$

$$= \begin{bmatrix} \sigma & \mathbf{w}^T \\ \mathbf{0} & \mathbf{B} \end{bmatrix} \begin{matrix} 1 \\ m-1 \end{matrix} \triangleq \mathbf{A}_1. \qquad (3.43)$$

$$\quad 1 \quad n-1$$

where $\mathbf{B} \triangleq \mathbf{U}_1^T \mathbf{AV}_1$ and $\mathbf{w}$ is a vector to be defined. The $\mathbf{0}$ in the (2,1) block above follows from the fact that $\mathbf{U}_1 \perp \mathbf{y}$, because $\boldsymbol{U}$ is orthonormal.

Now, we post-multiply both sides of (3.43) by the vector $\begin{bmatrix} \sigma \\ \mathbf{w} \end{bmatrix}$ and take 2-norms:

$$\left\| \mathbf{A}_1 \begin{bmatrix} \sigma \\ \mathbf{w} \end{bmatrix} \right\|_2^2 = \left\| \begin{bmatrix} \sigma & \mathbf{w}^T \\ \mathbf{0} & \mathbf{B} \end{bmatrix} \begin{bmatrix} \sigma \\ \mathbf{w} \end{bmatrix} \right\|_2^2 \geq (\sigma^2 + \mathbf{w}^T \mathbf{w})^2. \qquad (3.44)$$

This follows because the term on the extreme right is only the first element of the vector product of the middle term. But, as we have seen, matrix p-norms obey the following property:

$$||\mathbf{Ax}||_2 \leq ||\mathbf{A}||_2 \, ||\mathbf{x}||_2 \,. \qquad (3.45)$$

Therefore using (3.44) and (3.45), we have

$$||\mathbf{A}_1||_2^2 \left\| \begin{bmatrix} \sigma \\ \mathbf{w} \end{bmatrix} \right\|_2^2 \geq \left\| \mathbf{A}_1 \begin{bmatrix} \sigma \\ \mathbf{w} \end{bmatrix} \right\|_2^2 \geq (\sigma^2 + \mathbf{w}^T \mathbf{w})^2. \qquad (3.46)$$

Note that $\left\| \begin{bmatrix} \sigma \\ \mathbf{w} \end{bmatrix} \right\|_2^2 = \sigma^2 + \mathbf{w}^T \mathbf{w}$. Dividing (3.46) by this quantity, we obtain

$$||\mathbf{A}_1||_2^2 \geq \sigma^2 + \mathbf{w}^T \mathbf{w}. \qquad (3.47)$$

But, we defined $\sigma = ||\mathbf{A}||_2$. Therefore, the following must hold:

$$\sigma = ||\mathbf{A}||_2 = \left\| \mathbf{U}^T \mathbf{AV} \right\|_2 = ||\mathbf{A}_1||_2 \qquad (3.48)$$

where the equality on the right follows because the matrix 2-norm is invariant to matrix pre- and post-multiplication by an orthonormal matrix. By comparing (3.47) and (3.48), we have the result $\mathbf{w} = \mathbf{0}$.

Substituting this result back into (3.43), we now have

$$\mathbf{A}_1 = \begin{bmatrix} \sigma & \mathbf{0} \\ \mathbf{0} & \mathbf{B} \end{bmatrix}. \tag{3.49}$$

The whole process repeats using only the component $\mathbf{B}$, until $\mathbf{A}_n$ becomes diagonal. □

Problems

1. Prove that $||\boldsymbol{A}||_2 = \sigma_1$.
2. Prove that the trace of a projector is equal to its rank.
3. Prove that the SVD and the ED on a positive definite symmetric matrix are equivalent.
4. If $\boldsymbol{X}$ is rank deficient, prove that $\boldsymbol{X}^T\boldsymbol{X}$ is also rank deficient.
5. We are given a complete orthonormal matrix $\boldsymbol{Q} \in \mathbb{R}^{n \times n}$. Find $\boldsymbol{X}$ such that $\boldsymbol{Q}\boldsymbol{X}\boldsymbol{Q}^T$ is a projector onto span$[\boldsymbol{q}_1, \ldots, \boldsymbol{q}_r]$, where $r < n$.
6. Using Matlab® or other suitable language, construct a 6×4 matrix $\boldsymbol{A}$ with rank = 2.

   (a) Construct the projectors for each of the four fundamental subspaces of $\boldsymbol{A}$.
   (b) Explain how to test whether a vector $\boldsymbol{y}$ is in a specified subspace $\mathcal{S}$.
   (c) Construct a random vector $\boldsymbol{x}$ and project it onto the four subspaces, using the respective projectors from part a, to yield the result $\boldsymbol{y}$.
   (d) Verify that each $\boldsymbol{y}$ is indeed in the subspace corresponding to the respective $\boldsymbol{P}$.

7. The following questions have to do with the discussion on the use of the SVD in solving systems of equations such as $\boldsymbol{A}\boldsymbol{x} = \boldsymbol{b}$.

   (a) Consider the case where the matrix $\boldsymbol{A}$ is tall and full rank. What is the solution $\boldsymbol{x}$ such that $||\boldsymbol{A}\boldsymbol{x} - \boldsymbol{b}||_2^2$ is mimimised? Give an expression for this error. Consider the range of possibilities for $\boldsymbol{b}$ in relation to $R(\boldsymbol{A})$.
   (b) Now consider the case where $\boldsymbol{A}$ is short and full rank. What is the solution $\boldsymbol{x}$ such that $||\boldsymbol{x}||_2^2$ is minimum?

8. Consider a tall matrix $\boldsymbol{A} \in \mathbb{R}^{m \times n}$ with column rank $r < n - 1$, whose SVD is given by $\boldsymbol{A} = \boldsymbol{U}\boldsymbol{\Sigma}\boldsymbol{V}^T$.

   (a) Explain how the SVD can be changed with $\boldsymbol{A}$ remaining constant. Thus explain under what conditions the SVD is not unique.
   (b) Repeat the above for the case where $\boldsymbol{A}$ is short and $r < m - 1$.

9. Given the (tall) matrix $\boldsymbol{A}$ above, explain how to determine a vector $\boldsymbol{x}$ so that $\boldsymbol{x} \in R(\boldsymbol{A})_{\perp}$. Illustrate with an example.
10. We are given an $m \times k$ matrix $\boldsymbol{A}$ and a $k \times n$ matrix $\boldsymbol{B}$, where $m > k > n$, i.e. $\boldsymbol{A}$ is tall and $\boldsymbol{B}$ is short. We form $\boldsymbol{C} = \boldsymbol{AB}$.

    (a) Prove that the row spaces of $\boldsymbol{B}$ and $\boldsymbol{C}$ are equivalent. *Hint:* Consider the projector for each subspace.
    (b) Prove that the column spaces of $\boldsymbol{A}$ and $\boldsymbol{C}$ are equivalent.
    (c) Find a basis for $R(\boldsymbol{B}^T)$ in terms of the SVD of $\boldsymbol{C}$.
    (d) Using matlab®, provide an example showing that $R(\boldsymbol{C}^T) \equiv R(\boldsymbol{B}^T)$. Also demonstrate that these are both equivalent to the range of the basis determined in step 10 above.
    (e) Do the results change if $\boldsymbol{A}$, $\boldsymbol{B}$ or both become rank deficient?

11. We are given a very tall matrix $\boldsymbol{A}$ where $m \gg n$ and where m is very large. We are interested in determining only the matrix $\boldsymbol{V}$ of the SVD. Solving this directly by computation of the SVD is expensive, due to the fact that $\boldsymbol{U}$ is very large and therefore requires a lot of time and memory to calculate. Suggest a more practical approach for determining $\boldsymbol{V}$. Also show how to find the first r columns of $\boldsymbol{U}$, given $\boldsymbol{V}$. Using similar ideas, extend your method to the case where $\boldsymbol{A}$ is large and very short, and we are interested only in $\boldsymbol{U}$. Also show how to determine the first r columns of $\boldsymbol{V}$. *Hint:* Consider the matrices $\boldsymbol{A}^T\boldsymbol{A}$ and $\boldsymbol{A}\boldsymbol{A}^T$, respectively.
12. We are given a coloured noise process generated by filtering white noise. From this process, we generate a matrix $\boldsymbol{X} \in \mathbb{R}^{m\times n}$, where $m > n$ in the manner described in Sect. 2.3. The covariance matrix $\hat{\boldsymbol{R}}_1$ may be written as $\hat{\boldsymbol{R}}_1 = \frac{1}{m}\boldsymbol{X}^T\boldsymbol{X}$. Compare the eigenvalues and eigenvectors of $\hat{\boldsymbol{R}}_1$ with those of $\hat{\boldsymbol{R}}_2 \triangleq \frac{1}{n}\boldsymbol{X}\boldsymbol{X}^T$. What happens to the eigenvector waveforms as both m and n become large?
13. Consider a projector $\boldsymbol{P}$ onto some subspace $\mathcal{S}$. Prove that for any $\boldsymbol{x}$ of suitable length, $||\boldsymbol{x} - \boldsymbol{P}\boldsymbol{x}||_2^2 \leq ||\boldsymbol{x}||_2^2$. Also prove for the Frobenius norm case instead of the 2-norm case.

References

1. G.H. Golub, C.F. Van Loan, *Matrix Computations*, 3rd edn. (The Johns Hopkins University, Baltimore, MD, 1996)
2. A. Papoulis, *Random Variables and Stochastic Processes* (McGraw-Hill, New York, 1994)

Chapter 4
The Quadratic Form

4.1 The Quadratic Form and Positive Definiteness

Consider a square matrix $\boldsymbol{A} \in \mathbb{R}^{n \times n}$. Then the formulation

$$\boldsymbol{x}^T \boldsymbol{A}\boldsymbol{x}$$

is referred to as a *quadratic form* in $\boldsymbol{A}$, where $\boldsymbol{x} \in \mathbb{R}^n$. The matrix $\boldsymbol{A}$ is *positive definite* iff for any $\boldsymbol{0} \neq \boldsymbol{x} \in \mathbb{R}^n$,

$$\boldsymbol{x}^T \boldsymbol{A}\boldsymbol{x} > 0. \tag{4.1}$$

The matrix $\boldsymbol{A}$ is *positive semi-definite* if and only if, for any $\boldsymbol{x} \neq \boldsymbol{0}$, we have

$$\boldsymbol{x}^T \boldsymbol{A}\boldsymbol{x} \geq 0, \tag{4.2}$$

which, as we see later, includes the possibility that $\boldsymbol{A}$ is rank deficient.

It may be verified by direct multiplication that the quadratic form can also be expressed as

$$\boldsymbol{x}^T \boldsymbol{A}\boldsymbol{x} = \sum_{i=1}^{n} \sum_{j=1}^{n} a_{ij} x_i x_j. \tag{4.3}$$

It is only the symmetric part of $\boldsymbol{A}$ which is relevant in a quadratic form expression. To see this, we define the symmetric part $\boldsymbol{T}$ of $\boldsymbol{A}$ as $\boldsymbol{T} \stackrel{\Delta}{=} \frac{1}{2}[\boldsymbol{A} + \boldsymbol{A}^T]$, and the asymmetric part $\boldsymbol{S}$ of $\boldsymbol{A}$ as $\boldsymbol{S} \stackrel{\Delta}{=} \frac{1}{2}[\boldsymbol{A} - \boldsymbol{A}^T]$. Then we have the desired properties that $\boldsymbol{T}^T = \boldsymbol{T}$, $\boldsymbol{S} = -\boldsymbol{S}^T$, and $\boldsymbol{A} = \boldsymbol{T} + \boldsymbol{S}$. Note the diagonal elements of $\boldsymbol{S}$ must be zero.

J. Reilly, *Fundamentals of Linear Algebra for Signal Processing*,
https://doi.org/10.1007/978-3-031-68915-4_4

We can express (4.3) as

$$\boldsymbol{x}^T \boldsymbol{A}\boldsymbol{x} = \sum_{i=1}^{n}\sum_{j=1}^{n} t_{ij}x_i x_j + \sum_{i=1}^{n}\sum_{j=1}^{n} s_{ij}x_i x_j. \tag{4.4}$$

We now consider only the second term on the right in (4.4). Since $\boldsymbol{S} = -\boldsymbol{S}^T$, for any index pair $(i, j), i \neq j$, the quantity $s_{ij}x_i x_j$ exactly cancels the term $s_{ji}x_j x_i$ corresponding to the (j, i)th pair, because $s_{ij} = -s_{ji}$. Further, since the diagonal terms of $\boldsymbol{S}$ must be zero, this second term is always zero for any $\boldsymbol{x}$. Thus, when considering quadratic forms, it suffices to consider only the symmetric part $\boldsymbol{T}$ of the matrix; i.e. we have the result $\boldsymbol{x}^T\boldsymbol{A}\boldsymbol{x} = \boldsymbol{x}^T\boldsymbol{T}\boldsymbol{x}$. This is useful, since now we can substitute an arbitrary square matrix $\boldsymbol{A}$ in a quadratic form by its symmetric part $\boldsymbol{T}$, which is square and symmetric, meaning we can represent it by its eigendecomposition. As we will see, this substitution offers new interpretations and insight on the quadratic form.

This result generalises to the case where $\boldsymbol{A}$ is complex. It is left as an exercise to show that (i) $\boldsymbol{x}^H\boldsymbol{A}\boldsymbol{x} = \boldsymbol{x}^H\boldsymbol{T}\boldsymbol{x}$, where $\boldsymbol{T} \triangleq \frac{1}{2}[\boldsymbol{A} + \boldsymbol{A}^H]$, and (ii) that the quantity $\boldsymbol{x}^H\boldsymbol{A}\boldsymbol{x}$ is pure real.

Quadratic forms on positive definite matrices are used very frequently in least squares and adaptive filtering applications. Also as we see later, quadratic forms play a fundamental role in defining the multivariate Gaussian probability density function.

Theorem 4.1 *A matrix* $\boldsymbol{A}$ *is positive definite if and only if all eigenvalues of the symmetric part of* $\boldsymbol{A}$ *are positive.*

Proof Let the eigendecomposition on the symmetric part $\boldsymbol{T}$ of $\boldsymbol{A}$ be represented as $\boldsymbol{T} = \boldsymbol{V}\boldsymbol{\Lambda}\boldsymbol{V}^T$. Since only the symmetric part of $\boldsymbol{A}$ is relevant, the quadratic form on $\boldsymbol{A}$ may be expressed as $\boldsymbol{x}^T\boldsymbol{A}\boldsymbol{x} = \boldsymbol{x}^T\boldsymbol{T}\boldsymbol{x} = \boldsymbol{x}^T\boldsymbol{V}\boldsymbol{\Lambda}\boldsymbol{V}^T\boldsymbol{x}$. Let us define the variable $\boldsymbol{z}$ as $\boldsymbol{z} \triangleq \boldsymbol{V}^T\boldsymbol{x}$. As we have seen previously in Chap. 2, $\boldsymbol{z}$ is a rotation of $\boldsymbol{x}$ due to the fact $\boldsymbol{V}$ is orthonormal. Thus we have

$$\begin{aligned} \boldsymbol{x}^T\boldsymbol{A}\boldsymbol{x} &= \boldsymbol{z}^T\boldsymbol{\Lambda}\boldsymbol{z} \\ &= \sum_{i=1}^{n} z_i^2\lambda_i. \end{aligned} \tag{4.5}$$

This expression is greater than zero for arbitrary $\boldsymbol{x} \neq \boldsymbol{0}$ iff $\lambda_i > 0, i = 1, \ldots, n$. □

We also see from (4.5) that if the equality in the quadratic form is satisfied (i.e. $\boldsymbol{x}^T\boldsymbol{A}\boldsymbol{x} = 0$ for some $\boldsymbol{x} \neq \boldsymbol{0}$ and corresponding $\boldsymbol{z}$), then $\boldsymbol{A}$ has a non-empty nullspace, and at least one eigenvalue of $\boldsymbol{T}$ must be zero. Hence, if $\boldsymbol{A}$ is symmetric, then $\boldsymbol{A}$ being positive *semi-definite* implies that at least one eigenvalue of $\boldsymbol{A}$ must also be zero, which means that $\boldsymbol{A}$ is rank deficient.

4.2 The Locus of Points $\{z|z^T\Lambda z = 1\}$

Let $\boldsymbol{A}$ be positive definite. Here we investigate the set of points satisfying $\{z|z^T\Lambda z = 1\}$, where z is defined as above. We can write

$$\begin{aligned} z^T\Lambda z &= \sum_{i=1}^{n} z_i^2\lambda_i \\ &= \sum_{i=1}^{n} \frac{z_i^2}{\frac{1}{\lambda_i}}. \end{aligned} \tag{4.6}$$

When this quantity is equated to 1, we have the canonical form of an ellipse in the variables z_i, with principal axis lengths $\sqrt{\frac{1}{\lambda_i}}$. The principal axes are aligned along the corresponding elementary basis directions $\boldsymbol{e}_1, \boldsymbol{e}_2, \ldots, \boldsymbol{e}_n$.

Since $z = \boldsymbol{V}^T\boldsymbol{x}$ where $\boldsymbol{V}$ is orthonormal, the locus of points $\{\boldsymbol{x}|\boldsymbol{x}^T\boldsymbol{A}\boldsymbol{x} = 1\}$ is a rotated version of the ellipse in (4.6). This ellipse has the same principal axes lengths as before, but the ith principal axis now lines up along the ith eigenvector $\boldsymbol{v}_i$ of $\boldsymbol{A}$.

The locus of points $\{\boldsymbol{x}|\boldsymbol{x}^T\boldsymbol{A}\boldsymbol{x} = k,\ k > 0\}$ defines a scaled version of the ellipse above. In this case, the ith principal axis length is given by the quantity $\sqrt{\frac{k}{\lambda_1}}$.

Example A three-dimensional plot of $y = \boldsymbol{x}^T\boldsymbol{A}\boldsymbol{x}$ is shown plotted in Fig. 4.1 for $\boldsymbol{A}$ given by

$$\boldsymbol{A} = \begin{bmatrix} 2 & 1 \\ 1 & 2 \end{bmatrix}. \tag{4.7}$$

Note that this curve is elliptical in cross section in a plane $y = k$ as discussed above. A quick calculation verifies the eigenvalues of $\boldsymbol{A}$ are 3, 1 with corresponding eigenvectors $[1, 1]^T$ and $[1, -1]^T$. For $y = k = 1$, the lengths of the principal axes of the ellipse are then $1/\sqrt{3}$ and 1. It can be verified from the figure the principal axis lengths are indeed the lengths indicated and are lined up along the directions of the eigenvectors as required.

In practice, the quadratic form is often expressed using a matrix inverse (e.g. $\boldsymbol{\Sigma}^{-1}$, where $\boldsymbol{\Sigma}$ is a positive definite matrix), rather than in terms of the matrix itself. This is particularly true in the case of a multivariate Gaussian probability density function, as we see in the sequel. In this case, the quadratic form becomes $\boldsymbol{x}^T\boldsymbol{\Sigma}^{-1}\boldsymbol{x}$. Recall that the eigenvalues of $\boldsymbol{\Sigma}^{-1}$ are the reciprocals of those of $\boldsymbol{\Sigma}$. We express the eigendecomposition of $\boldsymbol{\Sigma}^{-1}$ as $\boldsymbol{V}\boldsymbol{\Lambda}^{-1}\boldsymbol{V}^T$ and (4.6) becomes

$$z^T\Lambda^{-1}z = \sum_{i=1}^{n} \frac{z_i^2}{\lambda_i},$$

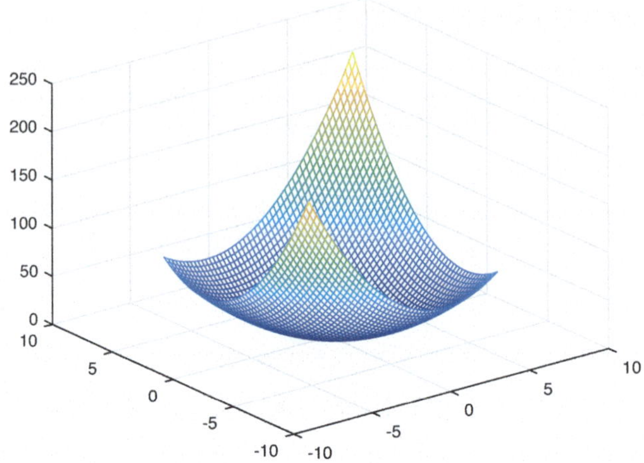

Fig. 4.1 Three-dimensional plot of the quadratic form $\boldsymbol{y} = \boldsymbol{x}^T \boldsymbol{A}\boldsymbol{x}$ with $\boldsymbol{A}$ given by (4.7)

where $z = \boldsymbol{V}^T\boldsymbol{x}$. When this expression is equated to 1, we again have an ellipse in the variables z_i. However, in this case, the semi-axis lengths are given as $\sqrt{\lambda_i}$ rather than $\sqrt{\frac{1}{\lambda_i}}$ as before. When the quadratic form is expressed in terms of $\boldsymbol{x}$ and $\boldsymbol{\Sigma}^{-1}$ rather than z and $\boldsymbol{\Lambda}^{-1}$, the ellipse becomes aligned along the eigenvectors rather than the coordinate axes.

In the case when the mean $\boldsymbol{\mu}$ of $\boldsymbol{x}$ is non-zero, the quadratic form in $\boldsymbol{A}$ is often replaced with the expression

$$(\boldsymbol{x} - \boldsymbol{\mu})^T \boldsymbol{A}(\boldsymbol{x} - \boldsymbol{\mu}).$$

This has the effect of translating the centre of the ellipse described by $\boldsymbol{x}^T\boldsymbol{A}\boldsymbol{x} = k$ from the point $\boldsymbol{\mu}$ to the origin.

Positive definiteness of $\boldsymbol{A}$ in the quadratic form $\boldsymbol{x}^T\boldsymbol{A}\boldsymbol{x}$ is the matrix analog to the scalar a being positive in the scalar expression ax^2. The scalar equation $y = ax^2$ is a parabola which faces upwards if a is positive. Likewise, the equation $y = \boldsymbol{x}^T\boldsymbol{A}\boldsymbol{x} = \sum_{i=1}^n z_i^2\lambda_i$, where $z = \boldsymbol{V}^T x$ as before, is a multidimensional parabola. The parabola faces upwards in all directions if $\boldsymbol{A}$ is positive definite. If $\boldsymbol{A}$ is not positive (semi) definite, then some eigenvalues are negative, and the curve faces down in the orientations corresponding to the negative eigenvalues and up in those corresponding to the positive eigenvalues.

Theorem 4.2 *A (square) symmetric matrix $\boldsymbol{A} \in \mathbb{R}^{n\times n}$ can be decomposed into the form $\boldsymbol{A} = \boldsymbol{B}\boldsymbol{B}^T$ if and only if $\boldsymbol{A}$ is positive definite or positive semi-definite.*

The matrix $\boldsymbol{B}$ is referred to as a *square root factor* of $\boldsymbol{A}$.

Proof (Necessary condition; i.e. if $\boldsymbol{A} = \boldsymbol{B}\boldsymbol{B}^T$, then $\boldsymbol{A}$ is positive definite.) Let us define $\boldsymbol{z}$ as $\boldsymbol{B}^T\boldsymbol{x}$. Then for any $\boldsymbol{x}$,

$$\begin{aligned} \boldsymbol{x}^T A\boldsymbol{x} &= \boldsymbol{x}^T \boldsymbol{B}\boldsymbol{B}^T \boldsymbol{x} \\ &= \boldsymbol{z}^T \boldsymbol{z} \\ &\geq 0. \end{aligned} \tag{4.8}$$

Conversely (sufficient condition): Since $\boldsymbol{A}$ is square and symmetric, we can write $\boldsymbol{A} = \boldsymbol{V}\boldsymbol{\Lambda}\boldsymbol{V}^T$. Since $\boldsymbol{A}$ is positive definite by hypothesis, all the eigenvalues are positive, and so we can write $\boldsymbol{A} = (\boldsymbol{V}\boldsymbol{\Lambda}^{1/2})(\boldsymbol{V}\boldsymbol{\Lambda}^{1/2})^T$. Therefore $\boldsymbol{B} = \boldsymbol{V}\boldsymbol{\Lambda}^{1/2}$ is a square root factor. However, we can also define $\boldsymbol{B}$ as $\boldsymbol{B} = \boldsymbol{V}\boldsymbol{\Lambda}^{1/2}\boldsymbol{Q}^T$ where $\boldsymbol{Q} \in \mathbb{R}^{m\times n}, m \geq n$ has orthonormal columns, such that $\boldsymbol{Q}^T\boldsymbol{Q} = \boldsymbol{I}$. Then $\boldsymbol{A} = \boldsymbol{V}\boldsymbol{\Lambda}^{1/2}\boldsymbol{Q}^T\boldsymbol{Q}\boldsymbol{\Lambda}^{1/2}\boldsymbol{V}^T = \boldsymbol{B}\boldsymbol{B}^T$. It is therefore clear that $\boldsymbol{B}$ is not unique, and it follows that the square root factor $\boldsymbol{B}$ of $\boldsymbol{A}$ is unique only up to an orthonormal ambiguity. □

The fact that $\boldsymbol{A}$ can be decomposed into two symmetric factors in this way is the fundamental idea behind the Cholesky factorisation, which is a major topic of the following chapter.

4.3 The Gaussian Multivariate Probability Density Function

Here, we very briefly introduce this topic so we can use this material for an example of the application of the Cholesky decomposition later in this course, and also in least square analysis to follow shortly. This topic is a good application of quadratic forms. More detail is provided in several books, e.g. [2, 3]. First we consider the univariate case of the Gaussian probability distribution function (*pdf*). The *pdf* $p(x)$ of a univariate Gaussian-distributed random variable x with mean μ and variance σ^2 is given as

$$p(x) = \frac{1}{\sqrt{2\pi\sigma^2}} \exp\left[-\frac{1}{2\sigma^2}(x-\mu)^2\right]. \tag{4.9}$$

This is the familiar bell-shaped curve. It is completely specified by two parameters—the mean μ which determines the position of the peak and the variance σ^2 which determines the width or spread of the curve.

Now consider the more interesting multidimensional case. Consider a Gaussian-distributed random vector $\boldsymbol{x} \in \mathbb{R}^n$ with mean $\boldsymbol{\mu}$ and covariance $\boldsymbol{\Sigma}$. In this case we can say $\boldsymbol{x} \sim N(\boldsymbol{\mu}, \boldsymbol{\Sigma})$, where $N(\cdot, \cdot)$ denotes a *normal* distribution, which is alternative designation for Gaussian distribution. The symbol $\sim$ indicates "is distributed as".

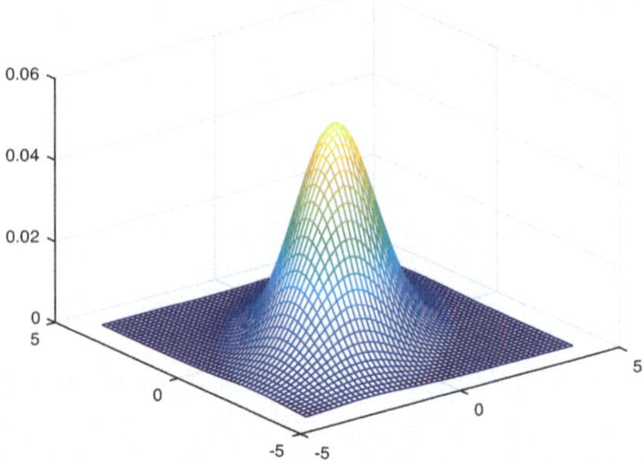

Fig. 4.2 A Gaussian probability density function with zero mean and covariance matrix [2 1; 1 2]

The multivariate *pdf* describing $\boldsymbol{x}$ is

$$p(\boldsymbol{x}) = (2\pi)^{-\frac{n}{2}} \mid \boldsymbol{\Sigma} \mid^{-\frac{1}{2}} \exp\left[-\frac{1}{2}(\boldsymbol{x}-\boldsymbol{\mu})^T\boldsymbol{\Sigma}^{-1}(\boldsymbol{x}-\boldsymbol{\mu})\right]. \tag{4.10}$$

We can see that the multivariate case collapses to the univariate case when the number of variables reduces to one. A plot of $p(\boldsymbol{x})$ vs. $\boldsymbol{x}$ for the case $n = 2$ is shown in Fig. 4.2, for a mean $\boldsymbol{\mu} = \boldsymbol{0}$ and covariance matrix $\boldsymbol{\Sigma} = \boldsymbol{\Sigma}_1$ defined as

$$\boldsymbol{\Sigma}_1 = \begin{bmatrix} 2 & 1 \\ 1 & 2 \end{bmatrix}. \tag{4.11}$$

The peak of the distribution occurs at the vector mean $\boldsymbol{\mu}$. Because the exponent in (4.10) is a quadratic form, the set of points satisfied by the equation $\left[\frac{1}{2}(\boldsymbol{x}-\boldsymbol{\mu})\boldsymbol{\Sigma}^{-1}(\boldsymbol{x}-\boldsymbol{\mu})\right] = k$ where k is a constant is an ellipse. Therefore this ellipse defines a contour of equal probability density. The interior of this ellipse defines a region into which an observation governed by this distribution will fall with a specified probability α, which is dependent on the value k. This probability level α is given as

$$\alpha = \int_{\mathcal{R}} (2\pi)^{-\frac{n}{2}} \mid \boldsymbol{\Sigma} \mid^{-\frac{1}{2}} \exp\left[-\frac{1}{2}(\boldsymbol{x}-\boldsymbol{\mu})\boldsymbol{\Sigma}^{-1}(\boldsymbol{x}-\boldsymbol{\mu})\right] d\boldsymbol{x}, \tag{4.12}$$

where $\mathcal{R}$ is the interior of the ellipse. Stated another way, an *ellipse* is the region in which any observation governed by the probability distribution (4.10) will fall with a specified probability level α. As k increases, the ellipse gets larger, and α increases. These ellipses are referred to as *joint confidence regions* (JCRs) at probability level α.

The covariance matrix $\boldsymbol{\Sigma}$ controls the shape of the ellipse. Because the quadratic form in this case is in $\boldsymbol{\Sigma}^{-1}$, the length of the ith principal axis is $\sqrt{2k\lambda_i}$, as previously explained. Therefore as the eigenvalues of $\boldsymbol{\Sigma}$ increase, the size of the JCRs increase (i.e. the variances of the distribution increase) for a given value of k.

We now investigate the relationship of the covariances between the variables (i.e. off-diagonal terms of the covariance matrix) and the shape of the Gaussian pdf. We have seen previously in Chap. 2 that covariance is a measure of dependence between individual random variables. We have also seen that as the off-diagonal covariance terms become larger, there is a larger disparity between the largest and smallest eigenvalues of the covariance matrix. Thus, as the covariances increase, the eigenvalues and thus the lengths of the semi-axes of the JCRs become more disparate; i.e. the JCRs of the Gaussian pdf become more elongated. This behaviour is illustrated in Fig. 4.3, which shows a multivariate Gaussian pdf for a mean $\boldsymbol{\mu} = \mathbf{0}$ and for a covariance matrix $\boldsymbol{\Sigma} = \boldsymbol{\Sigma}_2$ given as

$$\boldsymbol{\Sigma}_2 = \begin{bmatrix} 2 & 1.9 \\ 1.9 & 2 \end{bmatrix}. \tag{4.13}$$

Note that in this case, the covariance elements of $\boldsymbol{\Sigma}_2$ have increased substantially relative to those of $\boldsymbol{\Sigma}_1$ in Fig. 4.2, although the variances themselves (the main diagonal elements) have remained unchanged. By examining the pdf of Fig. 4.3, we see that the joint confidence ellipsoid has become elongated, as expected. (For $\boldsymbol{\Sigma}_1$ of Fig. 4.2 the eigenvalues are (3, 1), and for $\boldsymbol{\Sigma}_2$ of Fig. 4.3, the eigenvalues are $(3.9, 0.1)$). This elongation results in the conditional probability $p(x_1|x_2)$ for Fig. 4.3 having a much smaller variance (spread) than that for Fig. 4.2; i.e. when the covariances are larger, knowledge of one variable tells us more about the other. This is how the probability density function incorporates the information contained in the covariances between the variables. With regard to Gaussian probability density functions, the following concepts, (1) larger correlations between the variables, (2) larger disparity between the eigenvalues, (3) elongated joint confidence regions, and

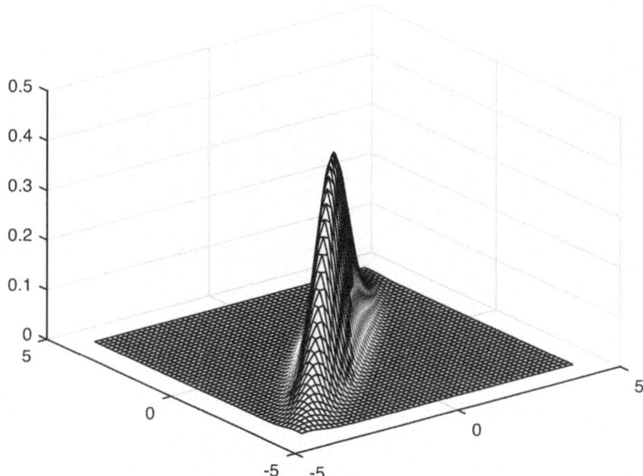

Fig. 4.3 A Gaussian *pdf* with larger covariance elements. The respective covariance matrix is given by (4.13)

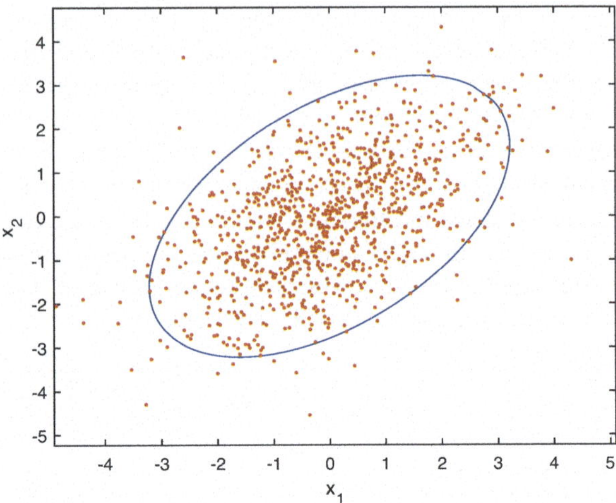

Fig. 4.4 The 95% JCR corresponding to the pdf of Fig. 4.2. The JCR boundary is shown as the blue ellipse, whereas the red dots constitute 1000 samples from this distribution. Note that the JCR is indeed elliptical and that roughly 5% of the samples fall outside the JCR, as expected

(4) lower variances of the conditional probabilities, are all closely inter-related and are effectively different manifestations of correlations between the variables.

An example of a two-dimensional JCR for the pdf of Fig. 4.2 is shown in Fig. 4.4, where the blue elliptical contour delimits the 95% joint confidence region of this distribution. There are 1000 red dots shown, each of which are samples drawn from this same distribution. It is evident that approximately 5% or 50 samples fall outside the JCR.

4.4 The Rayleigh Quotient

The *Rayleigh quotient* is a simple mathematical structure that has a great deal of interesting uses. The Rayleigh quotient $r(\boldsymbol{x})$ is defined as

$$r(\boldsymbol{x}) = \frac{\boldsymbol{x}^T \boldsymbol{A} \boldsymbol{x}}{\boldsymbol{x}^T \boldsymbol{x}}. \tag{4.14}$$

It is easily verified that if $\boldsymbol{x}$ is the ith eigenvector $\boldsymbol{v}_i$ of $\boldsymbol{A}$ (not necessarily normalised to unit norm), then $r(\boldsymbol{x}) = \lambda_i$:

$$\begin{aligned} \frac{\boldsymbol{v}_i^T \boldsymbol{A} \boldsymbol{v}_i}{\boldsymbol{v}_i^T \boldsymbol{v}_i} &= \frac{\lambda_i \boldsymbol{v}^T \boldsymbol{v}}{\boldsymbol{v}^T \boldsymbol{v}} \\ &= \lambda_i. \end{aligned} \tag{4.15}$$

In fact, it can be shown by differentiating $r(\boldsymbol{x})$ with respect to $\boldsymbol{x}$ that $\boldsymbol{x} = \boldsymbol{v}_i$ is a stationary point of $r(\boldsymbol{x})$.

4.5 Methods for Computing a Single Eigenpair

4.5.1 The Rayleigh Quotient Method

The Rayleigh quotient discussed above leads naturally to an iterative method for computing an eigenvalue/eigenvector pair of a square symmetric matrix $\boldsymbol{A}$ for the case where the respective eigenvalue is distinct. If $\boldsymbol{x}$ is an approximate eigenvector, then $r(\boldsymbol{x})$ gives us a reasonable approximation to the corresponding eigenvalue. Further, the inverse perturbation theory of Golub and Van Loan [1] says that if u is an eigenvalue, then the solution z to $(\boldsymbol{A} - u\boldsymbol{I})z = \boldsymbol{b}$, where $\boldsymbol{b}$ is an approximate eigenvector, gives us a better estimate of the eigenvector. These two ideas lead to the following *Rayleigh quotient* technique for calculating an eigenvector/eigenvalue pair:

initialise $k = 0$ and set $\boldsymbol{x}_0$ to an appropriate value; set $||\boldsymbol{x}_0||_2 = 1$.
iterate in k until convergence:

1. $\mu_k = r(\boldsymbol{x}_k)$ (new estimate of eigenvalue)
2. Solve $(\boldsymbol{A} - \mu_k\boldsymbol{I})z_{k+1} = \boldsymbol{x}_k$ for z_{k+1} (new estimate of corresponding eigenvector)
3. $\boldsymbol{x}_{k+1} = z_{k+1}/||z_{k+1}||_2$

end

This procedure exhibits cubic convergence to the eigenvector. At convergence, μ is an eigenvalue, and z is the corresponding eigenvector. Therefore, the matrix $(\boldsymbol{A} - \mu\boldsymbol{I})$ is singular, and z is in its nullspace. The solution z becomes extremely large, and the system of equations $(\boldsymbol{A} - \mu\boldsymbol{I})z = \boldsymbol{x}$ is satisfied only because of numerical error, since $\boldsymbol{x}$ should normally be $\boldsymbol{0}$. Nevertheless, accurate values of the eigenvalue and eigenvector are obtained.

4.5.2 The Power Method

This is a technique for computing the dominant eigenvalue λ_1 (i.e. the largest in absolute value) where λ_1 is unique. If $\boldsymbol{A}$ is not symmetric, we must also assume its eigenvectors are linearly independent.

We start with an initial guess $\boldsymbol{x}(0)$ of the eigenvector. Because the eigenvectors are linearly independent, we can express $\boldsymbol{x}(0)$ in the eigenvector basis $\boldsymbol{V} = [\boldsymbol{v}_1, \boldsymbol{v}_2 \ldots, \boldsymbol{v}_n]$ as

$$\boldsymbol{x}(0) = \sum_{j=1}^{n} c_j \boldsymbol{v}_j, \tag{4.16}$$

where the c_j are the coefficients of $\boldsymbol{x}(0)$ in the basis $\boldsymbol{V}$. We post-multiply $\boldsymbol{A}$ by $\boldsymbol{x}(0)$ to get $\boldsymbol{x}(1) = \boldsymbol{A}\boldsymbol{x}(0)$. Substituting (4.16) for $\boldsymbol{x}(0)$, we have

$$\boldsymbol{x}(1) = \sum_{j=1}^{n} c_j \boldsymbol{A}\boldsymbol{v}_j = \sum_{j=1}^{n} c_j \lambda_j \boldsymbol{v}_j.$$

We continue iteratively to post-multiply in this manner, and at the kth iteration, we have $\boldsymbol{x}(k) = \boldsymbol{A}\boldsymbol{x}(k-1)$ or

$$\boldsymbol{x}(k) = \sum_{j=1}^{n} c_j (\lambda_j)^k \boldsymbol{v}_j. \tag{4.17}$$

We multiply the above by $\left(\frac{\lambda_1}{\lambda_1}\right)^k = 1$ to obtain

$$\boldsymbol{x}(k) = \left(\lambda_1\right)^k \sum_{j=1}^{n} c_j \left(\frac{\lambda_j}{\lambda_1}\right)^k \boldsymbol{v}_j \longrightarrow \left(\lambda_1\right)^k c_1 \boldsymbol{v}_1, \tag{4.18}$$

as k becomes large. The result on the right follows because the terms $\left(\frac{\lambda_j}{\lambda_1}\right)^k \rightarrow 0,\ j \neq 1$, as k becomes large, since λ_1 is largest in absolute value. When (4.18) is satisfied, the method has converged, and we have $\boldsymbol{x}(k+1) = \lambda_1 \boldsymbol{x}(k)$. At this point, λ_1 is revealed, and $\boldsymbol{v}_1 = \boldsymbol{x}(k+1)$.

There is a practical matter remaining and that is from (4.18), we see that $\boldsymbol{x}(k)$ can become very large or very small in norm as k increases, depending whether $|\lambda_1|$ is greater than or less than 1, leading respectively to either floating point over- or underflow. This situation is easily remedied by replacing $\boldsymbol{x}(k)$ with $\frac{\boldsymbol{x}(k)}{||\boldsymbol{x}(k)||_2}$ at each iteration. Other scaling options which use an ∞- or 1-norm are more computationally efficient than the 2-norm and will also work well.

The schema for the power method is therefore given as follows:

1. Initialise $\boldsymbol{x}(0)$ to some suitable value. Often setting all the elements to one is a good choice. Initialise $k = 0$.
2. $\boldsymbol{x}(k+1) = \boldsymbol{A}\boldsymbol{x}(k)$.
3. normalise $\boldsymbol{x}(k+1)$ by replacing $\boldsymbol{x}(k+1)$ with $\frac{\boldsymbol{x}(k+1)}{||\boldsymbol{x}(k+1)||}$, where $||\cdot||$ is a suitable norm.
4. if $\frac{||\boldsymbol{x}(k+1)-\boldsymbol{x}(k)||}{||\boldsymbol{x}(k+1)||} < \epsilon$, where ϵ is some suitably small threshold,
   (a) $\lambda_1 = \frac{||\boldsymbol{x}(k+1)||}{||\boldsymbol{x}(k)||}$
   (b) $\boldsymbol{v}_1 = \boldsymbol{x}(k+1)$.
   (c) return.
5. $k = k + 1$
6. go to step 2.

Deflation It turns out the power method can be applied sequentially to calculate additional eigenpairs $(\lambda_2, \boldsymbol{v}_2)$, $(\lambda_3, \boldsymbol{v}_3)\dots$ using a straightforward deflation procedure. We rewrite the relation $\boldsymbol{A} = \boldsymbol{V}\boldsymbol{\Lambda}\boldsymbol{V}^T$ using the outer product rule for matrix multiplication using the fact that $\boldsymbol{\Lambda}$ is diagonal, in the form

$$\boldsymbol{A} = \sum_{j=1}^{n} \lambda_i \boldsymbol{v}_i \boldsymbol{v}_i^T.$$

The power method as described above allows us to specify λ_1 and $\boldsymbol{v}_1$ of the first term in this expansion. We can therefore define a deflated version $\boldsymbol{A}^{def}$ which removes the first term as

$$\boldsymbol{A}^{def} = \boldsymbol{A} - \lambda_1 \boldsymbol{v}_1 \boldsymbol{v}_1^T.$$

It is noted that the largest eigenvalue of $\boldsymbol{A}^{def}$ is now λ_2. Therefore the power method described above may be applied to $\boldsymbol{A}^{def}$ to yield λ_2 and $\boldsymbol{v}_2$, provided λ_2 is distinct. This process as described may then be applied sequentially to determine additional eigenpairs of $\boldsymbol{A}$.

One caveat regarding this sequential power method is that if the dimension of $\boldsymbol{A}$ is large, then small errors, due to floating point error, etc., in the early eigenpair estimates can compound, leaving the eigenpair estimates of the later stages inaccurate. There are better methods of calculating the complete eigendecomposition that use the QR decomposition (to be described later in Sect. 6.4).

It is straightforward to show that the *smallest* eigenpair may be determined using the power method on the matrix $\boldsymbol{A}^{-1}$ instead of $\boldsymbol{A}$. In this case, the method is referred to as the *inverse power method*. The proof is left as an exercise. *Hint:* What are the eigenvalues of $\boldsymbol{A}^{-1}$ relative to $\boldsymbol{A}$? The convergence of the inverse power method can be accelerated using the method described in Problem 2 at the end of this chapter.

Appendix

Alternate Differentiation of the Quadratic Form

We see that the quadratic form is a scalar. To differentiate a scalar with respect to the vector $\boldsymbol{x}$, we differentiate with respect to each element of $\boldsymbol{x}$ in turn and then assemble all the results back into a vector. We proceed as follows:

We write the quadratic form as

$$\boldsymbol{x}^T \boldsymbol{A}\boldsymbol{x} = \sum_{i=1}^{n}\sum_{j=1}^{n} x_i x_j a_{ij}. \tag{4.19}$$

When differentiating the above with respect to a particular element x_k, we need only consider the terms when either index i or j equals k. Therefore,

$$\frac{d}{dx_k}\boldsymbol{x}^T\boldsymbol{A}\boldsymbol{x} = \frac{d}{dx_k}\left[\sum_{\substack{j=1\\j\neq k}}^{n} x_k x_j a_{kj} + \sum_{\substack{i=1\\i\neq k}}^{n} x_i x_k a_{ik} + x_k^2 a_{kk}\right] \tag{4.20}$$

where the first term of (4.20) corresponds to holding i constant at the value k and the second corresponds to holding j constant at k. Care must be taken to include the term $x_k^2 a_{kk}$ corresponding to $i = j = k$ only once; therefore, it is excluded in the first two terms and added in separately. Equation (4.20) evaluates to

$$\begin{aligned}\frac{d}{dx_k}\boldsymbol{x}^T\boldsymbol{A}\boldsymbol{x} &= \sum_{j\neq k} x_j a_{kj} + \sum_{i\neq k} x_i a_{ik} + 2x_k a_{kk}\\ &= \sum_{j} x_j a_{kj} + \sum_{i} x_i a_{ik}\\ &= [\boldsymbol{A}\boldsymbol{x}]_k + \left[\boldsymbol{A}^T\boldsymbol{x}\right]_k\\ &= \left[(\boldsymbol{A}+\boldsymbol{A}^T)\boldsymbol{x}\right]_k\\ &= [2\boldsymbol{T}\boldsymbol{x}]_k\end{aligned}$$

where $(\cdot)_k$ denotes k^{th} element of the argument. By assembling these individual terms corresponding to $k = 1, \ldots, n$ back into a vector, we have the result that

$$\frac{d}{d\boldsymbol{x}}\boldsymbol{x}^T\boldsymbol{A}\boldsymbol{x} = 2\boldsymbol{T}\boldsymbol{x}.$$

It is interesting to find the stationary points of the quadratic form subject to a norm constraint; i.e. we seek the solution to

$$\max_{||\boldsymbol{x}||_2^2=1} \boldsymbol{x}^T\boldsymbol{A}\boldsymbol{x}.$$

To solve this, we form the Lagrangian

$$\boldsymbol{x}^T\boldsymbol{A}\boldsymbol{x} + \lambda\left(1 - \boldsymbol{x}^T\boldsymbol{x}\right).$$

Differentiating and setting the result to zero (realising that $d/d\boldsymbol{x}\ \left(\boldsymbol{x}^T\boldsymbol{x}\right) = 2\boldsymbol{x}$) give

$$\boldsymbol{T}\boldsymbol{x} = \lambda\boldsymbol{x}.$$

Thus, the eigenvectors are stationary points of the quadratic form, and the $\boldsymbol{x}$ which gives the maximum (or minimum), subject to a norm constraint, is the maximum (minimum) eigenvector of $\boldsymbol{A}$.

Problems

1. Consider the following generalisation of the power method, which determines the vectors $\boldsymbol{u}_1$, $\boldsymbol{v}_1$ and the quantity σ_1 of the SVD of a matrix $\boldsymbol{A}$:

   (a) Initialise $\boldsymbol{x}^{(0)}$ to some suitable value, assign $i = 0$:
   (b) Repeat until convergence

   - $\boldsymbol{y}^{(i+1)} = \boldsymbol{A}\boldsymbol{x}^{(i)}$
   - $\boldsymbol{x}^{(i+1)} = \boldsymbol{A}^T\boldsymbol{y}^{(i+1)}$
   - Normalise: $\boldsymbol{x}^{(i+1)} \leftarrow \frac{\boldsymbol{x}^{(i+1)}}{||\boldsymbol{x}^{(i+1)}||_2}$
   - $i \leftarrow i + 1$

   After convergence, $\boldsymbol{u}_1 = \boldsymbol{y}$, $\boldsymbol{v}_1 = \boldsymbol{x}$ and $\sigma_1 = \sqrt{\frac{||\boldsymbol{x}^{(i+1)}||_2}{||\boldsymbol{x}^{(i)}||_2}}$ (values taken before normalisation).

   (a) Prove the method works.
   (b) Explain how to modify the method to find the second-largest SVD component.
   (c) Why is the normalisation step necessary?
   (d) What happens if $\sigma_1 = \sigma_2$?

2. Consider the inverse power method for computing the smallest eigenpair of a matrix $\boldsymbol{A}$. Show that convergence can be significantly accelerated by replacing $\boldsymbol{A}$ with $\boldsymbol{A} - \gamma\boldsymbol{I}$, where γ is an estimate of the smallest eigenvalue, before inversion of $\boldsymbol{A}$.
3. Consider the multivariate quadratic equation given as

$$\frac{1}{2}\boldsymbol{x}^T\boldsymbol{A}\boldsymbol{x} + \boldsymbol{b}^T\boldsymbol{x} + c = 0,$$

   where $\boldsymbol{x} \in \mathbb{R}^n$ and $\boldsymbol{A}$ is positive definite.

   (a) Develop a closed-form solution for $\boldsymbol{x}$.
   (b) Explain what happens if $\boldsymbol{A}$ is mixed-definite or negative definite.
   (c) Explain what happens when $\boldsymbol{A}$ is positive semi-definite?

4. Prove that the diagonal elements of a positive definite symmetric matrix must be positive.
5. (a) The χ^2 distribution with n degrees of freedom is the probability density function of the quantity $\sum_{i=1}^{n} x_i^2$, where the x_i are independent, identically distributed (*iid*) Gaussian random variables with zero mean and unit vari-

ance. Prove that the quantity $(\boldsymbol{y} - \boldsymbol{\mu})^T \boldsymbol{\Sigma}^{-1}(\boldsymbol{y} - \boldsymbol{\mu})$ is χ^2 distributed with n degrees of freedom, where $\boldsymbol{y} \in \mathbb{R}^n$ is a Gaussian random vector with mean $\boldsymbol{\mu}$ and covariance matrix $\boldsymbol{\Sigma}$.

(b) What is the distribution of the quantity $z = \boldsymbol{G}^{-1}(\boldsymbol{y} - \boldsymbol{\mu})$, where $\boldsymbol{G}$ is the Cholesky factor[1] of $\boldsymbol{\Sigma}$?

6. Sketch the joint confidence region of the Gaussian probability distribution for $\boldsymbol{\mu} = [1, 1]^T$ and $\boldsymbol{\Sigma}$ given by (4.11) at the $\alpha = 95\%$ level, showing all relevant values.

References

1. G.H. Golub, C.F. Van Loan, *Matrix Computations*, 3rd edn. (The Johns Hopkins University, Baltimore, MD, 1996)
2. L.L. Scharf, C. Demeure, *Statistical Signal Processing: Detection, Estimation, and Time Series Analysis* (Prentice Hall, Englewood Cliffs, 1991)
3. H.L. Van Trees, *Detection, Estimation, and Modulation Theory, Part I: Detection, Estimation, and Linear Modulation Theory* (Wiley, New York, 2004)

[1] The Cholesky factor $\boldsymbol{G}$ of a square, symmetric, positive definite matrix $\boldsymbol{\Sigma}$ is lower triangular such that $\boldsymbol{G}\boldsymbol{G}^T = \boldsymbol{\Sigma}$.

Chapter 5
Gaussian Elimination and Associated Numerical Issues

5.1 Floating Point Arithmetic Systems

In this Section we discuss floating point number systems according to the IEEE 754 standard for floating point numbers [1]. A real number x can be represented in floating point form (denoted $fl(x)$) as

$$fl(x) = s \cdot f \cdot b^k \tag{5.1}$$

where

$$\begin{aligned} s &= \text{ sign bit} = \pm 1 \\ f &= \text{ fractional part of } x \text{ of length } t \text{ bits} \\ b &= \text{machine base} = 2 \text{ for binary systems} \\ k &= \text{ exponent} \end{aligned}$$

Note that the operation $fl(x)$(i.e. conversion from a real number x to its floating point representation) maps a real number x into a set of *discrete* points on the real number line. These points are determined by (5.1). This mapping has the property that the separation between points is proportional to $|x|$. Because the operation $fl(x)$ maps a continuous range of numbers into a discrete set, there is error associated with the representation $fl(x)$.

In the conversion process, the exponent is adjusted so that the most significant bit (MSB) of the fractional part is 1 and so that the binary point is immediately to the right of the MSB. For example, the binary number

$$x = .000010011110101101 1 \tag{5.2}$$

J. Reilly, *Fundamentals of Linear Algebra for Signal Processing*,
https://doi.org/10.1007/978-3-031-68915-4_5

could be represented as a floating point number with $t = 9$ bits as

$$1.00111101 \times 2^{-5}.$$

Since it is known that the MSB of the fractional part is a one, it does not need to be physically present in the actual floating-point number representation. This way, we get an extra bit, "for free". This means the number x in (5.2) may be represented as

$$\underbrace{00111101}_{f} \times 2^{-5}.$$

$\uparrow$ leading 1 assumed present

This above form only takes 8 bits instead of 9 to represent $fl(x)$ with the same precision.

The range of possible real numbers which can be mapped into the representation $|fl(x)|$ is:

$$1.00\ldots00 \times 2^{L} \leq |\mathrm{fl}(x)| \leq \overbrace{1.111111\ldots1}^{|\leftarrow t \text{ bits } \rightarrow|} \times 2^{U}$$

where L and U are the minimum and maximum values of the exponent, respectively. Note that any arithmetic operation which produces a result outside of these bounds results in a floating point overflow or underflow error.

Note that because the leading one in the most significant bit position is absent and assumed present, the representation of the number zero requires special attention. This is usually accomplished by reserving a special value of the exponent field for the number zero.

5.1.1 *Machine Epsilon u*

Since the operation $fl(x)$ maps the set of real numbers into a discrete set, the quantity $fl(x)$ involves error. The quantity *machine epsilon*, represented by the symbol u is the maximum relative error possible in $fl(x)$.

If we have a real number y and an approximation $\hat{y}$ to it, then the relative error ϵ_r in the representation of y is given as

$$\epsilon_r = \frac{|y - \hat{y}|}{|y|}.$$

We see that ϵ_r is the normalised version of the *absolute error* given by $|y - \hat{y}|$. This normalisation is important, since, e.g. if y is a time measurement of say one millisecond, then an error of $1/2$ a millisecond would usually be considered a sizeable error. On the other hand, if y were measured in years, an error of $1/2$ a millisecond would usually be considered very small indeed, yet the absolute error is the same in each case. Relative error, due to its normalisation, removes this difficulty.

The relative error ϵ_r in the quantity $fl(x)$ is given by

$$\epsilon_r = \frac{|fl(x) - x|}{|x|}. \tag{5.3}$$

But u is the maximum relative error. In the following, because both $fl(x)$ and x may be assumed to have the same exponent, we consider only the fractional part of $fl(x)$. The maximum value of ϵ_r is achieved when the numerator of (5.3) is maximum and the denominator is minimum. Therefore

$$\begin{aligned} u = \max \epsilon_r &= \frac{\max |fl(x) - x|}{\min |x|} \\ &= \frac{0.\overbrace{00\ldots0}^{\leftarrow t \text{ bits} \rightarrow}\,1111111111\ldots}{1} \\ &\rightarrow 2^{1-t} \end{aligned}$$

if the machine chops. By "chopping", we mean the machine constructs the fractional part of $fl(x)$ by retaining only the most significant t bits, and truncating the rest. If the machine rounds, then the relative error is one half that due to chopping; hence

$$u = 2^{-t}.$$

Thus, the number $fl(x)$ may be represented as

$$fl(x) = x(1 + \epsilon), \text{ where } |\epsilon| \leq u.$$

It is also noteworthy that if we perform operations on a sequence of n floating point numbers, then the worst-case error accumulates; i.e. if we evaluate $\sum_{i=1}^{n} fl(x_i)$, then it is possible in the worst case that each of the $fl(x_i)$ are subject to the maximum relative error of the same sign, in which case the maximum relative error of the sum becomes nu. This result holds for both addition and subtraction operations. It may also be shown that the same result also holds (to a first order approximation) for both multiplication and division operations.

In an actual computer implementation, s is a single bit (usually 0 to indicate a positive number, and 1 to represent a negative number). In single precision, the total length of a floating point number is typically 32 bits. Of these, 8 are used for the exponent k, one for s, leaving 23 for the fractional part f (t = 24 bits effective precision). This means for single precision arithmetic with chopping, $u = 2^{-23} = 1.19 \times 10^{-7}$.

5.1.2 *Absolute Value Notation*

It turns out that in order to perform error analysis on floating- point matrix computations, we need the absolute value notation:

If $\boldsymbol{A}$ and $\boldsymbol{B}$ are in $\mathbb{R}^{m\times n}$ then

$$\boldsymbol{B} = |\boldsymbol{A}| \Rightarrow b_{ij} = |a_{ij}|, \quad i = 1:m, \quad j = 1:n$$

$$\text{also } \boldsymbol{B} \leq \boldsymbol{A} \Rightarrow b_{ij} \leq a_{ij}, \quad i = 1:m, \quad j = 1:n$$

This notation is used often throughout this chapter.

From our discussion on floating point numbers, we then have

$$|fl(\boldsymbol{A}) - \boldsymbol{A}| \leq u|\boldsymbol{A}|. \tag{5.4}$$

5.1.3 *Catastrophic Cancellation*

Significant reduction in precision may result when subtracting two nearly equal floating-point numbers. If the fractional part of two numbers A and B are identical in their first r digits ($r \leq t$), then fl$(A - B)$ has only $t - r$ bits significance; i.e. we have lost r bits of significance in representing the difference. As r approaches t, the difference has very few significant bits in its fractional part. This reduction in precision is referred to as *catastrophic cancellation*.

We can demonstrate this phenomenon by example as follows: Let A and B be two numbers whose fractional parts are identical in their first $r = 7$ digits. Then for the case $t = 10$ and $b = 2$ (binary arithmetic)

$$\text{frac}(A) = \begin{matrix} |\leftarrow r_{\text{bits}} \rightarrow| \\ 1011011\ 101 \end{matrix}$$

$$\text{frac}(B) = \begin{matrix} |\leftarrow r_{\text{bits}} \rightarrow| \\ 1011011\ 001 \end{matrix}$$

where frac$(\cdot)$ is the fractional part of the number. Because the numbers are nearly equal, it may be assumed that their exponents have the same value. Then, we see that the difference frac$(A - B)$ is $(100)_2$, which has only $t - r = 3$ bits significance. We have lost 7 bits of significance in representing the difference, which results in a drastic increase in u. Thus the difference can be in significant error.

Another example of catastrophic cancellation is as follows: Find roots of the quadratic equation

$$x^2 + 1958.63x + 0.00253 = 0$$

Solution

$$x = \frac{-b \pm \sqrt{b^2 - 4ac}}{2a} \tag{5.5}$$

$$\textit{computed roots}: x_1 = -1958.62998,\ x_2 = -0.00000150$$

$$\textit{true roots}: x_1 = -1958.6299,\ x_2 = -0.0000012917$$

There are obviously serious problems with the accuracy of x_2, which corresponds to the "+" sign in (5.5) above. In this case, since $b^2 >> 4ac$, $\sqrt{b^2 - 4ac} \simeq b$. Hence, we are subtracting two nearly equal numbers when calculating x_2, which results in catastrophic cancellation.

Another example of catastrophic cancellation is in evaluating the inner product of two nearly orthogonal vectors. In this case the inner product $\boldsymbol{x}^T\boldsymbol{y} = \sum_{i=1}^{n} fl(x_i)y_i \approx 0$. So we can express the inner product as

$$\boldsymbol{x}^T\boldsymbol{y} = \sum_{i\in\mathcal{P}} fl(x_i)y_i - \sum_{i\in\mathcal{N}} fl(x_i)y_i \approx 0, \tag{5.6}$$

where $\mathcal{P}$, ($\mathcal{N}$) is the set of indexes representing positive (negative) terms. Thus all products in the first sum are positive, and those in the second sum are negative. Since each product $fl(x_i)y_i$ can be arbitrarily larger in absolute value than the overall result which is $\approx$0, the absolute value of each summation term above can be large, yet the final result is close to zero. Thus (5.6) inherently involves the subtraction of two nearly equal numbers, thus giving rise to catastrophic cancellation.

In the case where the vectors are not closely orthogonal, then many of the products (5.6) all have the same sign and the effect of catastrophic cancellation is suppressed, and so there is little if any reduction in the number of effective significant bits. In this case, the relative error in the inner product can be expressed in the form

$$\frac{\left|fl(\mathbf{x}^T\mathbf{y}) - \mathbf{x}^T\mathbf{y}\right|}{\left|\mathbf{x}^T\mathbf{y}\right|} \leq nu, \tag{5.7}$$

or

$$\left|fl(\mathbf{x}^T\mathbf{y}) - \mathbf{x}^T\mathbf{y}\right| \leq nu\left|\mathbf{x}^T\mathbf{y}\right|. \tag{5.8}$$

This is exactly the worst-case relative error we would expect when representing a sum of n floating point numbers. However, when the vectors become close to orthogonality, the number of effective significant bits becomes reduced due to catastrophic cancellation and so the bound of (5.8) no longer applies as is. It is shown [2] that in this case,

$$\left| fl(\boldsymbol{x}^T \mathbf{y}) - \boldsymbol{x}^T \mathbf{y} \right| \leq nu|\boldsymbol{x}|^T |\mathbf{y}| + O(u^2), \tag{5.9}$$

where absolute value notation has been applied; i.e. $|\boldsymbol{x}|$ and $|\boldsymbol{y}|$ denote the absolute value of the elements of the vectors. The notation $O(u^2)$, read "order u squared", indicates the presence of terms in u^2 and higher, which can be ignored due to the fact they may be considered small in comparison to the first-order term in u. Hence (5.9) tells us that if $|\boldsymbol{x}^T \mathbf{y}| \ll |\boldsymbol{x}|^T |\mathbf{y}|$, which happens when $\boldsymbol{x}$ is nearly orthogonal to $\boldsymbol{y}$, then the relative error in $f\,l(\boldsymbol{x}^T \mathbf{y})$ may be much larger than that indicated by (5.8). This is due to the catastrophic cancellation implicitly expressed in the form of (5.6).

Fix If the partial products are accumulated in a *double precision* register (length of fractional part = $2t$), little error results. This is because multiplication of two t-digit numbers can be stored exactly in a $2t$ digit mantissa. Hence, roundoff only occurs when converting to single precision, and the result is significant to approximately t bits significance in single precision.

5.2 Gaussian Elimination

In this section, we discuss the concept of Gaussian elimination in some detail. But first, we present a very quick review, by example, of the elementary approach to Gaussian elimination. Given the system of equations

$$\boldsymbol{A}\boldsymbol{x} = \boldsymbol{b}$$

where $\boldsymbol{A} \in \mathbb{R}^{3 \times 3}$ is nonsingular. The above system can be expanded into the form

$$\begin{bmatrix} a_{11} & a_{12} & a_{13} \\ a_{21} & a_{22} & a_{23} \\ a_{31} & a_{32} & a_{33} \end{bmatrix} \begin{bmatrix} x_1 \\ x_2 \\ x_3 \end{bmatrix} = \begin{bmatrix} b_1 \\ b_2 \\ b_3 \end{bmatrix}.$$

To solve the system, we transform $\boldsymbol{A}$ into the following upper triangular form $\boldsymbol{U}$ by Gaussian elimination:

$$\begin{bmatrix} a_{11} & a_{12} & a_{13} \\ & a'_{22} & a'_{23} \\ & & a''_{33} \end{bmatrix} \begin{bmatrix} x_1 \\ x_2 \\ x_3 \end{bmatrix} = \begin{bmatrix} b_1 \\ b'_2 \\ b''_3 \end{bmatrix}, \tag{5.10}$$

using a sequence of elementary row operations, as follows:

$$\text{row } 2' := \text{row } 2 - \frac{a_{21}}{a_{11}} \text{ row } 1,$$

$$\text{row } 3' := \text{row } 3 - \frac{a_{31}}{a_{11}} \text{ row } 1,$$

and

$$\text{row } 3'' := \text{row } 3' - \frac{a'_{32}}{a'_{22}} \text{ row } 2'. \tag{5.11}$$

The prime indicates the respective quantity has been changed. Each elementary operation preserves the solution of the original system of equations and is designed to place a zero in the appropriate place below the main diagonal of $\boldsymbol{A}$. It is assumed throughout that the denominator of the multipliers above (these are referred to as the *pivot elements*) are non-zero.

The result is expressed in the upper triangular system of equations in $\boldsymbol{U}$:

$$\boldsymbol{U}\boldsymbol{x} = \boldsymbol{b}'.$$

Once $\boldsymbol{A}$ has been triangularised, the solution $\boldsymbol{x}$ is obtained by applying *backwards substitution* to the system (5.10). With this procedure, x_n is first determined from the last row of (5.10). Then x_{n-1} may be determined, knowing x_n, from the second-last row, etc. The algorithm may be summarised by the following schema:

$$\begin{aligned}
\text{for } i &= n, \ldots, 1 \\
x_i &:= b_i \\
\text{for } j &= i+1, \ldots, n \\
x_i &:= x_i - u_{ij} x_j \\
x_i &:= \frac{x_i}{u_{ii}} \\
\text{end}
\end{aligned}$$

5.2.1 *What About the Accuracy of Back Substitution?*

With operations on floating point numbers, we must be concerned about the accuracy of the result, since the floating point numbers themselves contain error. We want to know if it is possible that the small errors in the floating point representation of real numbers can lead to large errors in the computed result $\hat{\boldsymbol{x}}$. In this vein, we can show [2] that the $\hat{\boldsymbol{x}}$ obtained by back substitution satisfies the expression

$$(\boldsymbol{U} + \boldsymbol{E})\hat{\boldsymbol{x}} = \boldsymbol{b}',$$

where

$$|\boldsymbol{E}| \leq nu|\boldsymbol{U}| + O(u^2),$$

and where absolute value notation has been used. Here u is the machine epsilon as before. The above equation says that $\hat{\boldsymbol{x}}$ is the *exact* solution to a perturbed system. If the perturbations are on the order of error induced by floating point error, then the underlying algorithm is as good as can be expected in the face of floating point error. We see that all elements of $\boldsymbol{E}$ are of $O(nu)$, which is exactly the worst-case error expected in $\boldsymbol{U}$ due to floating point error alone, with errors accumulating over n floating point numbers. This is the best that can be done with floating point systems. It is worthy of note that if elements of $\boldsymbol{E}$ have a larger magnitude, then the error in the solution can be large, such as in the case with Gaussian elimination without pivoting, as we see later. However in the case at hand, we can conclude that back substitution *is stable*. By a numerically stable algorithm, we mean one that produces relatively small errors in its output values for small errors in the input values.

The total number of floating point operations (FLOPS) required for Gaussian elimination of a matrix $\boldsymbol{A} \in \mathbb{R}^{n \times n}$ may be shown to be $O(\frac{2n^3}{3})$ (one "FLOP" is one floating point operation; i.e. a floating point add, subtract, multiply, or divide). It is easily shown that backward substitution requires $O(n^2)$ flops. Thus, the number of operations required to solve $\boldsymbol{Ax} = \boldsymbol{b}$ is dominated by the Gaussian elimination process for moderate and large n.

5.2.2 The LU Decomposition

Suppose we can find lower and upper $n \times n$ triangular matrices $\boldsymbol{L}$ (with ones along the main diagonal) and $\boldsymbol{U}$, respectively, such that

$$\boldsymbol{A} = \boldsymbol{LU}.$$

This decomposition of $\boldsymbol{A}$ is referred to as the *LU decomposition*. To solve the system $\boldsymbol{Ax} = \boldsymbol{b}$, or $\boldsymbol{LUx} = \boldsymbol{b}$, we define the variable $\boldsymbol{z}$ as $\boldsymbol{z} = \boldsymbol{Ux}$ and then

$$\begin{aligned} \text{solve} \quad \boldsymbol{Lz} &= \boldsymbol{b} \text{ for } \boldsymbol{z} \\ \text{and then} \quad \boldsymbol{Ux} &= \boldsymbol{z} \text{ for } \boldsymbol{x}. \end{aligned}$$

Since both systems are triangular, they are easy to solve. The first system requires only forward elimination and the second only back substitution. Forward elimination is the analogous process to backward substitution, but since it is performed on a lower triangular system, the unknowns are solved in ascending order for forward elimination (i.e. $x_1, x_2, \ldots, x_n$) instead of descending order $(x_n, x_{n-1}, \ldots, x_1)$ as in backward substitution. Forward substitution requires an equal number of flops as back substitution and is just as stable. Thus, once the LU factorisation is complete, the solution of the system is easy: the total number of flops required to solve $\boldsymbol{Ax} = \boldsymbol{b}$ is $2n^2$. The details of the computation of the LU factorisation and the number of FLOPS required is discussed later.

We are lead to several interesting questions:

1. How does one perform the LU decomposition?
2. How much computational effort is required to perform the LU decomposition?
3. What is the relationship of LU decomposition, if any, to Gaussian elimination?
4. Is the LU decomposition process numerically stable?

The answer to these questions is provided in the following sections.

5.2.3 Gaussian Elimination

Here we choose to describe the Gaussian elimination process as a sequence of matrix multiplications, rather than the algorithmic description given above. Not only is the resulting matrix description more compact, but it leads to theoretical insights that are not possible otherwise. In this vein, Gaussian elimination may be described as a sequence of Gauss transformations $\boldsymbol{M}_{n-1} \dots \boldsymbol{M}_1 \in \mathbb{R}^{n \times n}$ operating sequentially on $\boldsymbol{A} \in \mathbb{R}^{n \times n}$ such that

$$\boldsymbol{M}_{n-1} \dots \boldsymbol{M}_2 \boldsymbol{M}_1 \boldsymbol{A} = \boldsymbol{U}. \tag{5.12}$$

where $\boldsymbol{U}$ is the desired upper triangular result. We illustrate the process with an example using the matrix $\boldsymbol{A}$ defined as

$$\boldsymbol{A} = \begin{bmatrix} 2 & -1 & 2 & 3 \\ 1 & 1 & 4 & 1 \\ -2 & -3 & -1 & -2 \\ 1 & 1 & -4 & 3 \end{bmatrix}. \tag{5.13}$$

The first stage wipes out all elements of $\boldsymbol{A}$ below the main diagonal in the first column. This procedure can be implemented by pre-multiplication with the matrix $\boldsymbol{M}_1$ of the specific form as seen on the right below (for clarity, a blank entry indicates a 0):

$$\boldsymbol{A}^{(1)} = \boldsymbol{M}_1 \boldsymbol{A} = \begin{bmatrix} 1 & & & \\ -0.5 & 1 & & \\ 1.0 & & 1 & \\ -0.5 & & & 1 \end{bmatrix} \begin{bmatrix} 2 & -1 & 2 & 3 \\ 1 & 1 & 4 & 1 \\ -2 & -3 & -1 & -2 \\ 1 & 1 & -4 & 3 \end{bmatrix}$$

$$= \begin{bmatrix} 2 & -1 & 2 & 3 \\ 0 & 1.5 & 3 & -0.5 \\ 0 & -4 & 1 & 1 \\ 0 & 1.5 & -5 & 1.5 \end{bmatrix} = \boldsymbol{A}^{(1)}.$$

Notice that the elements of the first column of $\boldsymbol{M}_1$ are precisely the multipliers used in the algorithm description of the algorithm. The specific structure of $\boldsymbol{M}_1$ is such that it introduces zeros below the main diagonal in the first column of $\boldsymbol{A}^{(1)}$ as desired.

We now proceed to annihilate the second column of $\boldsymbol{A}^{(1)}$ below the main diagonal by pre-multiplying by a matrix $\boldsymbol{M}_2$ to get $\boldsymbol{A}^{(2)}$

$$\boldsymbol{M}_2\boldsymbol{A}^{(1)} = \begin{bmatrix} 1 & & & \\ & 1 & & \\ & 2.67 & 1 & \\ & -1 & & 1 \end{bmatrix} \begin{bmatrix} 2 & -1 & 2 & 3 \\ 0 & 1.5 & 3 & -0.5 \\ 0 & -4 & 1 & 1 \\ 0 & 1.5 & -5 & 1.5 \end{bmatrix}$$

$$= \begin{bmatrix} 2 & -1 & 2 & 3 \\ 0 & 1.5 & 3 & -0.5 \\ 0 & 0 & 9 & -0.33 \\ 0 & 0 & -8 & 2 \end{bmatrix} = \boldsymbol{A}^{(2)}.$$

Thus, we see that $\boldsymbol{A}^{(2)}$ has zeros below the main diagonal in the first two columns. We can now clear out the third column by pre-multiplication with a matrix $\boldsymbol{M}_3$ to yield the desired upper triangular form, as desired:

$$\boldsymbol{M}_3\boldsymbol{A}^{(2)} = \begin{bmatrix} 1 & & & \\ & 1 & & \\ & & 1 & \\ & & 0.89 & 1 \end{bmatrix} \begin{bmatrix} 2 & -1 & 2 & 3 \\ 0 & 1.5 & 3 & -0.5 \\ 0 & 0 & 9 & -0.33 \\ 0 & 0 & -8 & 2 \end{bmatrix}$$

$$= \begin{bmatrix} 2 & -1 & 2 & 3 \\ 0 & 1.5 & 3 & -0.5 \\ 0 & 0 & 9 & -0.33 \\ 0 & 0 & 0 & 1.70 \end{bmatrix} = \boldsymbol{A}^{(3)} \triangleq \boldsymbol{U}. \tag{5.14}$$

Thus Gaussian elimination can be effected by a series of matrix multiplications $\boldsymbol{M}_{n-1} \ldots, \boldsymbol{M}_1$, provided the $\boldsymbol{M}$'s have the particular structure that has been indicated. The matrices $\boldsymbol{M}_k$ are referred to as *Gauss transformations*.

We now generalise this process. Suppose at the kth iteration, $k < n$, we have already determined Gauss transforms $\boldsymbol{M}_{k-1} \ldots, \boldsymbol{M}_1$ so that the resulting matrix

$\boldsymbol{A}^{(k-1)}$ has the form

$$\boldsymbol{A}^{(k-1)} = \boldsymbol{M}_{k-1} \ldots \boldsymbol{M}_1 \boldsymbol{A} = \underbrace{\begin{bmatrix} \boldsymbol{A}_{11}^{(k-1)} & \boldsymbol{A}_{12}^{(k-1)} \\ \boldsymbol{0} & \boldsymbol{A}_{22}^{(k-1)} \end{bmatrix}}_{k-1 \quad n-k+1} \begin{matrix} k-1 \\ n-k+1 \end{matrix} \tag{5.15}$$

where $\boldsymbol{A}_{11}^{(k-1)}$ is upper triangular. The fact that $\boldsymbol{A}_{11}^{(k-1)}$ is upper triangular means that the first $k-1$ columns are zero below the main diagonal, as indicated. The kth stage of Gaussian elimination proceeds one step to make the first column of $\boldsymbol{A}_{22}^{(k-1)}$ zero below the upper left element (i.e. pivot element as we see later), to yield the matrix $\boldsymbol{A}^{(k)}$.

By comparison to the example above, it may be seen that the matrix $\boldsymbol{M}_k$ at the kth iteration has the following form:

$$\boldsymbol{M}_k = \boldsymbol{I} - \boldsymbol{\alpha}^{(k)} \boldsymbol{e}_k{}^T \tag{5.16}$$

where

$\boldsymbol{I}$ is the $n \times n$ identity matrix

$\boldsymbol{e}_k$ is the kth column of $\mathbf{I}$

i.e. $\boldsymbol{e}_k^T = (0, \ldots, 0, 1, 0, \ldots, 0)$

$\uparrow k^{\text{th}}$ position

and

$$\boldsymbol{\alpha}^{(k)} \triangleq (0, \ldots, 0, l_{k+1,k}, \ldots, l_{n,k})^T, \tag{5.17}$$

where

$$l_{ik} = \frac{a_{ik}^{(k-1)}}{a_{kk}^{(k-1)}}, \qquad i = k+1, \ldots, n. \tag{5.18}$$

Note that the terms $l_{ik} = \frac{a_{ik}^{(k-1)}}{a_{kk}^{(k-1)}}$ above are the multipliers required to introduce the required zeros, as in (5.11).

By evaluating (5.16), we see that $\boldsymbol{M}_k$ has the following structure:

$$\boldsymbol{M}_k = \begin{bmatrix} 1 & & & & & \boldsymbol{0} \\ & \ddots & & & & \\ & & 1 & & & \\ & & -l_{k+1,k} & 1 & & \\ & & \vdots & & \ddots & \\ \boldsymbol{0} & & -l_{n,k} & \cdots & & 1 \end{bmatrix} \begin{matrix} \\ \\ \leftarrow k^{\text{th}} \text{ row} \\ \\ \\ \\ \end{matrix} \tag{5.19}$$

$$\uparrow \; k^{\text{th}} \text{ column}$$

where it will be noted that this definition of $\boldsymbol{M}_k$ has the same structure as those used in the examples.

The Pivot Element The quantity $a_{kk}^{(k-1)}$, which is the upper left-hand element of $\boldsymbol{A}_{22}^{(k-1)}$, is the *pivot element* for the kth stage, which is assumed to be non-zero. This element plays a strategically significant role in the Gaussian elimination process, due to the fact it appears in the denominator of all the l_{ik} terms in (5.18). We will see that small pivot values lead to large elements in $\boldsymbol{U}$ and $\boldsymbol{L}$ and therefore have the potential to lead to large errors in the solution $\boldsymbol{x}$ due to catastrophic cancellation.

5.2.4 Recovery of the LU Factors from Gaussian Elimination

We now discuss the relationship between Gaussian elimination and the LU decomposition. Specifically, we investigate how to determine the $\boldsymbol{L}$ and $\boldsymbol{U}$ factors of $\boldsymbol{A}$ directly from the Gaussian elimination process, without the need for explicit additional computations. We note the Gaussian elimination process produces

$$\boldsymbol{M}_{n-1} \dots \boldsymbol{M}_1 \boldsymbol{A} = \boldsymbol{U} \tag{5.20}$$

where $\boldsymbol{U}$ is the upper triangular matrix resulting from the Gaussian elimination process. Each $\boldsymbol{M}_k$ is unit lower triangular (ULT) (ULT means lower triangular with one's on the main diagonal), and it is easily verified that the product of ULT matrices is also ULT. Therefore, we define a ULT matrix $\boldsymbol{L}^{-1}$ as

$$\boldsymbol{M}_{n-1} \dots \boldsymbol{M}_1 = \boldsymbol{L}^{-1} \tag{5.21}$$

From (5.20), we then have $\boldsymbol{L}^{-1}\boldsymbol{A} = \boldsymbol{U}$. But since the inverse of a ULT matrix is also ULT, then

$$\boldsymbol{A} = \boldsymbol{L}\boldsymbol{U}, \tag{5.22}$$

which is the product of lower and upper triangular factors as desired. We have therefore completed the relationship between LU decomposition and Gaussian elimination. $\boldsymbol{U}$ is simply the upper triangular matrix resulting from Gaussian elimination, and the ULT matrix $\boldsymbol{L}$ is the inverse of the product of the $\boldsymbol{M}_k$'s.

Efficient Recovery of L $\boldsymbol{L}$ can be recovered from the $\boldsymbol{M}_k$'s in a very efficient manner without any explicit computation. The reason is that the $\boldsymbol{M}_k$'s have a very simple structure which can be exploited to our advantage. We note from (5.21) that

$$\boldsymbol{L} = \boldsymbol{M}_1^{-1} \dots \boldsymbol{M}_{n-1}^{-1}. \tag{5.23}$$

Therefore, we formulate $\boldsymbol{L}$ efficiently in two steps: we first examine the relationship between $\boldsymbol{M}_k^{-1}$ and $\boldsymbol{M}_k, k = 1, \dots, (n-1)$ and then investigate the structure of $\prod_{k=1}^{n-1} \boldsymbol{M}_k^{-1}$.

The Structure of $\boldsymbol{M}_k^{-1}$ We note that

$$\boldsymbol{M}_k \boldsymbol{A}^{(k-1)} = \boldsymbol{A}^{(k)} \tag{5.24}$$

The matrix $\boldsymbol{A}^{(k)}$ is formed from $\boldsymbol{A}^{(k-1)}$ by implicitly performing a *subtraction operation* as indicated by the structure of $\boldsymbol{M}_k$:

$$\boldsymbol{M}_k = \boldsymbol{I} - \boldsymbol{\alpha}^{(k)} \boldsymbol{e}_k^T. \tag{5.25}$$

The matrix $\boldsymbol{M}_k^{-1}$ must operate on $\boldsymbol{A}^{(k)}$ to restore $\boldsymbol{A}^{(k-1)}$. Thus, we may surmise that $\boldsymbol{M}_k^{-1}$ can be constructed by performing some form of an *addition operation* on $\boldsymbol{A}^{(k)}$. In this vein, consider $\boldsymbol{M}_k^{-1}$ of the following form:

$$\boldsymbol{M}_k^{-1} = \boldsymbol{I} + \boldsymbol{\alpha}^{(k)} \boldsymbol{e}_k^T. \tag{5.26}$$

We may prove this form is indeed the desired inverse, as follows. Using the definition of $\boldsymbol{M}_k^{-1}$ from (5.26), we have

$$\begin{aligned}
\boldsymbol{M}_k^{-1} \boldsymbol{M}_k &= (\boldsymbol{I} + \boldsymbol{\alpha}^{(k)} \boldsymbol{e}_k{}^T)(\boldsymbol{I} - \boldsymbol{\alpha}^{(k)} \boldsymbol{e}_k{}^T) \\
&= \boldsymbol{I} - \boldsymbol{\alpha}^{(k)} \boldsymbol{e}_k{}^T + \boldsymbol{\alpha}^{(k)} \boldsymbol{e}_k{}^T - \boldsymbol{\alpha}^{(k)} \underbrace{\boldsymbol{e}_k{}^T \boldsymbol{\alpha}^{(k)}}_{0} \boldsymbol{e}_k{}^T \\
&= \boldsymbol{I}.
\end{aligned} \tag{5.27}$$

Thus $\boldsymbol{M}_k^{-1}$ from (5.26) is indeed the inverse of $\boldsymbol{M}_k$. We can show that the underbrace term in (5.27) is zero by considering that $\boldsymbol{\alpha}^{(k)}$ from (5.17) has nonzero elements only for those indices which are greater than k (i.e. below the main diagonal position). The only nonzero element of $\boldsymbol{e}_k^T$ is in the kth (diagonal) position. Therefore, $\boldsymbol{e}_k^T \boldsymbol{\alpha}^{(k)} = 0$ as indicated. We therefore see that by looking at the structure of $\boldsymbol{M}_k$ carefully, we can perform the inversion operation simply by inverting a set of signs!

As an example, consider the matrix $\boldsymbol{M}_1$ in (5.14). By applying (5.26) to calculate $\boldsymbol{M}_1^{-1}$ it easily verified that

$$\begin{bmatrix} 1 & & & \\ -0.5 & 1 & & \\ 1.0 & & 1 & \\ -0.5 & & & 1 \end{bmatrix}\begin{bmatrix} 1 & & & \\ 0.5 & 1 & & \\ -1.0 & & 1 & \\ 0.5 & & & 1 \end{bmatrix} = \boldsymbol{I}. \tag{5.28}$$

The Structure of $\boldsymbol{L} = \prod_k \boldsymbol{M}_k^{-1}$ From (5.21) we have

$$\begin{aligned} \boldsymbol{L} &= (\boldsymbol{M}_{n-1}, \ldots, \boldsymbol{M}_1)^{-1} \\ &= \boldsymbol{M}_1^{-1}, \ldots, \boldsymbol{M}_{n-1}^{-1} \\ &= \left(\boldsymbol{I} + \boldsymbol{\alpha}^{1)}\boldsymbol{e}_1^T\right), \ldots, \left(\boldsymbol{I} + \boldsymbol{\alpha}^{(n-2)}\boldsymbol{e}_{n-2}^T\right)\left(\boldsymbol{I} + \boldsymbol{\alpha}^{(n-1)}\boldsymbol{e}_{n-1}^T\right), \end{aligned} \tag{5.29}$$

where (5.29) follows from (5.26). For ease of presentation, we first consider the product of only the last two factors of (5.29), which can be written as

$$\left(\boldsymbol{I} + \boldsymbol{\alpha}^{(n-2)}\boldsymbol{e}_{n-2}^T\right)\left(\boldsymbol{I} + \boldsymbol{\alpha}^{(n-1)}\boldsymbol{e}_{n-1}^T\right) \tag{5.30}$$

$$= \boldsymbol{I} + \boldsymbol{\alpha}^{(n-2)}\boldsymbol{e}_{n-2}^T + \boldsymbol{\alpha}^{(n-1)}\boldsymbol{e}_{n-1}^T + \boldsymbol{\alpha}^{(n-2)}\underbrace{\boldsymbol{e}_{n-2}^T\boldsymbol{\alpha}^{(n-1)}}_{0}\boldsymbol{e}_{n-1}^T \tag{5.31}$$

$$= \boldsymbol{I} + \sum_{i=n-2}^{n-1} \boldsymbol{\alpha}_i\boldsymbol{e}_i^T. \tag{5.32}$$

The term in the underbrace in (5.31) is also zero in this case due to the same reasoning as that used in (5.27). We now note that introducing the additional factor $\left(\boldsymbol{I} + \boldsymbol{\alpha}^{(n-3)}\boldsymbol{e}_{n-3}^T\right)$ corresponding to the third-last factor in (5.29) has only the effect of changing the lower limit of the sum in (5.32) from $(n-2)$ to $(n-3)$. The introduction of this additional factor does create cross-product terms as in the underbrace in (5.31), but these are also equal to zero for the same reason.

Therefore, after including all the factors from 1 to $n-1$, we have

$$\boldsymbol{L} = \boldsymbol{I} + \sum_{i=1}^{n-1} \boldsymbol{\alpha}^{(i)}\boldsymbol{e}_i^T. \tag{5.33}$$

Each term in (5.33) is a square matrix whose only non-zero entries are below the main diagonal in its distinct, respective column. Thus, the addition operation in (5.33) can be performed with no arithmetic operations. The addition of $\boldsymbol{I}$ in (5.33) puts 1's on the main diagonal to complete the formulation of $\boldsymbol{L}$.

We illustrate this procedure by following up on our earlier example. The quantities $\boldsymbol{\alpha}_k$ from the example are given by extracting the elements of the respective $\boldsymbol{M}_k$ below the diagonal in the kth column (after changing sign) as $[0.5, \ -1, \ 0.5]^T$, $[-2.667, \ 1]^T$ and -0.8889 for $k = 1, 2, 3$ respectively. Thus according to (5.33), $\boldsymbol{L}$ is given as

$$\boldsymbol{L} = \begin{bmatrix} 1 & & & \\ 0.5 & 1 & & \\ -1 & -2.67 & 1 & \\ 0.5 & 1 & -0.89 & 1 \end{bmatrix}. \tag{5.34}$$

We see that multiplying the matrices $\boldsymbol{L}$ from above and $\boldsymbol{U}$ from (5.14) indeed yields the original $\boldsymbol{A}$ given by (5.13), as shown:

$$\begin{bmatrix} 1 & & & \\ 0.5 & 1 & & \\ -1 & -2.67 & 1 & \\ 0.5 & 1 & -0.89 & 1 \end{bmatrix} \begin{bmatrix} 2 & -1 & 2 & 3 \\ & 1.5 & 3 & -0.5 \\ & & 9 & -0.33 \\ & & & 1.70 \end{bmatrix} = \boldsymbol{A} = \begin{bmatrix} 2 & -1 & 2 & 3 \\ 1 & 1 & 4 & 1 \\ -2 & -3 & -1 & -2 \\ 1 & 1 & -4 & 3 \end{bmatrix}.$$

Discussion

1. Note that in performing the sequence of Gauss transformations, we are performing *exactly the same arithmetic operations* as with elementary Gaussian elimination.
2. LU decomposition is a "high-level" description of Gaussian elimination. Matrix-level descriptions highlight connections between algorithms that may appear quite different at the scalar level.
3. The Gaussian elimination process requires $O(\frac{2n^3}{3})$ flops. This is the lowest number of any triangularisation technique for square matrices with no specific structure.

Evaluation of Determinants Our example provides an avenue for efficient evaluation of determinants. Recall $\det(\boldsymbol{AB}) = \det(\boldsymbol{A})\det(\boldsymbol{B})$. Here we have $\det(\boldsymbol{L}) = 1$ (since the determinant of a square, triangular matrix is the product of its diagonal elements which are all 1 in this case); also note that $\boldsymbol{U}$ is triangular so its determinant is the product of its diagonal elements. Therefore $\det(\boldsymbol{A})$, for $\boldsymbol{A}$ given from (5.13), is $\det(\boldsymbol{L})\det(\boldsymbol{U}) = \prod u_{ii} = 56$. This is a far more efficient method for evaluating determinants than the cofactor expansion method briefly mentioned in Chap. 1. The number of FLOPS required for the cofactor expansion method is proportional to $n!$, whereas this technique founded on the Gaussian elimination process requires only $\approx \frac{2n^3}{3}$ operations.

5.3 Numerical Properties of Gaussian Elimination

The numerical properties of Gaussian elimination can be described by the following statement [2]:

> Let $\hat{L}\hat{U}$ be the computed LU decomposition of $A \in \mathbb{R}^{n\times n}$. Then $\hat{y}$ is the computed solution to $\hat{L}\hat{y} = b$, and $\hat{x}$ the computed solution to $\hat{U}\hat{x} = \hat{y}$. Then,
>
> $$(A + E)\hat{x} = b,$$
>
> where
>
> $$|E| \leq nu\,[3|A| + 5|L||U|] + O(u^2). \tag{5.35}$$

This analysis, as in the back substitution case, shows that $\hat{x}$ exactly satisfies a perturbed system. The question is whether the perturbation $|E|$ is always small. If $|E|$ is of the order induced by floating point representation alone (i.e. $\mathcal{O}(nu)$), we may conclude that Gaussian elimination yields a solution which is as accurate as possible in the face of floating point error. But unlike the back substitution case, further inspection reveals that (5.35) does not allow such an optimistic outlook. It may happen during the course of the Gaussian elimination procedure that the term $|L||U|$ may become large, if small pivot elements are encountered, causing $|E|$ to become large, as we consider in the following.

By referring to (5.18), we can see that if any pivot $a_{kk}^{(k-1)}$ is small in magnitude, then the kth column of M_k is large in magnitude. Because M_k pre-multiplies $A^{(k-1)}$, large elements in M_k will result in large elements in the block $A_{22}^{(k)}$ of (5.15). The result is that both U and L will have large elements as k varies over its range from $1, \ldots n-1$. Hence, $|\ E\ |$ in (5.35) is "large", resulting in an inaccurate solution.

The fact that large $|\ L\ |$ and $|\ U\ |$ lead to an unstable solution can also be explained in a different way as follows. Consider two different LU decompositions on the same matrix A:

1. $A = LU$ (large pivots)
2. $A = \Lambda R$ (small pivots)

Two different LU decompositions on the same matrix can exist, because it is possible to interchange rows and columns of the $A_{22}^{(k-1)}$ block to place elements with either large or small magnitude as desired into the pivot position, using row and column interchanges, as described in more detail later. Generally, the elements l_{ij} and u_{ij} of L and U, respectively, are small, whereas the elements λ_{ij} and r_{ij} of $\mathbf{\Lambda}$ and $\mathbf{R}$ are large in magnitude. Consider the (i, j)th element a_{ij} of A computed according to the two different decompositions. We have

$$a_{ij} = \mathbf{l}_i{}^T\mathbf{u}_j \qquad \begin{cases} \mathbf{l}_i{}^T = i\text{th row of } L \\ \mathbf{u}_j = j\text{th column of } U \end{cases} \tag{5.36}$$

and

$$a_{ij} = \boldsymbol{\lambda}_i{}^T \mathbf{r}_j \qquad \{ \text{ likewise.} \tag{5.37}$$

Let us assume that the pivots in the second case are small enough so that

$$|\lambda_{ij}| \text{ and } |r_{ij}| \gg |a_{ij}|, \qquad (i, j) \in [1, \ldots n]. \tag{5.38}$$

Thus (5.37) can be written in the form

$$a_{ij} = P - N \tag{5.39}$$

where P, (N) is the sum of all terms in (5.37) which are positive (negative). We see that (5.38) implies that both $|P|, |N| \gg |a_{ij}|$. Thus when the pivots are sufficiently small in magnitude, from (5.39) we see that two nearly equal numbers are being subtracted, which leads to *catastrophic cancellation*, and ensuing numerical instability.

We note, however, that the P and N terms corresponding to (5.36) do not satisfy (5.38), and as a result, little or no catastrophic cancellation arises from the computation of (5.39). In this case the resulting system is stable.

Thus, for stability, *large pivots* are required. Otherwise, even well-conditioned systems can have large error in the solution, when computed using Gaussian elimination.

As an example of the effects of small pivots, consider the matrix $\boldsymbol{A}$ given by

$$\boldsymbol{A} = \begin{bmatrix} -0.2725 & -2.0518 & 0.5080 & 1.1275 \\ 1.0984 & -0.3538 & 0.2820 & 0.3502 \\ -0.2779 & -0.8236 & 0.0335 & -0.2991 \\ 0.7015 & -1.5771 & -1.3337 & 0.0229 \end{bmatrix},$$

which has been designed so that a single pivot element of very small magnitude on the order of 10^{-13} appears in the $(2, 2)$ position after the first stage of Gaussian elimination. The $\boldsymbol{L}$ and $\boldsymbol{U}$ matrices which result after completing the Gaussian elimination process without pivoting contain elements with very large magnitude, on the order of 10^{12}. The computed solution $\boldsymbol{x}$ obtained using the LU decomposition without pivoting, for $\boldsymbol{b} = [-0.6888, 10.0022, -1.3670, -2.1863]^T$ (to four decimal places), is given as

$$\boldsymbol{x} = \begin{bmatrix} 1.0826 \\ 0.9882 \\ 1.0622 \\ 0.9704 \end{bmatrix},$$

whereas the true solution is $[1, 1, 1, 1]^T$. The relative error in this computed solution is 0.1082, which may be regarded as significant, depending on the application.

On the other hand, the solution obtained using the Matlab® linear equation solver, which does use pivoting, yields the true solution within a relative error of approximately u, which is 2.2204×10^{-16} on the Matlab® platform used to obtain these results.

5.3.1 Pivoting

We can greatly improve the numerical stability of Gaussian elimination using a *pivoting* process, the objective of which is to place the element with the largest magnitude in the $\boldsymbol{A}_{22}$ block at the kth stage into the pivot position, which is the upper left corner of the $\boldsymbol{A}_{22}$ block. In this way, the Gaussian elimination is as stable as possible. This largest element can be moved into the pivot position through a row and column interchanges. In a manner similar to the Gaussian elimination description, we wish to express these row and column interchanges using matrix operations. It is readily verified that row interchanges can be accomplished using the following matrix multiplication:

$$\boldsymbol{A}^{(ij)} = \boldsymbol{P}^{(ij)}\boldsymbol{A}$$

where $\boldsymbol{A}^{(ij)}$ is the matrix $\boldsymbol{A}$ with rows i and j interchanged and $\boldsymbol{P}^{(ij)}$ (referred to as a *permutation* matrix) is the identity matrix with rows i and j interchanged. Likewise, to interchange columns k and l, the following matrix multiplication is performed:

$$\boldsymbol{A}^{(kl)} = \boldsymbol{A}\boldsymbol{\Pi}^{(kl)}$$

where the permutation matrix $\boldsymbol{\Pi}^{(kl)}$ is the identity matrix with columns k and l interchanged.

Full pivoting, where both row and column interchanges are performed, is stable yet expensive, since arithmetic comparisons are almost as costly as flops, and many comparisons are required to search through the entire $\boldsymbol{A}_{22}$ block at each stage to search for the element with the largest magnitude. Note that both row and column permutations take place to swap the respective element into the pivot position. However, the number of comparisons can be drastically reduced if only row permutations take place. That is, the element with the largest magnitude in the leading column of the $\boldsymbol{A}_{22}$ block is permuted into the pivot position using only row interchanges. The result, which is known as *partial pivoting*, is almost as stable.

Note that the row- and column-interchange operations will destroy the integrity of the system of the original system of equations $\boldsymbol{A}\boldsymbol{x} = \boldsymbol{b}$. In effect, the matrix $\boldsymbol{A}$ has been replaced by the quantity $\boldsymbol{P}\boldsymbol{A}\boldsymbol{\Pi}$, where $\boldsymbol{P} = \boldsymbol{P}_{n-1}, \ldots, \boldsymbol{P}_1$, and $\boldsymbol{\Pi} = \boldsymbol{\Pi}_1, \ldots, \boldsymbol{\Pi}_n - 1$, where $\boldsymbol{P}_i$ ($\boldsymbol{\Pi}_i$) is the respective row (column) permutation matrix at the ith stage of the decomposition. Therefore, a system of equations which is equivalent to the original can be written in the form

$$[\boldsymbol{P}\boldsymbol{A}\boldsymbol{\Pi}][\boldsymbol{\Pi}^T\boldsymbol{x}] = \boldsymbol{P}\boldsymbol{b},$$

where we have made use of the fact that $\boldsymbol{\Pi}\,\boldsymbol{\Pi}^T = \boldsymbol{P}\boldsymbol{P}^T = \boldsymbol{I}$.[1] Thus, for every row interchange, we also exchange corresponding elements of $\boldsymbol{b}$, and for every column interchange, we exchange corresponding elements of $\boldsymbol{x}$.

5.3.2 *The Cholesky Decomposition [2]*

We now consider several modifications to the LU decomposition, which ultimately lead up to the Cholesky decomposition. These modifications are 1) the LDM decomposition, 2) the LDL decomposition on symmetric matrices, and 3) the LDL decomposition on positive definite symmetric matrices. The Cholesky decomposition is relevant only for square symmetric positive-definite matrices and is an important concept in its own right. Several examples of the use of the Cholesky decomposition are provided at the end of the section.

The LDM Factorisation If no zero pivots are encountered during the Gaussian elimination process, then we can factor $\boldsymbol{U}$ so that

$$\boldsymbol{U} = \boldsymbol{D}\boldsymbol{M}^T$$

where $\boldsymbol{D}$ is diagonal and $\boldsymbol{M}$ is unit lower triangular (ULT). It is then apparent that

$$\boldsymbol{A} = \boldsymbol{L}\boldsymbol{D}\boldsymbol{M}^T.$$

The matrix $\boldsymbol{M}$ is calculated simply by dividing each row of $\boldsymbol{U}$ by its diagonal element d_{ii} and then taking the transpose of the result. As expected, the errors involved in solving a system of equations according to this factorisation behave in the same manner as with ordinary Gaussian elimination, as in (5.35). Thus, pivoting is required for this case, to prevent growth in computed versions of $|\boldsymbol{L}|$, $|\boldsymbol{D}|$ or $|\boldsymbol{M}|$.

It is straightforward to show [2] for a *symmetric* nonsingular matrix $\boldsymbol{A} \in \mathbb{R}^{n\times n}$, the factors $\boldsymbol{L}$ and $\boldsymbol{M}$ are identical and thus we may write $\boldsymbol{A} = \boldsymbol{L}\boldsymbol{D}\boldsymbol{L}^T$. This means that for a symmetric matrix $\boldsymbol{A}$, the LU factorisation requires only $\frac{n^3}{3}$ flops, instead of $\frac{2}{3}n^3$ as for the general case. This is because only the lower factor need be computed.

Now consider the case where $\boldsymbol{A}$ is *positive definite*. Define the symmetric part $\boldsymbol{T}$ and the asymmetric part $\boldsymbol{S}$ of $\boldsymbol{A}$, respectively, as

$$\mathbf{T} = \frac{\boldsymbol{A} + \boldsymbol{A}^T}{2}, \quad \boldsymbol{S} = \frac{\boldsymbol{A} - \boldsymbol{A}^T}{2}.$$

It is shown [2] that the computed solution $\hat{\boldsymbol{x}}$ to a positive definite system of equations (not necessarily symmetric) satisfies

$$(\boldsymbol{A} + \boldsymbol{E})\hat{\boldsymbol{x}} = \boldsymbol{b}$$

[1] The proof is straightforward and is left as an exercise.

where

$$||\boldsymbol{E}||_F \leq u\left[3n||\boldsymbol{A}||_F + 5cn^2\left(||\boldsymbol{T}||_2 + \left|\left|\boldsymbol{S}\boldsymbol{T}^{-1}\boldsymbol{S}\right|\right|_2\right)\right] + O(u^2) \tag{5.40}$$

where c is a constant of modest size. Equation (5.40) is a significant result, since it implies that when $\boldsymbol{A}$ is symmetric and positive definite, $||\boldsymbol{S}\boldsymbol{T}^{-1}\boldsymbol{S}||_2$ is zero and the $\boldsymbol{E}$ matrix for the bound (5.40) is close to the error introduced by floating-point representation alone. Also, since it is independent of the factors $\boldsymbol{L}$, $\boldsymbol{D}$, or $\boldsymbol{M}$, there is no opportunity for the elements to become large in absolute value during the Gaussian elimination process, reducing the prospects for catastrophic cancellation. Thus, the bound (5.40) is stable *without pivoting*. Pivoting is generally not required because positive definite matrices tend to have larger magnitude elements along the main diagonal, and hence the element with the largest magnitude is already in the pivot position during the Gaussian elimination process.

Incorporating the discussion for the symmetric and positive-definite cases together, we have the following.

The Cholesky Decomposition Itself For $\boldsymbol{A} \in \mathbb{R}^{n\times n}$ symmetric and positive definite, there exists a lower triangular matrix $\boldsymbol{G} \in \mathbb{R}^{n\times n}$ with positive diagonal entries, such that $\boldsymbol{A} = \boldsymbol{G}\boldsymbol{G}^T$.

Proof Consider $\boldsymbol{A}$ which is positive definite and symmetric. (Note that covariance matrices fall into this class.) Therefore, $\boldsymbol{x}^T\boldsymbol{A}\boldsymbol{x} > 0, \boldsymbol{0} \neq \boldsymbol{x} \in \mathbb{R}^{n\times n}$, and hence $\boldsymbol{x}^T\boldsymbol{L}\boldsymbol{D}\boldsymbol{L}^T\boldsymbol{x} > 0$. If $\boldsymbol{A}$ is positive definite, then $\boldsymbol{L}$ is full rank; let $\boldsymbol{y} \triangleq \boldsymbol{L}^T\boldsymbol{x}$. Then, $\boldsymbol{y}^T\boldsymbol{D}\boldsymbol{y} > 0$, *if and only if* all elements of $\boldsymbol{D}$ are positive. Therefore, if $\boldsymbol{A}$ is positive definite, then $d_{ii} > 0, i = 1\ldots, n$.

Because $\boldsymbol{A}$ is symmetric, then $\boldsymbol{A} = \boldsymbol{L}\boldsymbol{D}\boldsymbol{L}^T$. Because the d_{ii} are positive, then $\boldsymbol{G} = \boldsymbol{L}\,\mathrm{diag}(\sqrt{d_{11}}, \ldots, \sqrt{d_{nn}})$. Then $\boldsymbol{G}\boldsymbol{G}^T = \boldsymbol{A}$ as desired. □

Therefore, in solving the system $\boldsymbol{A}\boldsymbol{x} = \boldsymbol{b}$, where $\boldsymbol{A}$ is symmetric and positive definite (e.g. for the case where $\boldsymbol{A}$ is a sample covariance matrix), the Cholesky decomposition requires fewer flops than regular LU decomposition, since a properly designed algorithm can take advantage of the fact the two factors are transposes of each other. Further, the factorisation does not require pivoting. Both these points result in significantly reduced execution times.

Computation of the Cholesky Decomposition An algorithm for computing the Cholesky decomposition, which offers a faster computation time over the conventional LU decomposition, is developed simply by direct comparison, e.g., in the 3×3 case, we have

$$\begin{bmatrix} g_{11} & & \\ g_{21} & g_{22} & \\ g_{31} & g_{32} & g_{33} \end{bmatrix} \begin{bmatrix} g_{11} & g_{21} & g_{31} \\ & g_{22} & g_{32} \\ & & g_{33} \end{bmatrix} = \underbrace{\begin{bmatrix} a_{11} & a_{12} & a_{13} \\ a_{12} & a_{22} & a_{23} \\ a_{13} & a_{23} & a_{33} \end{bmatrix}}_{\substack{\text{symmetric,} \\ \text{positive definite.}}}$$

By following a proper order, each element of $\boldsymbol{G}$ may be determined in sequence, simply by comparing a particular element a_{ij} of $\boldsymbol{A}$ with the inner product $\boldsymbol{g}_i^T \boldsymbol{g}_j$, where $\boldsymbol{g}_i^T$ is taken to be the ith row of $\boldsymbol{G}$. First, we may determine g_{11} by comparison with a_{11}. Then, all remaining elements of the first column of $\boldsymbol{G}$ may be determined once g_{11} is known. Then, g_{22} can be determined, and the process repeats. For example,

$$g_{11}{}^2 = a_{11} \rightarrow g_{11} = \sqrt{a_{11}}$$

Also,

$$g_{i1} = \frac{a_{i1}}{g_{11}} \quad i = 2, \ldots, n.$$

Thus, all elements in first column of G can be solved. Now, consider the second column. First, we solve g_{22}:

$$g_{21}^2 + g_{22}^2 = a_{22}$$

Thus,

$$g_{22} = (a_{22} - g_{21}^2)^{\frac{1}{2}}$$

where the term in the round brackets is positive if $\boldsymbol{A}$ is positive definite. Once g_{22} is determined, all remaining elements in the second column may be found by comparison with corresponding element in the second column of $\boldsymbol{A}$. The third and remaining columns are solved in a similar way. If the process works its way in turn through columns $1, \ldots, n$, each element in $\boldsymbol{G}$ is found by solving a single equation in one unknown. Determining each diagonal element involves finding a square root of a particular quantity. This quantity is always positive if $\boldsymbol{A}$ is positive definite.

If $\boldsymbol{A} = \boldsymbol{G}\boldsymbol{G}^T$, then to solve the system $\boldsymbol{A}\boldsymbol{x} = \boldsymbol{b}$ we first solve

$$\boldsymbol{G}\boldsymbol{z} = \boldsymbol{b} \text{ for } \boldsymbol{z}$$

then

$$\boldsymbol{G}^T\boldsymbol{x} = \boldsymbol{z} \text{ for } \boldsymbol{x}$$

If the positive square root is always taken in the computation of the Cholesky factorisation, then the Cholesky factorisation is unique. The Cholesky decomposition $\boldsymbol{A} = \boldsymbol{G}\boldsymbol{G}^T$ is a matrix analog of a scalar square-root operation. Note however that a square root factor of a positive definite matrix is not unique. The matrix $\boldsymbol{G}\boldsymbol{Q}^T$, where $\boldsymbol{Q}^T$ is any orthonormal matrix of dimension $n \times k$, where $k \geq n$ is also a square root factor. Another square root matrix of $\boldsymbol{A}$ is given as $\boldsymbol{V} \cdot \text{diag}(\lambda_1, \lambda_2, \ldots, \lambda_n)^{1/2}$, where the $\boldsymbol{v}_i$ and λ_i are the eigenvectors and eigenvalues of $\boldsymbol{A}$, respectively. The uniqueness of the Cholesky factor is a result of it being lower triangular with positive diagonal elements. The advantage of the Cholesky factorisation is that it is easier to compute compared to other square root factors.

5.3.3 *Application of the Cholesky Decomposition*

Generating a Vector Process with a Prescribed Covariance We may use the Cholesky decomposition to generate a random vector process $\boldsymbol{x} \in \mathbb{R}^n$ with a desired covariance matrix $\boldsymbol{\Sigma} \in \mathbb{R}^{n \times n}$. Since $\boldsymbol{\Sigma}$ must be symmetric and positive definite, let

$$\boldsymbol{\Sigma} = \boldsymbol{G}\boldsymbol{G}^T$$

be the Cholesky factorisation of $\boldsymbol{\Sigma}$. Let $\boldsymbol{w} \in \mathbb{R}^n$ be a white random vector with zero mean such that $E(\boldsymbol{w}\boldsymbol{w}^T) = \boldsymbol{I}$. Such $\boldsymbol{w}$'s are easily generated by random number generators on the computer, such as the command "randn" in Matlab®.

Then, define $\boldsymbol{x}$ as

$$\boldsymbol{x} = \boldsymbol{G}\boldsymbol{w}$$

The vector process $\boldsymbol{x}$ has the desired covariance matrix because

$$\begin{aligned} E(\boldsymbol{x}\boldsymbol{x}^T) &= E(\boldsymbol{G}\boldsymbol{w}\boldsymbol{w}^T\boldsymbol{G}^T) \\ &= \boldsymbol{G}E(\boldsymbol{w}\boldsymbol{w}^T)\boldsymbol{G}^T \\ &= \boldsymbol{G}\boldsymbol{G}^T \\ &= \boldsymbol{\Sigma}. \end{aligned}$$

where in the second line, we have used the fact that the inner expectation is identical to $\boldsymbol{I}$. This procedure is particularly useful for computer simulations when it is desired to create a random vector process with a specified covariance matrix.

Whitening a Noise Process Consider the MUSIC example discussed in Chap. 2. In this case we observe the vector process

$$\boldsymbol{x}_i = \boldsymbol{S}(\boldsymbol{\theta})\boldsymbol{a}_i + \boldsymbol{n}_i \tag{5.41}$$

where in this case we assume the noise covariance matrix is $\boldsymbol{\Sigma}$, which is assumed known. As a general rule, and as illustrated in Chap. 7 on least squares analysis, estimation of parameters in non-white noise results in increased variances, relative to the case where the noise is white. We therefore wish to whiten the noise before the estimation process begins. In this vein, let $\boldsymbol{G}$ be the Cholesky factorisation of $\boldsymbol{\Sigma}$

such that $\boldsymbol{G}\boldsymbol{G}^T = \boldsymbol{\Sigma}$. Pre-multiply both sides of (5.41) above by $\boldsymbol{G}^{-1}$:

$$\boldsymbol{G}^{-1}\boldsymbol{x}_i = \boldsymbol{G}^{-1}\boldsymbol{S}(\boldsymbol{\theta})\boldsymbol{a}_i + \boldsymbol{G}^{-1}\boldsymbol{n}_i \tag{5.42}$$

The noise component is now $\boldsymbol{G}^{-1}\boldsymbol{n}_i$. The corresponding noise covariance matrix is

$$\begin{aligned} E(\boldsymbol{G}^{-1}\boldsymbol{n}_i\boldsymbol{n}_i^T\boldsymbol{G}^{-T}) &= \boldsymbol{G}^{-1}E(\boldsymbol{n}\boldsymbol{n}^T)\boldsymbol{G}^{-T} \\ &= \boldsymbol{G}^{-1}\boldsymbol{\Sigma}\boldsymbol{G}^{-T} \\ &= \boldsymbol{G}^{-1}\boldsymbol{G}\boldsymbol{G}^T\boldsymbol{G}^{-T} \\ &= \boldsymbol{I} \end{aligned} \tag{5.43}$$

Thus, by pre-multiplying the original signal $\boldsymbol{x}$ by the inverse Cholesky factor of the noise, the resulting noise becomes *white* with unit variance. Since the noise covariance matrix is diagonal, the elements of $\boldsymbol{n}$ become uncorrelated due to the whitening operation.

We note that as a consequence of this whitening process, the signal component has also been transformed by $\boldsymbol{G}^{-1}$. Therefore, in the specific case of the MUSIC algorithm, we must therefore substitute $\boldsymbol{G}^{-1}\boldsymbol{S}$ for $\boldsymbol{S}$ to achieve correct results.

Note that in both the above cases, any square root matrix $\boldsymbol{B}$ such that $\boldsymbol{B}\boldsymbol{B}^T = \boldsymbol{\Sigma}$ will achieve the same effects. However, the Cholesky factor is typically the easiest one to compute.

5.4 The Sensitivity of Linear Systems

Up to now, we have quantified the effective error in the matrix $\boldsymbol{A}$ due to the effects of potential catastrophic cancellation and to the inherent error in floating point number systems. Examples of these effects are the $\boldsymbol{E}$-matrices as specified in (5.40) and (5.35).

But the capability to quantify error does not address the complete problem. We also need to know how sensitive the computed solution $\hat{\boldsymbol{x}}$ is to error in the quantities $\boldsymbol{A}$ and $\boldsymbol{b}$. In this respect, in this section, we develop the idea of the matrix *condition number* $\kappa(\boldsymbol{A})$ of a matrix $\boldsymbol{A}$.

Consider the system of linear equations

$$\boldsymbol{A}\boldsymbol{x} = \boldsymbol{b} \tag{5.44}$$

where $\boldsymbol{A} \in \mathbb{R}^{n\times n}$ is nonsingular and $\boldsymbol{b} \in \mathbb{R}^n$. How do perturbations in $\boldsymbol{A}$ or $\boldsymbol{b}$ affect the solution $\boldsymbol{x}$?

To gain insight, we consider several situations where small perturbations can induce large errors in $\boldsymbol{x}$. For the first example, we perform the singular value decomposition on $\boldsymbol{A}$:

$$\boldsymbol{A} = \boldsymbol{U}\boldsymbol{\Sigma}\boldsymbol{V}^T. \tag{5.45}$$

Let us now consider a perturbed version $\tilde{\boldsymbol{A}}$ (following the method of (5.35) or (5.40)), where $\tilde{\boldsymbol{A}} = \boldsymbol{A} + \epsilon\boldsymbol{E}$, and as before $\boldsymbol{E}$ is an error matrix and ϵ controls the magnitude of error. In this example, let $\boldsymbol{E}$ be taken as the outer product $\boldsymbol{E} = \boldsymbol{u}_n\boldsymbol{v}_n^T$. Then, the singular value decomposition of $\tilde{\boldsymbol{A}}$ is identical to that for $\boldsymbol{A}$, except the transformed σ_n, denoted $\tilde{\sigma_n}$, is replaced with $\sigma_n + \epsilon$.

Because $\boldsymbol{x} = \boldsymbol{A}^{-1}\boldsymbol{b}$, we have

$$\boldsymbol{x} = \boldsymbol{V}\boldsymbol{\Sigma}^{-1}\boldsymbol{U}^T\boldsymbol{b},$$

or, using the outer product representation for matrix multiplication, we have

$$\boldsymbol{x} = \sum_{i=1}^{n} \boldsymbol{v}_i \frac{\boldsymbol{u}_i{}^T\boldsymbol{b}}{\sigma_i}. \tag{5.46}$$

If we assume σ_n to be small in comparison to σ_1 and that σ_n and epsilon are of the same order of magnitude, then $\tilde{\sigma}_n$ contains large relative error. Further, since the term for $i = n$ in (5.46) (i.e. the one which has been significantly perturbed) contributes strongly to $\boldsymbol{x}$ (since $\tilde{\sigma}_n$ is small), the computed version $\hat{\boldsymbol{x}}$ of $\boldsymbol{x}$ is strongly perturbed. Therefore a small change in $\boldsymbol{A}$ of the type described can result in large changes in the solution $\boldsymbol{x}$.

The "approximation theorem" of Chap. 3 provides an additional viewpoint. Here, we see that the smallest singular value σ_n is the 2-norm distance of $\boldsymbol{A}$ from the set of singular matrices. Consider the matrix $\boldsymbol{A}_{n-1}$ defined in the theorem. Then $\boldsymbol{A}_{n-1}$ is the closest singular matrix in the 2-norm sense to $\boldsymbol{A}$, and this 2-norm distance is σ_n. Thus, if σ_n is small, then $\boldsymbol{A}$ is close to singularity, in which case the solution can vary arbitrarily. Thus as σ_n becomes smaller, the computed $\boldsymbol{x}$ becomes more sensitive to changes in either $\boldsymbol{A}$ or $\boldsymbol{b}$.

These examples indicate that a small σ_n can cause large errors in $\boldsymbol{x}$. But we don't have a precise idea of what "small" means in this context. "Small" relative to what? The following section addresses this question.

5.4.1 Derivation of Condition Number

Consider the perturbed system where there are errors in both $\boldsymbol{A}$ and $\boldsymbol{b}$. Here, the notation is simpler if we denote the errors as $\delta\boldsymbol{A}$ and $\delta\boldsymbol{b}_1$, respectively. The perturbed system becomes

$$(\boldsymbol{A} + \delta\boldsymbol{A})(\boldsymbol{x} + \delta\boldsymbol{x}) = \boldsymbol{b} + \delta\boldsymbol{b}_1$$

We can write the above as

$$\boldsymbol{A}(\boldsymbol{x}+\delta \boldsymbol{x})=\boldsymbol{b}+\delta \boldsymbol{b}_1-\delta \boldsymbol{b}_2+\mathcal{O}(u)^2$$

where $\delta \boldsymbol{b}_2=\delta \boldsymbol{A}(\boldsymbol{x}+\delta \boldsymbol{x})=\delta \boldsymbol{A}\boldsymbol{x}+\mathcal{O}(u)^2$ to a first-order approximation, which is reasonable since the errors are assumed small. The error $\delta \boldsymbol{b}_2$ is the error in $\boldsymbol{A}$ transformed to appear as an error in $\boldsymbol{b}$. Defining $\delta \boldsymbol{b} \triangleq \delta \boldsymbol{b}_1-\delta \boldsymbol{b}_2$, we have

$$\boldsymbol{A}(\boldsymbol{x}+\delta \boldsymbol{x})=\boldsymbol{b}+\delta \boldsymbol{b}+\mathcal{O}(u)^2.$$

We therefore have the following two equations, where we have suppressed the error term:

$$\boldsymbol{A}\boldsymbol{x}=\boldsymbol{b} \Longrightarrow \boldsymbol{x}=\boldsymbol{A}^{-1}\boldsymbol{b} \tag{5.47}$$

$$\boldsymbol{A}\delta \boldsymbol{x}=\delta \boldsymbol{b} \Longrightarrow \delta \boldsymbol{x}=\boldsymbol{A}^{-1}\delta \boldsymbol{b}. \tag{5.48}$$

The above leads to

$$\frac{||\delta \boldsymbol{x}||_2^2}{||\boldsymbol{x}||_2^2}=\frac{||\boldsymbol{A}^{-1}\delta \boldsymbol{b}||_2^2}{||\boldsymbol{A}^{-1}\boldsymbol{b}||_2^2}.$$

We now consider what is the worst possible relative error $\frac{||\delta \boldsymbol{x}||}{||\boldsymbol{x}||}$ in the solution $\boldsymbol{x}$ in the 2-norm sense. This occurs (for a fixed $||\delta \boldsymbol{b}||_2$ and $||\boldsymbol{b}||_2$) when the direction of $\delta \boldsymbol{b}$ from (5.48) is such that $||\delta \boldsymbol{x}||_2$ is maximum and, simultaneously, when $\boldsymbol{b}$ from (5.47) is in the direction such that the corresponding $||\boldsymbol{x}||_2$ is minimum.

Since the singular values of $\boldsymbol{A}^{-1}$ are the reciprocals of those of $\boldsymbol{A}$, the largest singular value of $\boldsymbol{A}^{-1}$ is $1/\sigma_n$, and the smallest is $1/\sigma_1$. Likewise, the $\boldsymbol{u}$-vector associated the largest singular value is $\boldsymbol{u}_n$, and is $\boldsymbol{u}_1$ if associated with the smallest singular value. With this in mind, it is straightforward to show from the ellipsoidal interpretation of the SVD of $\boldsymbol{A}^{-1}$ that the maximum of $||\delta \boldsymbol{x}||_2=||\boldsymbol{A}^{-1}\delta \boldsymbol{b}||_2$ with respect to $\delta \boldsymbol{b}$, for $||\delta \boldsymbol{b}||_2$ held constant, occurs when $\delta \boldsymbol{b}$ has the form $\delta \boldsymbol{b}=\boldsymbol{u}_n||\delta \boldsymbol{b}||_2$. Then

$$\begin{aligned}
\max_{\delta \boldsymbol{b}} ||\boldsymbol{A}^{-1}\delta \boldsymbol{b}||_2 &= \max_{\delta \boldsymbol{b}} ||\boldsymbol{V}\boldsymbol{\Sigma}^{-1}\boldsymbol{U}^T\delta \boldsymbol{b}||_2 \\
&= \max_{\delta \boldsymbol{b}} ||\boldsymbol{\Sigma}^{-1}\boldsymbol{U}^T\delta \boldsymbol{b}||_2 \\
&= \left|\left|\boldsymbol{\Sigma}^{-1}\boldsymbol{U}^T\ \boldsymbol{u}_n||\delta \boldsymbol{b}||_2\right|\right|_2
\end{aligned}$$

$$= \left\| \boldsymbol{\Sigma}^{-1} \begin{bmatrix} 0 \\ \vdots \\ 0 \\ ||\delta \boldsymbol{b}||_2 \end{bmatrix} \right\|_2$$

$$= \frac{1}{\sigma_n} ||\delta \boldsymbol{b}||_2. \tag{5.49}$$

In the second line, we have used the fact that the 2-norm is invariant to multiplication by the orthonormal matrix $\boldsymbol{V}$. The third line follows from the fact that the maximum occurs when $\delta \boldsymbol{b} = ||\delta \boldsymbol{b}||_2 \boldsymbol{u}_n$, and so from the orthonormality of $\boldsymbol{U}$, the quantity $\boldsymbol{U}^T \delta \boldsymbol{b}$ is a vector of zeros except for the last element, which is equal to $||\delta \boldsymbol{b}||_2$.

Using analogous logic, we see that the minimum of $||\boldsymbol{A}^{-1}\boldsymbol{b}||_2$ in (5.47) for fixed $||\boldsymbol{b}||_2$ occurs when the direction of $\boldsymbol{b}$ aligns with $\boldsymbol{u}_1$. In this case, following the same process as in (5.49), except replacing the maximum with minimum, we have

$$\min_{\boldsymbol{b}} ||\boldsymbol{A}^{-1}\boldsymbol{b}||_2 = \frac{1}{\sigma_1} ||\boldsymbol{b}||_2. \tag{5.50}$$

We can now use (5.50) and (5.49) as worst-case values in (5.47) and (5.48), respectively, to evaluate the worst-case upper bound on the relative error $\frac{||\delta \boldsymbol{x}||_2}{||\boldsymbol{x}||_2}$ in $\boldsymbol{x}$. We have

$$\frac{||\delta \boldsymbol{x}||_2}{||\boldsymbol{x}||_2} \leq \frac{\sigma_1}{\sigma_n} \frac{||\delta \boldsymbol{b}||_2}{||\boldsymbol{b}||_2}. \tag{5.51}$$

The quantity $\frac{||\delta \boldsymbol{b}||_2}{||\boldsymbol{b}||_2}$ in (5.51) may be interpreted as the relative error in $\boldsymbol{A}$ and $\boldsymbol{b}$. This relative error is magnified by the factor $\frac{\sigma_1}{\sigma_n}$ to give the worst-case relative error in the solution $\boldsymbol{x}$. The ratio $\frac{\sigma_1}{\sigma_n}$ is an important quantity in matrix analysis and is referred to as the *condition number* of the matrix $\boldsymbol{A}$ and is given the symbol $\kappa_2(\boldsymbol{A})$. The subscript 2 refers to the 2-norm used in the derivation in this case. In fact, the condition number may be derived using any suitable norm, as discussed in the Appendix of this chapter.

The analysis for this section gives an interpretation of the meaning of the condition number $\kappa_2(\boldsymbol{A})$. It also indicates in what directions $\boldsymbol{b}$ and $\delta \boldsymbol{b}$ must point to result in the maximum relative error in $\boldsymbol{x}$. We see for worst error performance, $\delta \boldsymbol{b}$ points along the direction of $\boldsymbol{u}_n$, and $\boldsymbol{b}$ points along $\boldsymbol{u}_1$. If the "SVD ellipsoid" is elongated, then there is a large disparity in the relative growth factors in $\delta \boldsymbol{x}$ and $\boldsymbol{x}$, and large relative error in $\boldsymbol{x}$ can result.

Properties of the Condition Number

1. $\kappa(\boldsymbol{A}) \geq 1$.
2. If $\kappa(\boldsymbol{A}) \sim 1$, we say the system is *well-conditioned*, and the error in the solution is of the same magnitude as that of $\boldsymbol{A}$ and $\boldsymbol{b}$.

3. If $\kappa(\boldsymbol{A})$ is large, then the system is *poorly conditioned*, and small errors in $\boldsymbol{A}$ or $\boldsymbol{b}$ could result in large errors in $\boldsymbol{x}$. In the practical case, the errors can be treated as random variables and hence are likely to have components along *all* the vectors $\boldsymbol{u}_i$, including $\boldsymbol{u}_n$. Thus in a practical situation with poor conditioning, error growth in the solution is almost certain to occur.
4. It is interesting to note that the *best-case* error growth factor is less than unity. This could happen if the directions for $\delta\boldsymbol{b}$ and $\boldsymbol{b}$ are reversed; i.e. $\delta\boldsymbol{b}$ aligns with $\boldsymbol{u}_1$ and $\boldsymbol{b}$ aligns with $\boldsymbol{u}_n$. However, in view of Item 3 above, this favourable scenario is unlikely to happen.
5. $\kappa(\boldsymbol{A})$ is norm-dependent. In the Appendix, an alternative derivation of condition number is presented, which is better suited to showing that any suitable norm can be used to evaluate it. The type of norm used is typically indicated by a subscript on the κ.

We still must consider how bad the condition number can be before it starts to seriously affect the accuracy of the solution for a given floating-point precision. In ordinary numerical systems, the errors in $\boldsymbol{A}$ or $\boldsymbol{b}$ result from the floating point representation of the numbers. The maximum relative error in the floating point number is u. The condition number $\kappa(\boldsymbol{A})$ is the worst-case factor by which this floating-point error is magnified in the solution. Thus, the relative error in the solution $\boldsymbol{x}$ is bounded from above by the quantity $O(u\kappa(\boldsymbol{A}))$. Therefore, if $\kappa(\boldsymbol{A}) \sim \frac{1}{u}$, then the relative error in the solution can approach unity, which means the result is meaningless. If $\kappa(\boldsymbol{A}) \sim \frac{10^{-r}}{u}$, then the relative error in the solution can be taken as 10^{-r}, and the solution is approximately correct to r decimal places.

Questions

What is the condition number of an orthonormal matrix?
What is the condition number of a singular matrix?

Example Here we consider a poorly conditioned 2×2 system of equations, given as

$$\begin{bmatrix} 1 & 1 \\ 0.92 & 1.08 \end{bmatrix} \begin{bmatrix} x_1 \\ x_2 \end{bmatrix} = \begin{bmatrix} 1 \\ 1 \end{bmatrix} \tag{5.52}$$

The two equations are shown plotted together in Fig. 5.1, where it may be seen they are close to being co-linear. The singular values of $\boldsymbol{A}$ are $[2.0016, 0.0799]$, to give a value of $\kappa(\boldsymbol{A}) = 25.0401$. The solution to the unperturbed system is $\boldsymbol{x} = [0.5, 0.5]$. The $\boldsymbol{U}$-matrix from the SVD of $\boldsymbol{A}$ is given as

$$\boldsymbol{U} = \begin{bmatrix} -0.7060 & -0.7082 \\ -0.7082 & 0.7060 \end{bmatrix}.$$

We now perturb the $\boldsymbol{b}$-vector in (5.52) in the direction corresponding to the worst-case error in $\boldsymbol{x}$, which is along the $\boldsymbol{u}_2$-axis. The perturbed value of $\boldsymbol{b}$ is given as

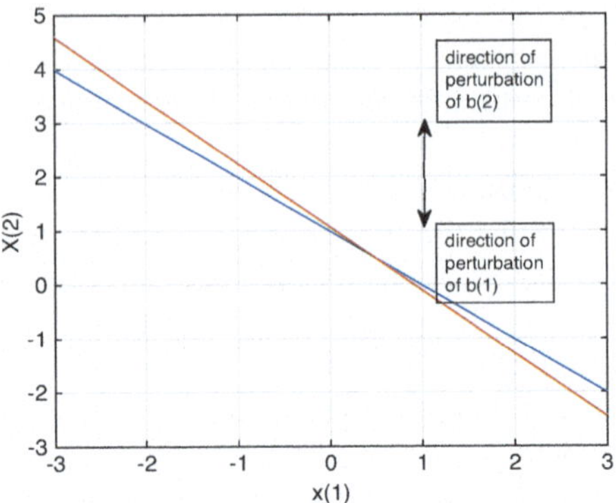

Fig. 5.1 Superimposed graphs of the equations for a 2×2 poorly conditioned system of equations. The value for $\boldsymbol{b}$ is $[1, 1]^T$, which is along the same direction as $\boldsymbol{u}_1$, whereas the perturbations $\delta\boldsymbol{b}$ in $\boldsymbol{b}$ are $[-0.01, 0.01]^T$, which is along the direction of $\boldsymbol{u}_2$. Visual inspection shows that this arrangement results in a relatively large shift in the point of intersection of the two lines, which corresponds to the solution of the system of equations

$\boldsymbol{b}_p = \boldsymbol{b} + 0.01\boldsymbol{u}_2$, where the value 0.01 was chosen to be the magnitude of the perturbation. The resulting directions in which $b(1)$ and $b(2)$ are perturbed are indicated in Fig. 5.1. We note that the unperturbed $\boldsymbol{b}$ already points approximately along the $\boldsymbol{u}_1$-direction, which is the direction corresponding to the worst-case error in the solution.

We now solve the perturbed system and compare the corresponding relative error in the solution with the upper bound given by (5.51), repeated here for convenience:

$$\frac{||\delta\boldsymbol{x}||_2}{||\boldsymbol{x}||_2} \leq \frac{\sigma_1}{\sigma_n}\frac{||\delta\boldsymbol{b}||_2}{||\boldsymbol{b}||_2}.$$

The value of $\frac{||\delta\boldsymbol{b}||_2}{||\boldsymbol{b}||_2}$ on the right, using the current values is 0.0070711. $\kappa(\boldsymbol{A}) = 25.0401$, so the worst-case relative error in $\boldsymbol{x}$ predicted by (5.51) is **0.17706**. The perturbed solution is $\boldsymbol{x} = [0.40807, 0.58485]^T$, which results in an actual relative error of [**0.17692**], which is seen to be very close to the worst-case upper bound predicted by (5.51), as it should be in this case.

An example of a well-conditioned system of equations is shown in Fig. 5.2, where it is seen in this case the two equations are almost orthogonal. The same experiment as above was conducted for this system of equations. In this case, $\kappa(\boldsymbol{A}) = 1.23721$, and $\frac{||\delta\boldsymbol{b}||_2}{||\boldsymbol{b}||_2} = 0.0070711$. The actual relative error in $\boldsymbol{x}$ is **0.0071661**, whereas the relative error of the worst-case upper bound is **0.0087479**. Thus we can see that in this case, due to the improved condition number, little growth in the error of the solution is evident, and the actual relative error in $\boldsymbol{x}$ is still close to the value predicted by the upper bound.

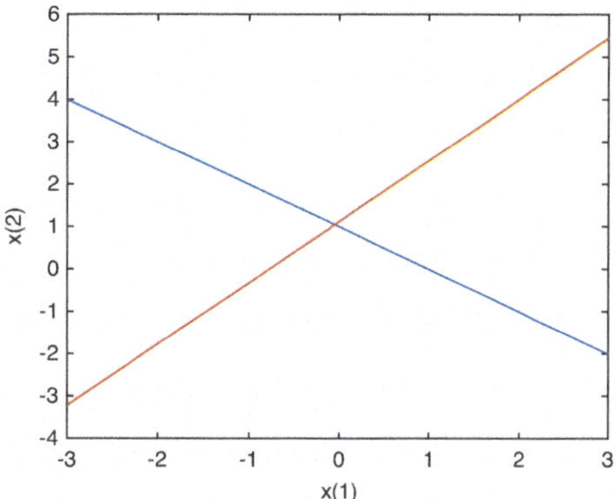

Fig. 5.2 Graphs corresponding to Fig. 5.1, but for a well-conditioned system of equations

5.5 The Interlacing Theorem and Condition Numbers [2]

Here we discuss a useful theorem which is useful in the context of condition numbers. We have a symmetric matrix $\boldsymbol{B}_n \in \mathbb{R}^{n\times n}$. Let $\boldsymbol{B}_r = \boldsymbol{B}(1 : r, 1 : r)$. Then

$$\lambda_{r+1}(\boldsymbol{B}_{r+1}) \leq \lambda_r(\boldsymbol{B}_r) \leq \lambda_r(\boldsymbol{B}_{r+1}) \leq \ldots \leq \lambda_2(\boldsymbol{B}_{r+1}) \leq \lambda_1(\boldsymbol{B}_r) \leq \lambda_1(\boldsymbol{B}_{r+1})$$

for $r = 1 : n - 1$. Here $\lambda_i(\boldsymbol{B})$ indicates the ith-largest eigenvalue of $\boldsymbol{B}$. To look at how this theorem is useful in the condition number context, we consider a square, symmetric $(n - 1) \times (n - 1)$ matrix $\boldsymbol{B}_{n-1}$. Its largest and smallest eigenvalues, respectively, are $\lambda_1(\boldsymbol{B}_{n-1})$ and $\lambda_{n-1}(\boldsymbol{B}_{n-1})$. We can denote the respective condition number as $\kappa(\boldsymbol{B}_{n-1})$. Now we add a column and row to form $\boldsymbol{B}_n$ (in such a way so that $\boldsymbol{B}_n$ remains symmetric). The largest and smallest eigenvalues of $\boldsymbol{B}_n$ are now $\lambda_1(\boldsymbol{B}_n)$ and $\lambda_n(\boldsymbol{B}_n)$. We can infer from the interlacing theorem that

$$\lambda_1(\boldsymbol{B}_n) \geq \lambda_1(\boldsymbol{B}_{n-1}), \text{ and}$$
$$\lambda_n(\boldsymbol{B}_n) \leq \lambda_{n-1}(\boldsymbol{B}_{n-1})$$

where we have set the value of r above equal to $n - 1$ in each case. These equations imply that $\kappa(\boldsymbol{B}_n) \geq \kappa(\boldsymbol{B}_{n-1})$. This means that increasing the size of a square symmetric matrix never decreases the condition number, and only under special conditions does it remain unchanged.

This treatment has special relevance when $\boldsymbol{B}$ is a covariance matrix, of the form $\boldsymbol{X}^T\boldsymbol{X}$. Adding an extra coumn to $\boldsymbol{X}$ increases the dimension of $\boldsymbol{B}$ by 1, in which case the condition number is most likely to increase. This fact has significant consequences in linear least squares estimation, which we discuss in Chaps. 7–9. Adding a column to $\boldsymbol{X}$ has the likely effect of making its columns more linearly dependent, thus increasing the condition number.

In Chap. 10, we discuss the concept of *regularisation*, which is a means of mitigating the effect of a poor condition number when solving a system of equations.

Appendix

Alternate Derivation of Condition Number

We now develop the idea of the *condition number*, which gives us a precise definition of the sensitivity of $\boldsymbol{x}$ to changes in $\boldsymbol{A}$ or $\boldsymbol{b}$ in Eq. (5.44). Now consider the perturbed system

$$(\mathbf{A} + \epsilon\mathbf{F})\mathbf{x}(\epsilon) = \mathbf{b} + \epsilon\mathbf{f} \tag{5.53}$$

where

ϵ is a small scalar
$\boldsymbol{F} \in \mathbb{R}^{n \times n}$ and $\boldsymbol{f} \in \mathbb{R}^n$ are errors
$\boldsymbol{x}(\epsilon)$ is the perturbed solution, such that $\boldsymbol{x}(0) = \boldsymbol{x}$.

We wish to place a lower bound on the relative error in $\boldsymbol{x}$ due to the perturbations. Since $\boldsymbol{A}$ is nonsingular, we can differentiate (5.53) implicitly wrt ϵ:

$$(\mathbf{A} + \epsilon\mathbf{F})\dot{\mathbf{x}}(\epsilon) + \mathbf{F}\mathbf{x}(\epsilon) = \mathbf{f} \tag{5.54}$$

For $\epsilon = 0$ we get

$$\dot{\mathbf{x}}(0) = \boldsymbol{A}^{-1}(\mathbf{f} - \mathbf{F}\mathbf{x}). \tag{5.55}$$

The Taylor series expansion for $\boldsymbol{x}(\epsilon)$ about $\epsilon = 0$ has the form

$$\mathbf{x}(\epsilon) = \mathbf{x} + \epsilon\dot{\mathbf{x}}(0) + O(\epsilon^2). \tag{5.56}$$

Substituting (5.55) into (5.56), we get

$$\mathbf{x}(\epsilon) - \mathbf{x} = \epsilon\mathbf{A}^{-1}(\mathbf{f} - \mathbf{F}\mathbf{x}) + O(\epsilon^2) \tag{5.57}$$

Hence by taking norms, we have

$$\begin{aligned} ||\boldsymbol{x}(\epsilon) - \boldsymbol{x}|| &= \left|\left|\epsilon\boldsymbol{A}^{-1}(\mathbf{f} - \mathbf{F}\mathbf{x}) + O(\epsilon^2)\right|\right| \\ &\leq \epsilon \left|\left|\boldsymbol{A}^{-1}(\mathbf{f} - \mathbf{F}\mathbf{x})\right|\right| + O(\epsilon^2) \end{aligned}$$

where the triangle inequality has been used, i.e. $||\boldsymbol{A}+\mathbf{b}|| \leq ||\boldsymbol{A}|| + ||\mathbf{b}||$. Using the property of p-norms, $||\boldsymbol{Ab}|| \leq ||\boldsymbol{A}||\,||\boldsymbol{b}||$, we have

$$\begin{aligned}||\boldsymbol{x}(\epsilon)-\boldsymbol{x}|| &\leq \epsilon \left|\left|\boldsymbol{A}^{-1}\right|\right| \, ||\{\mathbf{f}-\mathbf{Fx}\}|| + O(\epsilon^2)\\ &\leq \epsilon \left|\left|\boldsymbol{A}^{-1}\right|\right| \{||\mathbf{f}|| + ||\mathbf{Fx}||\} + O(\epsilon^2)\\ &\leq \epsilon \left|\left|\boldsymbol{A}^{-1}\right|\right| \{||\mathbf{f}|| + ||\mathbf{F}||\,||\mathbf{x}||\} + O(\epsilon^2).\end{aligned}$$

Therefore the relative error in $\boldsymbol{x}(\epsilon)$ can be expressed as

$$\begin{aligned}\frac{||\boldsymbol{x}(\boldsymbol{\epsilon})-\boldsymbol{x}||}{||\boldsymbol{x}||} &\leq \epsilon \left|\left|\boldsymbol{A}^{-1}\right|\right| \left\{\frac{||\mathbf{f}||}{||\boldsymbol{x}||} + ||\mathbf{F}||\right\} + O(\epsilon^2)\\ &= \epsilon \left|\left|\boldsymbol{A}^{-1}\right|\right| \, ||\boldsymbol{A}|| \left\{\frac{||\mathbf{f}||}{||\boldsymbol{A}||\,||\boldsymbol{x}||} + \frac{||\mathbf{F}||}{||\boldsymbol{A}||}\right\} + O(\epsilon^2).\end{aligned}$$

But since $\boldsymbol{Ax} = \boldsymbol{b}$, then $||\boldsymbol{b}|| \leq ||\boldsymbol{A}||\,||\boldsymbol{x}||$ and we have

$$\frac{||\boldsymbol{x}(\boldsymbol{\epsilon})-\boldsymbol{x}||}{||\boldsymbol{x}||} \leq \epsilon \left|\left|\boldsymbol{A}^{-1}\right|\right| \, ||\boldsymbol{A}|| \left\{\frac{||\mathbf{f}||}{||\mathbf{b}||} + \frac{||\mathbf{F}||}{||\boldsymbol{A}||}\right\} + O(\epsilon^2). \tag{5.58}$$

There are many interesting things about (5.58):

1. The left-hand side $= \frac{||\boldsymbol{x}(\epsilon)-\mathbf{x}||}{||\boldsymbol{x}||}$ is the *relative* error in $\boldsymbol{x}$ due to the perturbation.
2. $\epsilon \frac{||\mathbf{f}||}{||\mathbf{b}||}$ is the relative error in $\boldsymbol{b} \triangleq \rho_b$
3. $\epsilon \frac{||\mathbf{F}||}{||\boldsymbol{A}||}$ is the relative error in $\boldsymbol{A} \triangleq \rho_A$
4. $\left|\left|\boldsymbol{A}^{-1}\right|\right| \, ||\boldsymbol{A}||$ is defined as the *condition number* $\kappa(\boldsymbol{A})$ of $\boldsymbol{A}$.

From (5.58) we write

$$\frac{||\mathbf{x}(\boldsymbol{\epsilon})-\mathbf{x}||}{||\boldsymbol{x}||} \leq \kappa(\boldsymbol{A})(\rho_A + \rho_B) + O(\epsilon^2) \tag{5.59}$$

Thus we have the important result: eq. (5.59) says that, to a first-order approximation, the relative error in the computed solution $\boldsymbol{x}$ is bounded by the expression $\kappa(\boldsymbol{A})\times$ (relative error in $\boldsymbol{A}$ + relative error in $\boldsymbol{b}$). This is a rather intuitively satisfying result. Thus the condition number $\kappa(\boldsymbol{A})$ is the maximum amount the relative error in $\boldsymbol{A} + \boldsymbol{b}$ is magnified to give the relative error in the solution $\boldsymbol{x}$.

The condition number $\kappa(\boldsymbol{A})$ is norm-dependent. The most common norm is the 2-norm. In this case, $||\boldsymbol{A}||_2 = \sigma_1$. Further, since the singular values of $\boldsymbol{A}^{-1}$ are the reciprocals of those of $\boldsymbol{A}$, it is easy to verify that $\left|\left|\boldsymbol{A}^{-1}\right|\right|_2 = \sigma_n^{-1}$. Therefore, from the definition of condition number, we have

$$\kappa_2(\boldsymbol{A}) = \frac{\sigma_1}{\sigma_n} \tag{5.60}$$

Condition Number and Power Spectral Density

Theorem 5.1 *The condition number of a covariance matrix representing a random process is bounded from above by the ratio of the maximum to minimum value of the corresponding power spectrum of the process.*

Proof *(from [3])* Let $\boldsymbol{R} \in \mathbb{R}^{n\times n}$ be the covariance matrix of a stationary or wide-sense stationary random process $\boldsymbol{x}$, with corresponding eigenvectors $\boldsymbol{v}_i$ and eigenvalues λ_i. In this treatment, the eigenvectors do not necessarily have unit 2-norm. Consider the *Rayleigh quotient* discussed in Sect. 5.3:

$$\lambda_i = \frac{\boldsymbol{v}_i^T \boldsymbol{R}\boldsymbol{v}_i}{\boldsymbol{v}_i^T \boldsymbol{v}_i}. \tag{5.61}$$

The quadratic form in the numerator may be expressed in an expanded form as

$$\boldsymbol{v}_i^T \boldsymbol{R}\boldsymbol{v}_i = \sum_{k=1}^{n}\sum_{m=1}^{n} v_{ik} r(k-m) v_{im} \tag{5.62}$$

where v_{ik} denotes the kth element of the ith eigenvector $\boldsymbol{v}_i$ matrix $\boldsymbol{V}$ and $r(k-m)$ is the (k,m)th element of $\boldsymbol{R}$. Using the Wiener-Khintchine relation[2] we may write

$$r(k-m) = \frac{1}{2\pi}\int_{-\pi}^{\pi} S(\omega) e^{j\omega(k-m)} d\omega. \tag{5.63}$$

where $S(\omega)$ is the power spectral density of the process. Substituting (5.63) into (5.62) we have

$$\begin{aligned}\boldsymbol{v}_i^T \boldsymbol{R}\boldsymbol{v}_i &= \frac{1}{2\pi}\sum_{k=1}^{n}\sum_{m=1}^{n} v_{ik} v_{im}\int_{-\pi}^{\pi} S(\omega) e^{j\omega(k-m)} d\omega \\ &= \frac{1}{2\pi}\int_{-\pi}^{\pi} S(\omega) d\omega \sum_{k=1}^{n} v_{ik} e^{j\omega k}\sum_{m=1}^{n} v_{im} e^{-j\omega m}.\end{aligned} \tag{5.64}$$

At this point, we interpret the eigenvector $\boldsymbol{v}_i$ as a waveform in time. Let its corresponding Fourier transform $V_i(e^{j\omega})$ be given as

$$V_i(e^{j\omega}) = \sum_{k=1}^{n} v_{ik} e^{-j\omega k}. \tag{5.65}$$

[2] This relation states that the autocorrelation sequence $r(\cdot)$ and the power spectral density $S(\omega)$ are a Fourier transform pair [3].

We may therefore express (5.64) as

$$\boldsymbol{v}_i^T \boldsymbol{R} \boldsymbol{v}_i = \frac{1}{2\pi} \int_{-\pi}^{\pi} \mid V_i(e^{j\omega}) \mid^2 S(\omega) d\omega. \tag{5.66}$$

It may also be shown that

$$\boldsymbol{v}_i^T \boldsymbol{v}_i = \frac{1}{2\pi} \int_{-\pi}^{\pi} \mid V_i(e^{j\omega}) \mid^2 d\omega. \tag{5.67}$$

Substituting (5.66) and (5.67) into (5.61), we have

$$\lambda_i = \frac{\int_{-\pi}^{\pi} \mid V_i(e^{j\omega}) \mid^2 S(\omega) d\omega}{\int_{-\pi}^{\pi} \mid V_i(e^{j\omega}) \mid^2 d\omega}. \tag{5.68}$$

As an aside, (5.68) has an interesting interpretation in itself. The numerator may be regarded as the integral of the output power spectral density of a filter with coefficients $\boldsymbol{v}_i$, driven by the input process $\boldsymbol{x}$. The ith eigenvalue is this quantity normalised by the squared norm of $\boldsymbol{v}_i$.

Let $S_{\min}$ and $S_{\max}$ be the absolute minimum and maximum values of $S(\omega)$, respectively. Then it follows that

$$\int_{-\pi}^{\pi} \mid V_i(e^{j\omega}) \mid^2 S(\omega) d\omega \geq S_{\min} \int_{-\pi}^{\pi} \mid V_i(e^{j\omega}) \mid^2 d\omega \tag{5.69}$$

and

$$\int_{-\pi}^{\pi} \mid V_i(e^{j\omega}) \mid^2 S(\omega) d\omega \leq S_{\max} \int_{-\pi}^{\pi} \mid V_i(e^{j\omega}) \mid^2 d\omega \tag{5.70}$$

Hence, from (5.68) we can say that the eigenvalues λ_i are bounded by the maximum and minimum values of the spectrum $S(\omega)$ as follows:

$$S_{\min} \leq \lambda_i \leq S_{\max}, \qquad i = 1, \ldots, n. \tag{5.71}$$

Further, the condition number $\kappa(\boldsymbol{R})$ is bounded as

$$\kappa(\boldsymbol{R}) \leq \frac{S_{\max}}{S_{\min}}. \tag{5.72}$$

A consequence of (5.72) is that if a covariance matrix $\boldsymbol{R}$ is rank deficient, then there exist values of $\omega \in [-\pi, \pi]$ such that the power spectrum is zero. □

Problems

1. Suggest a sequence of modified Gauss transforms $N_i, i = n, n-1 \ldots, 2$, so that

$$D = NU$$

where D is diagonal, U is full-rank upper triangular, and $N = N_2, \ldots, N_{n-1}N_n$. All matrices are $n \times n$. Let $U_{i-1} = N_{i-1} \ldots N_n U$. Show that U_i differs from U_{i-1} only in the ith column. Use this fact to propose an efficient means of calculating N from the N_i. *Hint:* The N_i's are calculated in the order $n, n-1, \ldots, 2$.

2. Apply the regular Gaussian elimination process on the matrix on the augmented matrix $A_1 = [A \quad I]$ where A is invertible, to give the result $A_2 = [U \quad B_1]$. What is B_1?

   Apply a sequence of the modified Gaussian transforms as above on A_2 so that the U-partition becomes diagonal, to give $A_3 = [D \quad B_2]$. Then form $A_4 = D^{-1}A_3 = [I \quad B_3]$. What is B_3?

3. Let $A \in \mathbb{R}^{n\times n} = \begin{bmatrix} A_{11} & A_{12} \\ A_{21} & A_{22} \end{bmatrix}$, where $A_{11} \in \mathbb{R}^{k\times k}$ is nonsingular. Then $S = A_{22} - A_{21}A_{11}^{-1}A_{12}$ is called the *Schur complement* of A. Show that after k steps of the Gaussian elimination algorithm without pivoting, A_{22} has been replaced by S.

4. We have an observed data matrix X whose covariance matrix $R = X^T X$. Suggest a transform B on X so that the covariance matrix corresponding to XB equals I.

5. A white noise process is fed through a low-pass filter with cutoff frequency of f_o and a monotonic rolloff characteristic. The process is sampled in accordance with Nyquist's criterion at a frequency f_s only slightly larger than $2f_o$. The covariance matrix R_1 of this process is evaluated. Then, the sampling frequency is increased well above the value $2f_o$ and the resulting covariance matrix R_2 is again evaluated. Compare the condition number of R_1 with that of R_2 and explain your reasoning carefully. *Hint:* Consider section "Condition Number and Power Spectral Density".

6. Give bases for the row and column subspaces of A in terms of its L and U factors.

7. On the course website, you will find a .mat file named Ch5Prob7.mat. It contains a very poorly conditioned matrix A and a vector b.

   (a) What is the condition number of A in comparison to the matlab® machine epsilon? (These may be determined by the matlab® commands "cond" and "eps", respectively).

   (b) Solve the system $Ax = b$ and compare the computed result x with the true solution $x_o = [1, 1, 1, 1, 1]^T$. How does the relative error in x compare with the condition number bound?

(c) Calculate the relative residual errors $\frac{||Ax-b||}{||A||}$ and $\frac{||Ax_o-b||}{||A||}$ and explain your findings in comparison with the relative error in computed solution x.

(d) The matrix A has an approximate nullspace $\mathcal{N}(A)$. What is it? Add a vector $\delta x \in \mathcal{N}(A)$ to x_o to give the vector x_1 (where δx_1 has moderate to small norm) and again evaluate the relative residual error $\frac{||Ax_1-b||}{||A||}$. Explain your result.

8. For the system of equations given by (5.52), determine the directions for both b and δb so that $\frac{||\delta x||_2}{||x||_2}$ is minimum. Compare the actual value of the relative error obtained with these choices of b and δb to that corresponding to the best-case bound.
9. Prove that

$$||A||_2||(A^TA)^{-1}A^T||_2 = \kappa_2(A), \text{ and}$$
$$||A||_2^2||(A^TA)^{-1}||_2 = \kappa_2(A)^2.$$

Hint: Use the SVD.

10. How large should t (the number of bits in the fractional part of fl(x)) be to ensure the solution to a system of linear equations is (approximately) accurate to r decimal places when $K(A) = 10^6$?

References

1. M. Brain et al., An automatable formal semantics for IEEE-754 floating-point arithmetic. in *2015 IEEE 22nd Symposium on Computer Arithmetic* (2015), pp. 160–167. https://doi.org/10.1109/ARITH.2015.26
2. G.H. Golub, C.F. Van Loan, *Matrix Computations*, 3rd edn. (The Johns Hopkins University, Baltimore, 1996)
3. S. Haykin, *Adaptive Filter Theory Prentice Hall*, 4th edn. (Englewood Cliffs, 2001)

Chapter 6
The QR Decomposition

6.1 Overview of the QR Decomposition

Given $\boldsymbol{A} \in \mathbb{R}^{m \times n}$, then the QR decomposition may be defined as

$$\boldsymbol{A} = \boldsymbol{Q}\boldsymbol{R},$$

where $\boldsymbol{Q} \in \mathbb{R}^{m \times m}$ is orthonormal and $\boldsymbol{R} \in \mathbb{R}^{m \times n}$ is upper triangular.

Unless $\boldsymbol{A}$ is square, $\boldsymbol{R}$ must be padded with a block of zeros to maintain dimensional consistency. If $\boldsymbol{A}$ is tall, then the triangular portion of $\boldsymbol{R}$ is padded with an $(m - n) \times n$ block of zeros from below as in (6.1). When $\boldsymbol{A}$ is short, then an $m \times (n - m)$ is padded to the right. In most practical cases of interest, $m > n$, and our following discussion assumes this fact.

For $m \geq n$, we can partition the QR decomposition in the following manner:

$$\underset{n}{m\begin{bmatrix} \boldsymbol{A} \end{bmatrix}} = \underset{n \;\; m-n}{\begin{bmatrix} \boldsymbol{Q}_1 & \boldsymbol{Q}_2 \end{bmatrix}} \underset{n}{\begin{bmatrix} \boldsymbol{R}_1 \\ \boldsymbol{0} \end{bmatrix}} \begin{matrix} n \\ m-n \end{matrix} . \tag{6.1}$$

Then we have the following properties for the QR decomposition:

1. If $\boldsymbol{A} = \boldsymbol{Q}\boldsymbol{R}$ is a QR factorisation of a tall, full-rank matrix $\boldsymbol{A}$ as defined above, then

$$\text{span}(\boldsymbol{a}_1, \boldsymbol{a}_2, \ldots, \boldsymbol{a}_k) = \text{span}(\boldsymbol{q}_1, \boldsymbol{q}_2, \ldots, \boldsymbol{q}_k), \qquad k = 1, \ldots, n. \tag{6.2}$$

   This follows from the fact that since $\boldsymbol{R}$ is upper triangular, the column $\boldsymbol{a}_k$ is a linear combination of the columns $[\boldsymbol{q}_1, \ldots, \boldsymbol{q}_k]$ for $k = 1, \ldots, n$.

J. Reilly, *Fundamentals of Linear Algebra for Signal Processing*,
https://doi.org/10.1007/978-3-031-68915-4_6

2. Further to the above, if $\boldsymbol{A}$ is tall, then[1]

$$R(\boldsymbol{A}) = R(\boldsymbol{Q}_1)$$
$$R(\boldsymbol{A})_{\perp} = R(\boldsymbol{Q}_2),$$

and $\boldsymbol{A} = \boldsymbol{Q}_1\boldsymbol{R}_1$. If $\boldsymbol{R}_1$ has positive diagonal entries, then it is *unique*. Furthermore, $\boldsymbol{R}_1 = \boldsymbol{G}^T$, where $\boldsymbol{G}$ is the Cholesky factor of $\boldsymbol{A}^T\boldsymbol{A}$.
3. If $\boldsymbol{A} = \boldsymbol{Q}_1\boldsymbol{R}_1$ and $\boldsymbol{R}_1$ is nonsingular, then

$$\boldsymbol{A}\boldsymbol{R}_1^{-1} = \boldsymbol{Q}_1. \tag{6.3}$$

Hence, $\boldsymbol{R}_1^{-1}$ is a matrix which *orthonormalises* $\boldsymbol{A}$. In fact, this property not only holds for $\boldsymbol{R}_1^{-1}$ but also holds for any inverse square root factor of the matrix $\boldsymbol{A}^T\boldsymbol{A}$.

We note the QR decomposition does not possess the optimality property of the eigenvectors, as expressed by Lemma 2.1 of Sect. 2.5.

We now consider the *Gram–Schmidt* procedure which justifies the existence of the QR decomposition and provides a means of computing it.

6.2 Classical Gram–Schmidt (CGS)

We specify "classical" because later we consider other forms of the GS procedure. In the following, we assume $\boldsymbol{A}$ is full rank; extension to the rank-deficient case is considered later. The CGS method involves successively forming orthonormal columns $\boldsymbol{Q}$ from the columns of $\boldsymbol{A}$, beginning at the first column. The first column $\boldsymbol{q}_1$ is defined as shown in Fig. 6.1 as

$$\boldsymbol{q}_1 = \frac{\boldsymbol{a}_1}{||\boldsymbol{a}_1||_2}.$$

Thus, we see $\boldsymbol{q}_1$ is a vector of unit norm. The element r_{11} of $\boldsymbol{R}$ is given as $||\boldsymbol{a}_1||_2$.

Now consider the formation of the second column $\boldsymbol{q}_2$. The columns $\boldsymbol{a}_1$ and $\boldsymbol{a}_2$ of $\boldsymbol{A}$ are represented as shown in Fig. 6.1.

Because $\boldsymbol{R}_1$ is upper triangular, the column $\boldsymbol{a}_2$ is a linear combination of $\boldsymbol{q}_1$ and $\boldsymbol{q}_2$:

$$\boldsymbol{a}_2 \in \text{span}(\boldsymbol{q}_1, \boldsymbol{q}_2). \tag{6.4}$$

[1] Do not confuse the notation in this section: $R(\cdot)$ denotes *range*, and $\boldsymbol{R}$ denotes an upper triangular matrix.

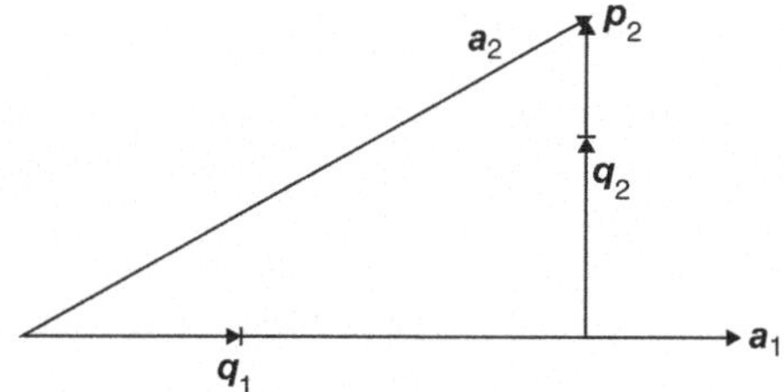

Fig. 6.1 Geometry of the Gram–Schmidt method for QR decomposition

Since $\boldsymbol{Q}$ is to be orthonormal, $\boldsymbol{q}_2$ must also satisfy

$$||\boldsymbol{q}_2||_2 = 1, \tag{6.5}$$

$$\boldsymbol{q}_2 \perp \boldsymbol{q}_1. \tag{6.6}$$

From Fig. 6.1, we may satisfy (6.4)–(6.6) by considering a vector $\boldsymbol{p}_2$, which is the projection of $\boldsymbol{a}_2$ onto the orthogonal complement subspace of $\boldsymbol{q}_1$. The vector $\boldsymbol{p}_2$ is thus defined as

$$\boldsymbol{p}_2 = \boldsymbol{P}_2^{\perp}\boldsymbol{a}_2 = (\boldsymbol{I} - \boldsymbol{q}_1\boldsymbol{q}_1^T)\boldsymbol{a}_2.$$

Then, the vector $\boldsymbol{q}_2$ is determined by normalising the 2-norm of $\boldsymbol{p}_2$:

$$\boldsymbol{q}_2 = \frac{\boldsymbol{p}_2}{||\boldsymbol{p}_2||_2}.$$

We define the matrix $\boldsymbol{Q}^{(2)} \triangleq [\boldsymbol{q}_1, \boldsymbol{q}_2]$. In this case, the second column $\boldsymbol{r}_2$ of $\boldsymbol{R}_1$ contains the coefficients of $\boldsymbol{a}_2$ relative to the basis $\boldsymbol{Q}^{(2)}$. Thus,

$$\boldsymbol{r}_2 = \left(\boldsymbol{Q}^{(2)}\right)^T \boldsymbol{a}_2.$$

To generalise, at the kth stage, we have the matrix $\boldsymbol{Q}^{(k-1)} = [\boldsymbol{q}_1, \ldots, \boldsymbol{q}_{k-1}]$. The vector $\boldsymbol{q}_k$ may be determined by finding the vector $\boldsymbol{p}_k$ which is the projection of $\boldsymbol{a}_k$ onto the orthogonal complement subspace $R(\boldsymbol{Q}^{(k-1)})^{\perp}$. Thus,

$$\boldsymbol{p}_k = (\boldsymbol{P}_k)^{\perp}\boldsymbol{a}_k = \left(\boldsymbol{I} - \boldsymbol{Q}^{(k-1)}\boldsymbol{Q}^{(k-1)T}\right)\boldsymbol{a}_k$$

and

$$\boldsymbol{q}_k = \frac{\boldsymbol{p}_k}{||\boldsymbol{p}_k||_2}.$$

We now define $\boldsymbol{Q}^{(k)} \triangleq [\boldsymbol{Q}^{(k-1)}, \boldsymbol{q}_k]$. The column $\boldsymbol{r}_k$ is then defined as

$$\boldsymbol{r}_k = \left(\boldsymbol{Q}^{(k)}\right)^T \boldsymbol{a}_k.$$

We then increment k by 1 and iterate the process until $k = n - 1$. At that point, the matrices $\boldsymbol{Q}$ and $\boldsymbol{R}$ are completely determined.

We note that the matrix $\boldsymbol{Q} \in \mathbb{R}^{m \times n}$ generated from the Gram–Schmidt process is not a full orthonormal matrix in the case when $m > n$. There are only n columns, which however are enough to determine an orthonormal basis for $R(\mathbf{A})$, and to be useful in solving least squares problems. This is in contrast to other methods, to be discussed later, which give the complete orthonormal matrix $\boldsymbol{Q} \in \mathbb{R}^{m \times m}$.

Unfortunately, the classical GS method is not numerically stable. This is because if columns of $\boldsymbol{A}$ are close to linear dependence, catastrophic cancellation occurs in computing the $\boldsymbol{p}_k$. This error is quickly compounded, because the resulting errored $\boldsymbol{q}_k$ are used in successive stages and quickly lose orthogonality. Nevertheless, the CGS process is useful as an analytical tool and for geometric interpretation.

6.2.1 Modified G–S Method for QR Decomposition

In this section we investigate an alternative method for computing the QR decomposition by the Gram–Schmidt process. It requires the same number of flops as classical GS but is stable. Thus, this modified method offers an effective method of computing the QR decomposition.

We are given a matrix $\boldsymbol{A} \in \mathbb{R}^{m \times n}$, $m > n$. In this case (as with classical Gram–Schmidt), the matrix $\boldsymbol{Q}$ which is obtained contains only n columns.

Because $\boldsymbol{R}$ is upper triangular, we can write successive columns of $\boldsymbol{A}$ as

$$\begin{aligned}
\boldsymbol{a}_1 &= r_{11}\boldsymbol{q}_1 \\
\boldsymbol{a}_2 &= r_{12}\boldsymbol{q}_1 + r_{22}\boldsymbol{q}_2 \\
\boldsymbol{a}_3 &= r_{13}\boldsymbol{q}_1 + r_{23}\boldsymbol{q}_2 + r_{33}\boldsymbol{q}_3 \\
\vdots \quad & \quad \vdots \qquad\quad \vdots \qquad\quad \vdots \qquad \ddots \\
\boldsymbol{a}_n &= r_{1n}\boldsymbol{q}_1 + r_{2n}\boldsymbol{q}_2 + r_{3n}\boldsymbol{q}_3 \cdots r_{nn}\boldsymbol{q}_n.
\end{aligned} \tag{6.7}$$

With classical G–S, the matrix $\boldsymbol{R}$ is computed column-wise. However, with this modified G–S procedure, we note $\boldsymbol{R}$ is computed row by row.

To describe the method, we note that the first column $\boldsymbol{q}_1$ of $\boldsymbol{Q}$ is given as $\boldsymbol{q}_1 = \boldsymbol{a}_1/r_{11}$ and $r_{11} = ||\boldsymbol{a}_1||_2$. Since $\boldsymbol{Q}^T\boldsymbol{A} = \boldsymbol{R}$, the elements r_{ij} of $\boldsymbol{R}$ equal $\boldsymbol{q}_i^T\boldsymbol{a}_j$, for $i < j$. Therefore the remaining elements $[r_{12}, \ldots, r_{1n}]^T$ of the first row $\boldsymbol{r}_1^T$ of $\boldsymbol{R}$ are given as

$$[r_{12}, \ldots, r_{1n}] = \boldsymbol{q}_1^T \boldsymbol{A}(:, 2:n),$$

where the Matlab® notation has been used. We see that the first column on the right in (6.7) is now completely determined. We can proceed to the second stage of the algorithm by forming a matrix $\boldsymbol{B}$ by subtracting this first column from both sides of (6.7):

$$\boldsymbol{B}^{(1)} = \boldsymbol{A} - \boldsymbol{q}_1 \boldsymbol{r}_1^T.$$

The first column of $\boldsymbol{B}^{(1)}$ is therefore zero. Then from (6.7), we have

$$\begin{aligned} \boldsymbol{b}_2 &= r_{22}\boldsymbol{q}_2 \\ \boldsymbol{b}_3 &= r_{23}\boldsymbol{q}_2 + r_{23}\boldsymbol{q}_3 \\ \vdots \quad & \quad \vdots \qquad\quad \vdots \quad \ddots \\ \boldsymbol{b}_n &= r_{2n}\boldsymbol{q}_2 + r_{3n}\boldsymbol{q}_3 \quad \ldots \; r_{nn}\boldsymbol{q}_n. \end{aligned} \tag{6.8}$$

From (6.8) it is evident that the column $\boldsymbol{q}_2$ and row $\boldsymbol{r}_2^T$ may be formed from $\boldsymbol{B}^{(1)}$ in exactly the same manner as $\boldsymbol{q}_1$ and $\boldsymbol{r}_1^T$ were formed from $\boldsymbol{A}$. The method proceeds n steps in this way until completion.

To formalise the process, assume we are at the kth stage of the decomposition. At this stage we determine kth column of $\boldsymbol{Q} = \boldsymbol{q}_k$ and the kth row of $\boldsymbol{R} = \boldsymbol{r}_k^T$. We define the matrix $\boldsymbol{A}^{(k)}$ in the following way:

$$\underbrace{\boldsymbol{A} - \textstyle\sum_{i=1}^{k-1} \boldsymbol{q}_i \boldsymbol{r}_i^T}_{\uparrow \text{ sum of outer products}} \quad = \quad [\underset{k-1}{\boldsymbol{0}}, \; \underset{n-k+1}{\boldsymbol{A}^{(k)}}].$$

We partition $\boldsymbol{A}^{(k)}$ as

$$\boldsymbol{A}^{(k)} = [\underset{1}{z} \; \underset{n-k}{\boldsymbol{B}^{(k)}}] \; m. \tag{6.9}$$

This situation above corresponds to having just subtracted out the $(k-1)$th column in (6.7). Then,

$$r_{kk} = ||z||_2$$

and

$$\boldsymbol{q}_k = \frac{z}{r_{kk}}.$$

The kth row of $\boldsymbol{R}$ may then be calculated as

$$\left[r_{k,k+1}, \ldots, r_{k,n}\right] = \boldsymbol{q}_k^T \boldsymbol{B}^{(k)}.$$

We now proceed to the $(k+1)$th stage by removing the component $\boldsymbol{q}_k$ from each column in $\boldsymbol{B}^{(k)}$:

$$\boldsymbol{A}^{(k+1)} = \boldsymbol{B}^{(k)} - \boldsymbol{q}_k \left[r_{k,k+1}, \ldots, r_{k,n} \right].$$

Then, increment k and go to (6.9).

This method, unlike the classical Gram Schmidt, is *stable*. The numerical stability results from the fact that errors in $\boldsymbol{q}_k$ at the kth stage are not compounded into succeeding stages. It also requires the same number of flops as classical G–S. It may therefore be observed that modified G–S is a very attractive method for computing the QR decomposition, since it has excellent stability properties, coupled with relatively few flops for its computation.

6.3 Householder Transformations

Householder transforms are generally the preferred method for performing the QR decomposition. They are fast to implement and have stable numerical properties.

6.3.1 Description of the Householder Algorithm

We have seen previously that the vector $\boldsymbol{x}_S = \boldsymbol{P}\boldsymbol{x}$ is the projection of $\boldsymbol{x}$ onto the range of $\boldsymbol{P}$ and that

$$\boldsymbol{x}_\perp = (\boldsymbol{I} - \boldsymbol{P})\boldsymbol{x}$$

is the projection of $\boldsymbol{x}$ onto the orthogonal complement subspace of $R(\boldsymbol{P})$. We now have a new variation of the projector matrix. Specifically, the matrix

$$\boldsymbol{H} = \boldsymbol{I} - 2\boldsymbol{P} \tag{6.10}$$

is a *reflection* matrix. The vector $\boldsymbol{x}_r = \boldsymbol{H}\boldsymbol{x}$ is a reflection of $\boldsymbol{x}$ in the orthogonal complement subspace of $R(\boldsymbol{P})$. This fact may be justified with the aid of Fig. 6.2. In this figure and in the sequel, we assume the matrix $\boldsymbol{P}$ is defined in terms of a single vector $\boldsymbol{v}$ as $\boldsymbol{P} = \boldsymbol{v}(\boldsymbol{v}^T\boldsymbol{v})^{-1}\boldsymbol{v}^T$. It is easily verified that the matrix $\boldsymbol{H}$ defined by (6.10) is orthonormal and symmetric.

The $\boldsymbol{H}$ matrices may be used to zero out selected components of a vector. For example, by choosing the vector $\boldsymbol{v}$ in the appropriate fashion, all elements of a vector $\boldsymbol{x}$ may be zeroed, except the first, x_1. This is done by choosing $\boldsymbol{v}$ so that the reflection of $\boldsymbol{x}$ in $\text{span}(\boldsymbol{v})^\perp$ lines up with the x_1-axis. Thus, in this manner, all elements of $\boldsymbol{x}$ are eliminated except the first.

We can use this property to perform a QR decomposition on a matrix $\boldsymbol{A}$. The method proceeds in a manner similar to Gaussian elimination, in that we eliminate

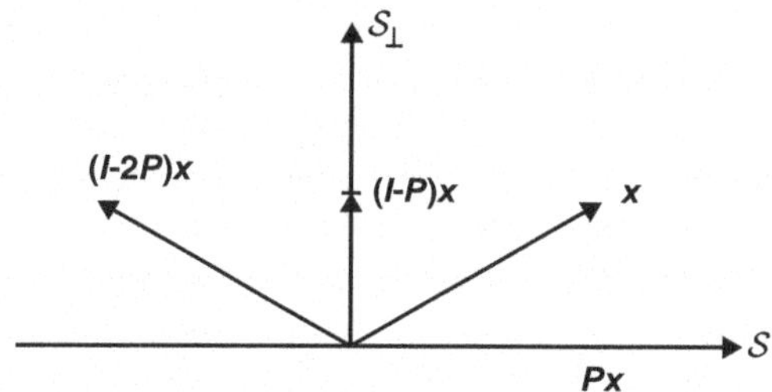

Fig. 6.2 Geometry of a reflector matrix

all elements below the main diagonal in each column successively. We can find a vector $\boldsymbol{v}_1$ so that

$$\boldsymbol{A}_1 = \boldsymbol{H}_1\boldsymbol{A},$$

where $\boldsymbol{H}_1$ is defined from $\boldsymbol{v}_1$ according to (6.10). The matrix $\boldsymbol{A}_1$ has zeros below the main diagonal in the first column, as desired. Then, we can find a new $\boldsymbol{v}_2$ so that

$$\boldsymbol{A}_2 = \boldsymbol{H}_2\boldsymbol{A}_1$$

has zeros below the main diagonal in both the first and second columns. This may be done by designing $\boldsymbol{H}_2$ so that the first column of $\boldsymbol{H}_2\boldsymbol{A}_1$ is the same as that of $\boldsymbol{A}_1$, and so that the second column of $\boldsymbol{H}_2\boldsymbol{A}_1$ is zero below the main diagonal.

The process continues for $n-1$ stages. At that stage we have

$$\boldsymbol{R} \stackrel{\Delta}{=} \boldsymbol{A}_{n-1} = \boldsymbol{H}_{n-1}\ldots\boldsymbol{H}_1\boldsymbol{A}, \tag{6.11}$$

where $\boldsymbol{R}$ is upper triangular.

Because the $\boldsymbol{H}$'s are orthonormal, $\prod_{i=1}^{n}\boldsymbol{H}_i$ is also orthonormal. Thus, from (6.11), we have

$$\boldsymbol{A} = \boldsymbol{Q}\boldsymbol{R},$$

where $\boldsymbol{Q}^T = \prod_{i=n}^{1}\boldsymbol{H}_i$, and thus the QR decomposition is complete.

Let us now consider the first stage of the Householder process. Extension to other stages is done later. How do we choose $\boldsymbol{P}$ (or more specifically $\boldsymbol{v}$) so that $\boldsymbol{y} = (\boldsymbol{I} - 2\boldsymbol{P})\boldsymbol{x}$ has zeros in every position except the first, for any $\boldsymbol{x} \in \mathbb{R}^n$? That is, how do we define $\boldsymbol{v}$ so that $\boldsymbol{y} = \boldsymbol{H}\boldsymbol{x}$ is a multiple of $\boldsymbol{e}_1$? Here goes:

$$\begin{aligned}\boldsymbol{H}\boldsymbol{x} &= (\boldsymbol{I} - 2\boldsymbol{P})\boldsymbol{x}\\ &= \left(\boldsymbol{I} - 2\boldsymbol{v}(\boldsymbol{v}^T\boldsymbol{v})^{-1}\boldsymbol{v}^T\right)\boldsymbol{x}\\ &= \boldsymbol{x} - \frac{2\boldsymbol{v}^T\boldsymbol{x}}{\boldsymbol{v}^T\boldsymbol{v}}\boldsymbol{v}.\end{aligned} \tag{6.12}$$

Householder made the observation that if $\boldsymbol{v}$ is to reflect the vector $\boldsymbol{x}$ onto the $\boldsymbol{e}_1$ -axis, then $\boldsymbol{v}$ must be in the same plane as that defined by $[\boldsymbol{x}, \boldsymbol{e}_1]$, or in other words, $\boldsymbol{v} \in \text{span}(\boldsymbol{x}, \boldsymbol{e}_1)$. Accordingly, we set $\boldsymbol{v} = \boldsymbol{x} + \alpha \boldsymbol{e}_1$, where α is a scalar to be determined. At this stage, this assignment may appear to be rather arbitrary, but as we see later, it leads to a simple and elegant result.

Substituting this definition for $\boldsymbol{v}$ into (6.12), where x_1 is the first element of $\boldsymbol{x}$, we get

$$\boldsymbol{v}^T \boldsymbol{x} = \boldsymbol{x}^T \boldsymbol{x} + \alpha x_1 \tag{6.13}$$

$$\boldsymbol{v}^T \boldsymbol{v} = \boldsymbol{x}^T \boldsymbol{x} + 2\alpha x_1 + \alpha^2. \tag{6.14}$$

Thus,

$$\begin{aligned} \boldsymbol{H}\boldsymbol{x} &= \boldsymbol{x} - \frac{2\boldsymbol{v}^T \boldsymbol{x}}{\boldsymbol{v}^T \boldsymbol{v}} [\boldsymbol{x} + \alpha \boldsymbol{e}_1] \\ &= \left[1 - \frac{2(\boldsymbol{x}^T \boldsymbol{x} + \alpha x_1)}{\boldsymbol{x}^T \boldsymbol{x} + 2\alpha x_1 + \alpha^2} \right] \boldsymbol{x} - 2\alpha \frac{\boldsymbol{v}^T \boldsymbol{x}}{\boldsymbol{v}^T \boldsymbol{v}} \boldsymbol{e}_1. \end{aligned} \tag{6.15}$$

To make $\boldsymbol{H}\boldsymbol{x}$ have zeros everywhere except in the first component, the first term above is forced to zero. If we set $\alpha = ||\boldsymbol{x}||_2$, then the first term is

$$\left[1 - \frac{2\left(||\boldsymbol{x}||_2^2 + ||\boldsymbol{x}||_2 x_1 \right)}{\left(||\boldsymbol{x}||_2^2 + 2\, ||\boldsymbol{x}||_2 x_1 + ||\boldsymbol{x}||_2^2 \right)} \right] = 0$$

as desired. Substituting this choice of α into (6.13) and (6.14), respectively, we see that $\boldsymbol{v}^T \boldsymbol{x} = ||\boldsymbol{x}||_2^2 + x_1\, ||\boldsymbol{x}||_2^2$, whereas $\boldsymbol{v}^T \boldsymbol{v} = 2\, ||\boldsymbol{x}||_2^2 + 2x_1\, ||\boldsymbol{x}||_2^2 = 2\boldsymbol{v}^T \boldsymbol{x}$. The last term in (6.15) then becomes

$$\boldsymbol{H}\boldsymbol{x} = -\, ||\boldsymbol{x}||_2\, \boldsymbol{e}_1. \tag{6.16}$$

Thus, we see that by defining $\boldsymbol{v} = \boldsymbol{x} + ||\boldsymbol{x}||_2\, \boldsymbol{e}_1$, then $\boldsymbol{H}\boldsymbol{x}$ has zeros everywhere except in the first position.

Note that we could also have achieved the same effect by setting $\alpha = -||\boldsymbol{x}||_2$ in (6.15). The choice of sign of α affects the numerical stability of the algorithm. If $\boldsymbol{x}$ is close to a multiple of $\boldsymbol{e}_1$, then $\boldsymbol{v} = \boldsymbol{x} - \text{sign}(x_1)||\boldsymbol{x}||_2 \boldsymbol{e}_1$ has small norm; hence large relative error can exist in factor $\beta \triangleq \frac{2}{\boldsymbol{v}^T \boldsymbol{v}}$. This difficulty can be avoided if the sign of α is chosen as the sign of x_1 (the first component of $\boldsymbol{x}$); that is,[2]

$$\boldsymbol{v} = \boldsymbol{x} + \text{sign}(x_1)||\boldsymbol{x}||_2 \boldsymbol{e}_1. \tag{6.17}$$

[2] sign(x) = +1 if x is positive and −1 if x is negative.

The corresponding matrix $\boldsymbol{H}$ is given from the second line of (6.12) as

$$\boldsymbol{H} = \boldsymbol{I} - \frac{2\boldsymbol{v}\boldsymbol{v}^T}{(\boldsymbol{v}^T\boldsymbol{v})}. \tag{6.18}$$

6.3.2 Example of Householder Elimination

Suppose $\boldsymbol{x} = (1, 1, 1)^T$.

What is $\boldsymbol{H}$ such that the only nonzero element of $\boldsymbol{H}\boldsymbol{x}$ is in the first position? That is, $\boldsymbol{H}\boldsymbol{x} \in \text{ span } \{\boldsymbol{e}_1\}$. The process is very simple.

Since $\boldsymbol{H}$ is uniquely determined by $\boldsymbol{v}$, we must find $\boldsymbol{v}$. From (6.17),

$$\mathbf{v} = \mathbf{x} + \alpha \mathbf{e}_1, \quad \text{where } \alpha = +||\boldsymbol{x}||_2$$

in this case. Thus, since $||x||_2 = \sqrt{3}$,

$$\boldsymbol{v} = \begin{bmatrix} 1 \\ 1 \\ 1 \end{bmatrix} + \begin{bmatrix} \sqrt{3} \\ 0 \\ 0 \end{bmatrix} = \begin{bmatrix} 1+\sqrt{3} \\ 1 \\ 1 \end{bmatrix},$$

and $\boldsymbol{H}$ is formed from (6.18) as

$$\begin{aligned}
\boldsymbol{H} &= \boldsymbol{I} - 2\frac{\boldsymbol{v}\boldsymbol{v}^T}{\boldsymbol{v}^T\boldsymbol{v}} \\
&= \boldsymbol{I} - \frac{2}{\boldsymbol{v}^T\boldsymbol{v}}\boldsymbol{v}\boldsymbol{v}^T \\
&= \boldsymbol{I} - 0.21132 \begin{bmatrix} (1+\sqrt{3})^2 & 1+\sqrt{3} & 1+\sqrt{3} \\ 1+\sqrt{3} & 1 & 1 \\ 1+\sqrt{3} & 1 & 1 \end{bmatrix} \\
&= \begin{bmatrix} -0.57734 & -0.57734 & -0.57734 \\ -0.57734 & 0.78868 & -0.21132 \\ -0.57734 & -0.21132 & 0.78868 \end{bmatrix}.
\end{aligned}$$

We see that

$$\boldsymbol{H}\boldsymbol{x} = \begin{bmatrix} -1.73202 \\ 0 \\ 0 \end{bmatrix},$$

which is exactly the way it is supposed to be. Note from this example $\boldsymbol{H}\boldsymbol{x}$ has the same 2-norm as $\boldsymbol{x}$. This is a consequence of (6.16), which itself follows from the orthonormality of $\boldsymbol{H}$.

6.3.3 Selective Elimination [1]

We have discussed the Householder procedure for annihilating all elements of a vector except the first. We now consider how the Householder procedure may be generalised to eliminate *any* contiguous block of vector components. This procedure is necessary in computing a QR decomposition, since only the elements below the main diagonal in a specific column are annihilated, while the remaining elements must remain intact.

Suppose we wish to eliminate all elements $x_k, \ldots, x_j$ of any $\boldsymbol{x} \in \mathbb{R}^n$, where

$$1 < k < j \leq n, \quad \boldsymbol{x} \in \mathbb{R}^n.$$

Then, the vector $\boldsymbol{v}$ for this case has the form:

$$\boldsymbol{v}^T = [0, \ldots, 0, \underbrace{x_k + \text{sign}(x_k)\alpha, x_{k+1}, \ldots, x_j}_{\substack{\text{this has same structure as a} \\ (j-k+1)\text{-dimensional Householder} \\ \text{vector as in (6.17)).}}} 0, \ldots, 0],$$

where $\alpha^2 = x_k^2 + \ldots + x_j^2$.

In this case, if we define $\boldsymbol{H}$ to have the form

$$\boldsymbol{H} = \text{diag}\left[\boldsymbol{I}_{k-1}, \overline{\mathbf{H}}, \boldsymbol{I}_{n-j}\right],$$

where $\overline{\mathbf{H}} = \boldsymbol{I} - 2\boldsymbol{v}\boldsymbol{v}^T/\boldsymbol{v}^T\boldsymbol{v}$ is the Householder matrix formed by $\boldsymbol{v}$'s nontrivial portion, then we have in this case

$$\boldsymbol{H}\boldsymbol{x} = [\underbrace{x_1, \ldots, x_{k-1}}_{\substack{\text{these elements} \\ \text{are unchanged}}}, -\text{sign}(x_k)\alpha, \underbrace{0, 0, \ldots 0}_{\substack{\text{0's in desired} \\ \text{positions}}}, \underbrace{x_{j+1}, \ldots, x_n}_{\substack{\text{these elements} \\ \text{also unchanged}}}]^T.$$

By using Householder matrices $\boldsymbol{H}$ constructed in this way, the complete QR decomposition may be effected, by choosing the block to be eliminated as that below the main diagonal in the respective column of $\boldsymbol{A}$.

6.3.4 Householder Numerical Properties [1]

Let $\beta = \frac{2}{\boldsymbol{v}^T\boldsymbol{v}}$ and $\hat{\mathbf{v}}$ and $\hat{\beta}$ be the computed versions of $\boldsymbol{v}$ and β, respectively.

Then,

$$\hat{\mathbf{H}} = \mathbf{I} - \hat{\beta}\hat{\mathbf{v}}\hat{\mathbf{v}}^T,$$

and it is shown [4] that

$$\left\|\boldsymbol{H} - \hat{\mathbf{H}}\right\| \leq 10u, \quad u = \text{ machine epsilon.}$$

The matrix $\boldsymbol{HA}$ has a block of zeros in a desired location. The floating point matrix $fl\left[\hat{\mathbf{H}}\mathbf{A}\right]$ satisfies

$$fl\left[\hat{\mathbf{H}}\mathbf{A}\right] = \boldsymbol{H}(\boldsymbol{A} + \boldsymbol{E}),$$

where

$||\boldsymbol{E}||_2 \leq cp^2u||A||_2$
c is a constant of order1
p is the number of elements which are zeroed.

Conclusion The computed Householder transformation process on a matrix $\boldsymbol{A}$ is an exact Householder transformation on a matrix close to $\boldsymbol{A}$. Thus, the Householder procedure is stable.

6.4 The QR Method for Computing the Eigendecomposition

In this section we use the QR decomposition to compute the eigendecomposition of a *square* matrix. While the method is applicable to both symmetric and nonsymmetric matrices, in this treatment we consider only the symmetric case, since that is the most relevant for the field of signal processing. A more general description is given in [3].

We first consider the *similarity transform* of a matrix $\boldsymbol{A}$. Let $\boldsymbol{A}$ be a square symmetric matrix. Then

$$\boldsymbol{Av} = \lambda\boldsymbol{v}$$

for any eigenpair $\boldsymbol{v}$, λ. We now substitute the matrix $\boldsymbol{C} = \boldsymbol{BAB}^{-1}$ for $\boldsymbol{A}$, where $\boldsymbol{B}$ is an invertible matrix of compatible size. $\boldsymbol{C}$ is referred to as a *similarity transform* of $\boldsymbol{A}$. Then $\boldsymbol{A} = \boldsymbol{B}^{-1}\boldsymbol{CB}$, and we have

$$\boldsymbol{B}^{-1}\boldsymbol{CBv} = \lambda\boldsymbol{v}.$$

Define $\boldsymbol{u} = \boldsymbol{B}\boldsymbol{v}$, and we have

$$\boldsymbol{C}\boldsymbol{u} = \lambda \boldsymbol{u}.$$

Thus we see that $\boldsymbol{C}$ has the same eigenvalues as $\boldsymbol{A}$, and the eigenvectors $\boldsymbol{u}$ of $\boldsymbol{C}$ are transformed versions of those of $\boldsymbol{A}$.

The QR method itself is iterative. We initialise by performing a QR decomposition on the square symmetric matrix $\boldsymbol{A}(0) = \boldsymbol{A}$, where the iteration index is indicated in the parentheses:

$$\boldsymbol{A}(0) = \boldsymbol{Q}(0)\boldsymbol{R}(0),$$

and form $\boldsymbol{A}(1) = \boldsymbol{R}(0)\boldsymbol{Q}(0)$, simply by reversing the factors. Substituting $\boldsymbol{R}(0) = \boldsymbol{Q}(0)^T\boldsymbol{A}(0)$, we have $\boldsymbol{A}(1) = \boldsymbol{Q}(0)^T\boldsymbol{A}(0)\boldsymbol{Q}(0)$, so $\boldsymbol{A}(1)$ is a similarity transform of $\boldsymbol{A}(0)$, and hence they have the same eigenvalues. We then perform a QR decomposition on $\boldsymbol{A}(1)$ to obtain

$$\boldsymbol{A}(1) = \boldsymbol{Q}(1)\boldsymbol{R}(1)$$

and again reverse the factors to form $\boldsymbol{A}(2) = \boldsymbol{R}(1)\boldsymbol{Q}(1)$. Since $\boldsymbol{R}(1) = \boldsymbol{Q}(1)^T\boldsymbol{A}(1)$, we have $\boldsymbol{A}(2) = \boldsymbol{Q}(1)^T\boldsymbol{A}(1)\boldsymbol{Q}(1) = \boldsymbol{Q}(1)^T\boldsymbol{Q}(0)\boldsymbol{A}(0)\boldsymbol{Q}(0)\boldsymbol{Q}(1)$. We continue iterating in this fashion. We see that at each iteration k, the matrix $\boldsymbol{A}(k)$ is a similarity transform of $\boldsymbol{A}(0)$ and that the matrix $\boldsymbol{Q}(k)$ is appended fore and aft of the previous product. At each successive stage of this process, $\boldsymbol{A}(k)$ becomes increasingly diagonal, and eventually, $\boldsymbol{A}(k)$ becomes completely diagonal for large enough k, thus revealing the eigenvalues. An extensive discussion on the convergence characteristics of the QR procedure is presented in Golub and Van Loan [1].

At convergence, we can write for k sufficiently large

$$\boldsymbol{A}(k) = \boldsymbol{Q}(k)^T\boldsymbol{Q}(k-1)^T \ldots \boldsymbol{Q}(0)^T\boldsymbol{A}\ \boldsymbol{Q}(0)\boldsymbol{Q}(1)\ldots\boldsymbol{Q}(k) = \boldsymbol{\Lambda},$$

where $\boldsymbol{\Lambda}$ is the diagonal eigenvalue matrix. It follows that

$$\boldsymbol{A} = \underbrace{\boldsymbol{Q}(0)\boldsymbol{Q}(1)\ldots\boldsymbol{Q}(k)}_{\boldsymbol{V}}\boldsymbol{\Lambda}\underbrace{\boldsymbol{Q}(k)^T\boldsymbol{Q}(k-1)^T\ldots\boldsymbol{Q}(0)^T}_{\boldsymbol{V}^T}.$$

By comparison with the general form of the eigendecomposition $\boldsymbol{A} = \boldsymbol{V}\boldsymbol{\Lambda}\boldsymbol{V}^T$ for a symmetric matrix, it is clear that the eigenvector matrix $\boldsymbol{V}$ is given by the product of the $\boldsymbol{Q}(k)$ as shown above.

The use of orthogonal matrices for the similarity transformations is desirable, since then the 2-norm of the matrix or vector is preserved. Otherwise, there is a potential for matrix or vector products to become large in specific directions, as shown in Fig. 3.1, Chap. 3. Large growth in elements can introduce the possibility for catastrophic cancellation to occur, in the manner discussed in Sect. 5.3.

6.4.1 Enhancements to the QR Method

We discuss two techniques that simultaneously reduce the computation time and improve the convergence rate of the QR method as described above. The first is conversion of $\boldsymbol{A}$ to tridiagonal form, which is applicable only in the symmetric case, and the second is the *shifted* QR method which is a modification of the basic method to accelerate convergence.

Tridiagonalisation The motivation of the tridiagonalisation approach is to introduce as many zeros into a transformed version of $\boldsymbol{A}$ as possible before the iterations begin. Assuming these zeros persist from one iteration to the next, the QR decomposition at each stage is easier because fewer elements need to be zeroed out at each step and further because the algorithm is executed on the resulting tridiagonal matrix, it exhibits faster convergence.

A square symmetric matrix can be converted to tridiagonal form with one similarity transform. (*Tridiagonal* means that only the main and first upper and lower diagonals are nonzero.) The process is quite straightforward. The original matrix $\boldsymbol{A}$ is replaced with $\boldsymbol{A}(0)$ as follows:

$$\boldsymbol{A}(0) = \boldsymbol{Q}_o \boldsymbol{A} \boldsymbol{Q}_o^T, \tag{6.19}$$

where $\boldsymbol{Q}_o$ is the product of Householder transforms required to eliminate all elements below the first lower diagonal in each column. It is left as an exercise to show that if pre-multiplication of a symmetric $\boldsymbol{A}$ by $\boldsymbol{Q}_o$ eliminates elements below the first lower diagonal in each column, then post-multiplication by $\boldsymbol{Q}_o^T$ eliminates elements to the right of the first upper diagonal in each row, thus maintaining symmetry. The QR method then proceeds on the transformed matrix $\boldsymbol{A}(0)$ instead of $\boldsymbol{A}$ and is much faster as a result.

One might well ask: If we can tridiagonalise $\boldsymbol{A}$ with one similarity transform, then why cannot we completely diagonalise $\boldsymbol{A}$ in one step? The answer lies in the fact that we can indeed find a $\boldsymbol{Q}_o$ so that $\boldsymbol{Q}_o\boldsymbol{A}$ has zeros in all positions below the main diagonal. The problem is that the post-multiplication by $\boldsymbol{Q}_o^T$ as in (6.19) overwrites the zeros that result from the pre-multiplication. As an example, we take the following symmetric Toeplitz matrix for $\boldsymbol{A}$:

$$\boldsymbol{A} = \begin{bmatrix} 4 & 3 & 2 & 1 \\ 3 & 4 & 3 & 2 \\ 2 & 3 & 4 & 3 \\ 1 & 2 & 3 & 4 \end{bmatrix}. \tag{6.20}$$

We eliminate the first column using a Householder matrix $\boldsymbol{H}$ given by

$$\boldsymbol{H} = \begin{bmatrix} -0.7303 & -0.5477 & -0.3651 & -0.1826 \\ -0.5477 & 0.8266 & -0.1156 & -0.0578 \\ -0.3651 & -0.1156 & 0.9229 & -0.0385 \\ -0.1826 & -0.0578 & -0.0385 & 0.9807 \end{bmatrix}.$$

The matrix $\boldsymbol{HA}$ is then

$$\boldsymbol{HA} = \begin{bmatrix} -5.4772 & -5.8424 & -5.1121 & -3.6515 \\ 0.0000 & 1.2010 & 0.7487 & 0.5276 \\ 0.0000 & 1.1340 & 2.4991 & 2.0184 \\ 0.0000 & 1.0670 & 2.2496 & 3.5092 \end{bmatrix}, \tag{6.21}$$

which has zeros below the main diagonal in the first column as desired. However, when we post-multiply by $\boldsymbol{H}^T$, we get

$$\boldsymbol{HAH}^T = \begin{bmatrix} 9.7333 & -1.0275 & -1.9022 & -2.0465 \\ -1.0275 & 0.8757 & 0.5318 & 0.4192 \\ -1.9022 & 0.5318 & 2.0977 & 1.8177 \\ -2.0465 & 0.4192 & 1.8177 & 3.2933 \end{bmatrix},$$

and thus it is apparent that the zeros introduced in (6.21) have been overwritten by the later post-multiplication by $\boldsymbol{H}^T$, and the procedure has not accomplished our objective to introduce as many zeros as possible into $\boldsymbol{A}(0)$. So now we accept the fact that the best we can do is to tridiagonalise. As a first step in this respect, we formulate a Householder matrix $\boldsymbol{H}$ to wipe out the elements below the first lower diagonal in the first column. This is given by

$$\boldsymbol{H} = \begin{bmatrix} 1.0000 & 0 & 0 & 0 \\ 0 & -0.8018 & -0.5345 & -0.2673 \\ 0 & -0.5345 & 0.8414 & -0.0793 \\ 0 & -0.2673 & -0.0793 & 0.9604 \end{bmatrix}.$$

Notice that the selective elimination procedure as in Sect. 6.3.3 has been used, since the element (1,1) is not affected in the tridiagonalisation procedure. This explains the identity structure in the first row and column of $\boldsymbol{H}$. The resulting matrix $\boldsymbol{HAH}^T$ has zeros outside of the first upper and lower diagonal. The second column below and the second row to the right of the respective diagonal are then zeroed out in a corresponding fashion. The overall result of the tridiagonalisation procedure is then given by

$$\boldsymbol{HAH}^T = \begin{bmatrix} 4.0000 & -3.7417 & -0.0000 & -0.0000 \\ -3.7417 & 8.2857 & -2.6030 & -0.0000 \\ -0.0000 & -2.6030 & 3.0396 & -0.2254 \\ -0.0000 & -0.0000 & -0.2254 & 0.6747 \end{bmatrix}.$$

We see that the only nonzero elements are indeed located along the three main diagonals, as desired. After tridiagonalisation, regular QR iterations as described above are applied, but now only one element in each column requires elimination, a fact that greatly speeds up both convergence and execution of the algorithm. The tridiagonal structure is maintained at each QR iteration.

The QR method may also be applied to nonsymmetric square matrices using exactly the same procedure as for the symmetric case. The only difference is that the initial matrix $\boldsymbol{A}(0)$ cannot be made tridiagonal, since it is not symmetric. In this case we make do with just zeroing out the elements below the first lower diagonal. This is the so-called upper Hessenburg form of the matrix. The overall process requires somewhat more computation than the symmetric case.

Shifted QR We can significantly improve the convergence rate of the QR algorithm using a shifting procedure. We replace the basic QR iteration described earlier with the following procedure:

$$\boldsymbol{A}(k) - \alpha_k \boldsymbol{I} = \boldsymbol{Q}(k)\boldsymbol{R}(k), \text{ and then } \boldsymbol{A}(k+1) = \boldsymbol{R}(k)\boldsymbol{Q}(k) + \alpha_k \boldsymbol{I},$$

where α_k is chosen to be close to the smallest eigenvalue. This variation of the QR iteration may be justified, because $\boldsymbol{A}(k)$ is similar to $\boldsymbol{A}(k+1)$, which we can show as follows:

$$\begin{aligned} \boldsymbol{Q}(k)^T \boldsymbol{A}(k) \boldsymbol{Q}(k) &= \boldsymbol{Q}(k)^T \left(\boldsymbol{Q}(k)\boldsymbol{R}(k) + \alpha_k \boldsymbol{I} \right) \boldsymbol{Q}(k) \\ &= \boldsymbol{R}(k)\boldsymbol{Q}(k) + \alpha_k \boldsymbol{I} \\ &= \boldsymbol{A}(k+1). \end{aligned}$$

Therefore the shift is justified in the QR procedure because the eigenvalues are not altered. In practice, an estimate of λ_{min} is provided by the element in the lower right corner of $\boldsymbol{A}(k)$. As indicated in [3], if α_k is sufficiently close to λ_{min}, the shifted QR algorithm can exhibit cubic convergence[3] to the eigenvalue.

The Jacobi method is yet another approach to computing the complete eigendecomposition using orthonormal similarity transforms, specifically for symmetric matrices. However it is not as commonly used as the QR method. The interested reader is referred to [2].

6.5 The Rank-Deficient QR Decomposition

Here we examine the structure of the QR decomposition when the matrix $\boldsymbol{A}$ is rank deficient. If $\boldsymbol{A} \in \mathbb{R}^{m \times n}, m > n, \text{rank}(\boldsymbol{A}) = r < n$, then we wish to preserve a

[3] By cubic convergence, we mean that if the error at iteration k is ϵ_k for suitably large k (where ϵ_k may be assumed small), then the error is $o(\epsilon_k^3)$ at iteration $k+1$. Thus the convergence is very fast.

fundamental property of the QR decomposition, i.e.

$$R(\boldsymbol{A}) = \text{span}\left[\boldsymbol{q}_1, \ldots, \boldsymbol{q}_r\right]. \tag{6.22}$$

Only in this case $\boldsymbol{Q}_r = \left[\boldsymbol{q}_1, \ldots, \boldsymbol{q}_r\right]$ can act as an orthonormal basis for $\boldsymbol{A}$.

We construct an example to show this is not always true. Suppose the rank 2 matrix $\boldsymbol{A}$ is defined as follows:

$$\boldsymbol{A} = \begin{bmatrix} -0.4437 & 0.1500 & -0.4119 \\ 0.4836 & -0.1635 & -1.5977 \\ 0.6345 & -0.2146 & 0.4580 \\ -0.2555 & 0.0864 & 0.5244 \end{bmatrix}.$$

Then the QR decomposition of $\boldsymbol{A}$ degenerates as follows:

$$\begin{aligned} \boldsymbol{A} &= \boldsymbol{Q}\boldsymbol{R} \\ &= \begin{bmatrix} -0.468 & 0.849 & 0.047 & 0.239 \\ 0.510 & 0.237 & -0.767 & 0.308 \\ 0.669 & 0.253 & 0.638 & 0.286 \\ -0.269 & -0.398 & 0.048 & 0.875 \end{bmatrix} \begin{bmatrix} 0.948 & -0.321 & -0.457 \\ 0 & \mathbf{0} & -0.822 \\ 0 & 0 & 1.524 \\ 0 & 0 & 0 \end{bmatrix}. \end{aligned} \tag{6.23}$$

Because of the zero in the $\boldsymbol{R}(2, 2)$ position, we see that $R(\boldsymbol{A}) \neq \text{span}[\boldsymbol{q}_1\boldsymbol{q}_2]$ as desired. Further, this QR decomposition is of no value in solving the LS problem, because $\boldsymbol{R}$ is not full rank. The problem in (6.23) is that there are no r columns (in this case two columns) of $\boldsymbol{Q}$ that can act as an orthonormal basis for $R(\boldsymbol{A})$.

We now show that column-permutation matrices $\boldsymbol{\Pi}$ can be applied at every stage so that the QR decomposition on $\boldsymbol{A}$ may be expressed in the form

$$\boldsymbol{Q}^T\boldsymbol{A}\boldsymbol{\Pi} = \underset{\quad r \quad n-r}{\begin{bmatrix} \boldsymbol{R}_{11} & \boldsymbol{R}_{12} \\ \mathbf{0} & \mathbf{0} \end{bmatrix}} \begin{matrix} r \\ m-r \end{matrix}, \tag{6.24}$$

where $\boldsymbol{R}_{11} \in \mathbb{R}^{r\times r}$ is upper triangular and nonsingular and $\boldsymbol{R}_{12}$ is a rectangular matrix. In this case it is clear that the rank-deficient QR decomposition in the form of (6.24) indeed satisfies (6.22), where $R(\boldsymbol{A}) = \text{span}[\boldsymbol{q}_1 \ldots, \boldsymbol{q}_r]$. The permutation matrix $\boldsymbol{\Pi}$ is determined in such a way so that at each stage i, $i = 1, \ldots, r$, the diagonal elements r_{ii} of $\boldsymbol{R}_{11}$ are as large in magnitude as possible, thus avoiding the degenerate form of (6.23). But what is the procedure to determine $\boldsymbol{\Pi}$?

To answer this, consider the ith stage, $i < r$, of the QR decomposition with column pivoting. Here, the first i columns have been annihilated below the main diagonal by an appropriate QR decomposition procedure, such as Householder. There exist an orthonormal $\boldsymbol{Q}$ and a permutation matrix $\boldsymbol{\Pi}$ so that

$$Q^T A \Pi = \begin{bmatrix} R_{11} & R_{12} \\ 0 & R_{22} \end{bmatrix} \begin{matrix} i \\ n-i \end{matrix} \qquad (6.25)$$
$$\quad i \quad n-i \; ,$$

where R_{11} is upper triangular.

The $(i+1)$th stage of the decomposition proceeds by first post-multiplying both sides of (6.25) by a permutation matrix $\Pi^{(i+1)}$ (to swap the desired column into the leading position of R_{22}, as discussed shortly) and then pre-multiplying both sides by an orthonormal matrix $Q^{(i+1)}$ such that the first column $r_{22}(1)$ of the R_{22} partition is annihilated below the first element. The superscript $(\cdot)^{(i+1)}$ on both Q and Π denotes the specific matrix at stage $i+1$. The Q in (6.25) at the $(i+1)$th stage is given by $Q^{(i+1)} \ldots Q^{(2)} Q^{(1)}$, and the Π in (6.25) is $\Pi^{(1)} \Pi^{(2)} \ldots \Pi^{(i+1)}$.

Since at the $(i+1)$th stage we wish to eliminate only the elements $i+1:n$ of the first column $r_{22}(1)$ of the R_{22} block, then according to the selective elimination procedure discussed earlier, the orthonormal matrix $Q^{(i+1)}$ to execute the $(i+1)th$ stage has the form

$$Q^{(i+1)} = \begin{bmatrix} I & 0 \\ 0 & \tilde{Q} \end{bmatrix} \begin{matrix} i \\ m-i \end{matrix} \; , \qquad (6.26)$$
$$\quad i \quad m-i$$

where the $\tilde{Q}$ above is the matrix which eliminates the desired elements of column $r_{22}(1)$. Since $\tilde{Q}$ is orthonormal, the element $r(i+1, i+1)$ in the top left (diagonal) position of R_{22} after the multiplication is complete is therefore equal to $||r_{22}(1)||_2$. It is now clear that to place the elements with the largest possible magnitudes along the diagonal of R, we must choose the permutation matrix $\Pi^{(i+1)}$ at the $(i+1)th$ stage so that the column of R_{22} with maximum 2-norm is swapped into the lead column position. This procedure ensures that the resulting QR decomposition will have the form of (6.24) as desired. Effectively, this procedure ensures that no zeros are introduced along the diagonal of R until after stage r.

We can write (6.25) at the completion of ith stage in the form

$$\underset{i \quad n-i}{\begin{bmatrix} \tilde{A}_1 & \tilde{A}_2 \end{bmatrix}} = \underset{i \quad m-i}{\begin{bmatrix} Q_1 & Q_2 \end{bmatrix}} \underset{i \quad n-i \; ,}{\begin{bmatrix} R_{11} & R_{12} \\ & 0 \; R_{22} \end{bmatrix}} \begin{matrix} i \\ m-i \end{matrix} \qquad (6.27)$$

where the tilde over the A-blocks indicates that the columns of A have been permutated as prescribed by $\Pi^{(i)}$. From our previous discussions, Q_1 is an orthonormal basis for $R(A_1)$, and Q_2 is an orthonormal basis for $R(A_1)_\perp$. It follows directly from the block multiplication in (6.27) that the elements of the column $r_{22}(k)$ of R_{22} are the coefficients of the column $\tilde{a}_2(k)$ in the basis Q_2. Thus, the column $r_{22}(1)$ which is annihilated after the permutation step at the $(i+1)$th stage corresponds to the column of $\tilde{A}_2$ which has the largest component in $R(\tilde{A}_1)_\perp$.

After the $(i+1)$th stage of the decomposition is complete, $\boldsymbol{\Pi}^{(i+1)}$ post-multiplies the previously accumulated product $\boldsymbol{\Pi}$, and $\boldsymbol{Q}^{(i+1)}$ pre-multiplies its previous accumulated product. After r stages, the QR decomposition in the form of (6.24) is complete. This form of the QR decomposition now has the correct structure to apply to the rank-deficient least squares problem, as discussed in Chapter 8. The final result of the rank-deficient QR decomposition may therefore be expressed as

$$\boldsymbol{Q}^T \boldsymbol{A}\boldsymbol{\Pi} = \underset{\quad r \quad n-r}{\begin{bmatrix} \boldsymbol{R}_{11} & \boldsymbol{R}_{12} \\ \boldsymbol{0} & \boldsymbol{0} \end{bmatrix}} \begin{matrix} r \\ m-r \end{matrix}, \tag{6.28}$$

where the formulation of $\boldsymbol{\Pi}$ and $\boldsymbol{Q}$ has been discussed above, and the first r columns of $\boldsymbol{Q}$ form a basis for $R(\boldsymbol{A})$ as desired.

Appendices

In this appendix, we examine two additional forms of QR decomposition—the Givens rotation and fast Givens rotation methods. They are both useful methods, particularly when only specific elements of $\boldsymbol{A}$ need to be eliminated.

Givens Rotations

We have seen so far in this chapter that the QR decomposition may be executed by the Gram–Schmidt and Householder procedures. We now discuss the QR decomposition by *Givens rotations*. A Givens transformation (rotation) is capable of annihilating a single zero in any position of interest. Givens rotations require a larger number of flops compared to Householder to compute a complete QR decomposition on a matrix $\boldsymbol{A}$. Nevertheless, they are very useful in some circumstances, because they can be used to eliminate only specific elements.

In this presentation we consider a Givens transformation $\boldsymbol{J}(i, k, \theta)$ to annihilate the (i, k)th element of the product $\boldsymbol{J}\boldsymbol{A}$ below the main diagonal; that is, for $i > k$. The matrix $\boldsymbol{J}$ has the form:

$$\boldsymbol{J} = \begin{bmatrix} 1 & & & & & & \\ & 1 & & & & & \\ & & c & & s & & \\ & & & \ddots & & & \\ & & -s & & c & & \\ & & & & & 1 & \\ & & & & & & 1 \end{bmatrix} \begin{matrix} \\ \\ k \\ \\ i \\ \\ \\ \end{matrix}, \tag{6.29}$$

where the marked columns are k and i respectively.

where $s = \sin(\theta)$ $c = \cos(\theta)$, and θ is an angle to be determined. $\boldsymbol{J}$ has the form of an identity matrix except for the (c, s) entries. These c, s entries occupy positions involving all combinations of the indices (i, k). The transformation $\boldsymbol{J}$ on $\boldsymbol{x}$ rotates $\boldsymbol{x}$ by θ radians in the $i - k$ plane.

A sequence of Givens rotations may be used to effect the QR decomposition. One rotation (pre-multiplication by $\boldsymbol{J}$) exists for every element to be eliminated; thus,

$$\underbrace{\boldsymbol{J}_{n,n-1} \cdots \boldsymbol{J}_{n2} \cdots \boldsymbol{J}_{32} \boldsymbol{J}_{n1} \cdots \boldsymbol{J}_{21}}_{\boldsymbol{Q}^T} \boldsymbol{A} = \underset{\text{upper triangular}}{\boldsymbol{R}}.$$

We therefore see that the QR decomposition may be effected by a sequence of Givens rotations. The resulting upper triangular product is $\boldsymbol{R}$, and the product of all the Givens transformations is $\boldsymbol{Q}^T$.

There are two conditions that must exist on each $\boldsymbol{J}$:

1. If $\boldsymbol{Q}$ is to be orthonormal, then each $\boldsymbol{J}$ must be orthonormal. (Because the product of orthonormal matrices is orthonormal.)
2. The (i, k)th element of $\boldsymbol{JA}$ must be zero.

The first condition is satisfied by the structure of $\boldsymbol{J}$ from (6.29):

$$\boldsymbol{J}^T\boldsymbol{J} = \begin{bmatrix} 1 & & & & & & \\ & 1 & & & & & \\ & & c & & -s & & \\ & & & \ddots & & & \\ & & s & & c & & \\ & & & & & 1 & \\ & & & & & & 1 \end{bmatrix} \begin{bmatrix} 1 & & & & & & \\ & 1 & & & & & \\ & & c & & s & & \\ & & & \ddots & & & \\ & & -s & & c & & \\ & & & & & 1 & \\ & & & & & & 1 \end{bmatrix} = \begin{bmatrix} 1 & & & & & \\ & 1 & & & 0 & \\ & & 1 & & & \\ & & & 1 & & \\ & 0 & & & 1 & \\ & & & & & 1 \end{bmatrix},$$

which holds for any θ. Condition 2 is satisfied by considering the following diagram:

$$\underset{\boldsymbol{J}}{\begin{bmatrix} 1 & & & & \\ & \ddots & & & \\ & & c \; s & & \\ & & -s \; c & & \\ & & & \ddots & \\ & & & & 1 \end{bmatrix}} \underset{\boldsymbol{A}}{\begin{bmatrix} a_{11} & & \cdots\cdots & & a_{1n} \\ \vdots & \cdots & a_{kk} \; a_{ki} & \cdots & \vdots \\ \vdots & \cdots & a_{ik} \; a_{ii} & \cdots & \vdots \\ a_{n1} & & \cdots\cdots & & a_{nn} \end{bmatrix}} \quad \begin{matrix} c = \cos(\theta) \\ s = \sin(\theta). \end{matrix}$$

The evaluation of the (i, k)th element of the product $\boldsymbol{JA}$ for $i > k$ must then be as follows:

$$-sa_{kk} + ca_{ik} = 0.$$

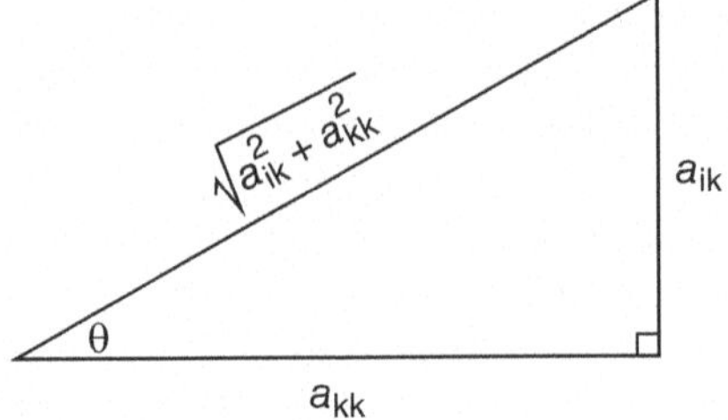

Fig. 6.3 Geometry relating to Givens rotations

Since the c and s are sines and cosines, respectively, the above relation corresponds to a trigonometric relationship involving the angle θ, as depicted in Fig. 6.3. It follows from the above that

$$\frac{s}{c} = \tan\theta = \frac{a_{ik}}{a_{kk}}.$$

We therefore have with the aid of Fig. 6.3

$$s = \frac{a_{ik}}{\left(\sqrt{a_{kk}^2 + a_{ik}^2}\right)}$$

$$c = \frac{a_{kk}}{\left(\sqrt{a_{kk}^2 + a_{ik}^2}\right)}.$$

Notice that θ is *not explicitly computed*. The matrix $\boldsymbol{J}(i, k, \theta)$ is now completely specified.

The following algorithm computes the c and s in the most stable numerical fashion and ensures that $\boldsymbol{J}(i, k))\boldsymbol{x}$ has a 0 in the (i, k)th position:

$$\text{If } x_k = 0 \text{ then } c := 1 \text{ and } s := 0 \text{ else}$$

$$\text{If} \quad |x_k| \geq |x_i|$$

$$\text{then } t := \tfrac{x_i}{x_k};\ s := \frac{1}{(1+t^2)^{\frac{1}{2}}};\quad c := st$$

$$\text{else } t := \tfrac{x_k}{x_i};\ c := \frac{1}{(1+t^2)^{\frac{1}{2}}};\quad s := ct.$$

This algorithm assures that $|t| \leq 1$. If $|t|$ becomes large, we may run into stability problems in calculating c and s.

It is easily verified that the following facts hold true when evaluating the product $\boldsymbol{J}(i,k)\boldsymbol{A}$:

1. The ith row of $\boldsymbol{JA} \leftarrow c\boldsymbol{a}_k^T + s\boldsymbol{a}_i^T$
2. The kth row of $\boldsymbol{JA} \leftarrow -s\boldsymbol{a}_k^T + c\boldsymbol{a}_i^T$
3. All other rows of $\boldsymbol{A}$ are unchanged

where $\boldsymbol{a}_i^T$ and $\boldsymbol{a}_k^T$ are the ith and kth rows of $\boldsymbol{A}$, respectively. Thus, only the ith and kth rows of the product $\boldsymbol{JA}$ are actually relevant in the Givens analysis.

The order in which elements are annihilated in the QR decomposition is critical. They must be annihilated below the main diagonal in the column ordering $1, \ldots, n$; otherwise, zeros written in previous stages will be overwritten in later stages. This is the same ordering that is necessary with Gaussian elimination, for the same reason.

Numerical Stability of QR Decomposition by Givens
QR decomposition by Givens rotation is of the same degree of stability as for Householder. Both are very stable and more so than Gaussian elimination for triangularisation.

"Fast" Givens Method for QR Decomposition

Even though the ordinary Givens method is stable, it is expensive to compute. In this section we discuss a modified Givens method which is almost as stable yet is considerably faster.

First, we present a quick review of "slow" Givens: Suppose we have a vector $\boldsymbol{x} = [x_1 \ldots x_i \ldots x_n]$, and we wish to annihilate the element x_i for the value $k = 1$. This is done using Givens rotations in the following way:

$$\begin{array}{c} \\ k \\ \\ i \\ \\ \\ \\ \end{array}\begin{array}{c} \begin{array}{cc} k & i \end{array} \\ \begin{bmatrix} c & \cdots & s & 0 & \cdots & 0 \\ \vdots & \ddots & & & & \\ -s & & c & & & \\ 0 & & & 1 & & \\ \vdots & & & & \ddots & \\ 0 & & & & & 1 \end{bmatrix} \end{array} \begin{bmatrix} x_1 \\ \vdots \\ x_i \\ \vdots \\ x_n \end{bmatrix} = \begin{bmatrix} x_1^{(1)} \\ \vdots \\ 0 \\ \vdots \\ x_n^{(1)} \end{bmatrix},$$

where the bracketed superscript indicates the corresponding element has been changed once, and

$$c = \frac{x_1}{\sqrt{x_1^2 + x_i^2}} \qquad s = \frac{x_i}{\sqrt{x_1^2 + x_i^2}}.$$

The relevant portion of this process may be represented at the 2×2 level as

$$\begin{bmatrix} c & s \\ -s & c \end{bmatrix} \begin{bmatrix} x_1 \\ x_i \end{bmatrix} = \begin{bmatrix} x_1' \\ 0 \end{bmatrix}. \tag{6.30}$$

Each element is eliminated in turn, using an appropriate Givens matrix $\boldsymbol{J}$, in the order of Gaussian Elimination, until an upper triangular matrix is obtained. Note that each element is eliminated by an *orthonormal matrix*.

The difficulty with the original Givens method is that, generally, none of the elements of the $\boldsymbol{J}$-matrix at the 2×2 level in (6.30) are 0 or 1. Thus, the update of a given element from (6.30) involves two multiplications and one add for each element in rows k and i. We now consider a faster form of Givens where the off-diagonal elements of the transformation matrix are replaced by ones. This reduces the number of explicit multiplications required for the evaluation of each altered element of the product from 2 to 1.

In this vein, let us consider *Fast Givens:* The idea here is to eliminate each element of $\boldsymbol{a}$ using a simplified transformation matrix, denoted as $\boldsymbol{M}$, to reduce the number of flops required over ordinary Givens. The result is that the $\boldsymbol{M}$ used for fast Givens is orthogonal but not orthonormal.

We can therefore speculate that we can triangularise $\boldsymbol{A}$ as

$$\boldsymbol{M}^T \boldsymbol{A} = \begin{bmatrix} \boldsymbol{S} \\ \boldsymbol{0} \end{bmatrix}, \tag{6.31}$$

where $\boldsymbol{A} \in \Re^{m \times n}, m > n$, $\boldsymbol{S} \in \Re^{n \times n}$ is upper triangular, and $\boldsymbol{M} \in \Re^{m \times m}$ has orthogonal but not orthonormal columns.

Hence

$$\boldsymbol{M}^T \boldsymbol{M} = \boldsymbol{D} = \operatorname{diag}(d_1 \dots d_m), \tag{6.32}$$

and $\boldsymbol{M}\boldsymbol{D}^{-\frac{1}{2}}$ is orthonormal.

We deal with the fast Givens problem at the 2×2 level. Let $\boldsymbol{x} = [x_1 \; x_2]^T$, and we define the matrix $\boldsymbol{M}_1$ as

$$\boldsymbol{M}_1 = \begin{bmatrix} \beta_1 & 1 \\ 1 & \alpha_1 \end{bmatrix}. \tag{6.33}$$

As with regular Givens, there are two conditions on $\boldsymbol{M}_1$ if the appropriate element of $\boldsymbol{M}^T \boldsymbol{A}$ is to be annihilated and $\boldsymbol{M}_1$ is to be orthogonal:

$$\text{(1)} \quad \boldsymbol{M}_1 \boldsymbol{x} = \begin{bmatrix} x_1' \\ 0 \end{bmatrix}$$

$$\text{(2)} \quad {\boldsymbol{M}_1}^T \boldsymbol{M}_1 = \operatorname{diag}(d_1, d_2).$$

To satisfy the first condition, we note using (6.33)

$$\boldsymbol{M}_1 \boldsymbol{x} = \begin{bmatrix} \beta_1 x_1 + x_2 \\ x_1 + \alpha_1 x_2 \end{bmatrix}$$

Therefore, for the second element of the product to be zero, we must have

$$\alpha_1 = \frac{-x_1}{x_2}. \tag{6.34}$$

To satisfy the second condition, we note

$$\boldsymbol{M}_1{}^T \boldsymbol{M}_1 = \begin{bmatrix} \beta_1 & 1 \\ 1 & \alpha_1 \end{bmatrix} \begin{bmatrix} \beta_1 & 1 \\ 1 & \alpha_1 \end{bmatrix} = \begin{bmatrix} \beta_1^2 + 1 & \beta_1 + \alpha_1 \\ \beta_1 + \alpha_1 & \alpha_1^2 + 1 \end{bmatrix}. \tag{6.35}$$

Hence, we must have $\beta_1 = -\alpha_1$. This completely defines the matrix $\boldsymbol{M}_1$. At the $m \times m$ level, the matrix M_1 has a form analogous to slow Givens:

$$\boldsymbol{M}_1(i,k) = \begin{bmatrix} 1 \cdots & 0 & \cdots & 0 & \cdots 0 \\ \vdots \ddots & & & & \\ 0 \cdots & \beta_1 & \cdots & 1 & \cdots 0 \\ & & \ddots & & \\ 0 \cdots & 1 & \cdots & \beta_1 & \cdots 0 \\ & & & & \ddots \\ 0 & & & & 1 \end{bmatrix} \begin{matrix} \\ \\ k \\ \\ i \\ \\ \\ \end{matrix} \tag{6.36}$$
$$\qquad\qquad k \qquad i \qquad .$$

Each element of $\boldsymbol{A}$ is eliminated in turn, just as with slow Givens. We note from (6.33) that

$$\boldsymbol{M}_1^T \boldsymbol{M}_1 = \begin{bmatrix} 1 + \beta_1^2 & 0 \\ 0 & 1 + \beta_1^2 \end{bmatrix};$$

hence, with each elimination, the rows of $\boldsymbol{M}^T \boldsymbol{A}$ grow by a factor of $1+\beta_1^2 = 1+\alpha_1^2$. But we see from (6.34) that if $|x_1| \gg |x_2|$, this growth factor can become large and lead to the potential of floating point overflow.

To control this growth, we consider a different form of $\boldsymbol{M}$:

$$\boldsymbol{M}_2 = \begin{bmatrix} 1 & \beta_2 \\ \alpha_2 & 1 \end{bmatrix}.$$

In this case, to satisfy the two conditions, it is easily verified that

$$\alpha_2 = \frac{-x_2}{x_1}$$

and $\beta_2 = -\alpha_2$. Hence, if $x_1 > x_2$, choose form $\boldsymbol{M}_2$, else choose $\boldsymbol{M}_1$. This way, the growth with each elimination is controlled to within a factor of 2. (With many eliminations, this still can be a problem).

The great advantage to the fast Givens approach is that the triangularisation may be accomplished using half the number of multiplications compared to slow Givens and may be done without square roots, which is good for VLSI implementations.

Flop Counts

The following table presents a flop count [1] for various methods of QR decomposition of a matrix $\boldsymbol{A} \in \mathbb{R}^{m \times n}$:

$$\begin{aligned} \text{Householder: } & 2n^2(m - n/3) \quad \text{1 flop} = \text{1 floating-pt. op. (add, mult, div or subt.)} \\ \text{slow Givens: } & 3n^2(m - n/3) \\ \text{fast Givens: } & 2n^2(m - n/3) \\ \text{Gram–Schmidt } & 2mn^2 \\ \text{by comparison, Gauss: } & \tfrac{2}{3}n^3, \end{aligned}$$

where, of course, the Gaussian elimination entry applies only to a non-orthogonal transformation of a *square* matrix. Thus, even though Householder and both forms of Givens are very stable and furthermore yield an orthonormal decomposition, they are significantly slower than Gaussian elimination.

Problems

1. Consider a QR decomposition on a matrix $\mathbf{A}^{m \times n}$, $m > n$, which has proceeded k steps, $k < n$ (i.e. k columns of $\mathbf{A}$ have been eliminated). A Householder method has been used. The situation may be depicted in the following equation:

$$m \begin{bmatrix} \boldsymbol{A}_1 & \boldsymbol{A}_2 \end{bmatrix}_{\; k \;\; n-k} = \begin{bmatrix} \boldsymbol{Q}_1 & \boldsymbol{Q}_2 \end{bmatrix}_{\; k \;\; m-k} \begin{bmatrix} \boldsymbol{R}_{11} & \boldsymbol{R}_{12} \\ \mathbf{0} & \boldsymbol{R}_{22} \end{bmatrix}_{\; k \;\; n-k} \begin{matrix} k \\ m-k \end{matrix},$$

where $\mathbf{Q}_1 \in \mathbb{R}^{m\times k}$, $\mathbf{R}_{11} \in \mathbb{R}^{k\times k}$ is upper triangular, and $\mathbf{R}_{22} \in \mathbb{R}^{(m-k)\times(n-k)}$. The matrix $\mathbf{A}$ may be partitioned as $\mathbf{A} = [\mathbf{A}_1 \ \ \mathbf{A}_2]$, where $\mathbf{A}_1$ has k columns.

(a) The matrix $\mathbf{Q}_1$ is an orthonormal basis for $R(\mathbf{A}_1)$. In terms of the $\mathbf{R}$-matrix, how can the quantity $||\mathbf{a}_i||_2$, $i \le k$ be determined?
(b) Let $s_i, i > k$ be the projection of $\boldsymbol{a}_i \in A_2$ onto $R(A_1)$. Determine the quantity $||\boldsymbol{s}_i||_2$ using only the $\mathbf{R}$-matrix.
(c) What is an orthonormal basis for $R(\mathbf{A}_1)_\perp$? Denote $\mathbf{p}_i$ as the projection of $\mathbf{a}_i$, $i > k$, onto $R(\mathbf{A}_1)_\perp$. Determine the quantity $||\mathbf{p}_i||_2$ using only the $\mathbf{R}$-matrix. Explain your reasoning carefully.

2. We are given a tall matrix $\boldsymbol{Q}_1$ with orthonormal columns. Explain how to find a $\boldsymbol{Q}_2$ so that $\boldsymbol{Q} = [\boldsymbol{Q}_1 \ \ \boldsymbol{Q}_2]$ is a complete orthonormal matrix.
3. Consider the QR decomposition on a tall rank-deficient matrix. Explain the characteristics of $\boldsymbol{R}$ in this case.
4. (a) Given that $\boldsymbol{P}$ is a projector onto a subspace $\mathcal{S}$, we have seen that the operation $(\boldsymbol{I} - 2\boldsymbol{P})\boldsymbol{x}$ reflects a vector $\boldsymbol{x}$ in the orthogonal complement subspace $\mathcal{S}_\perp$. What is the matrix which reflects $\boldsymbol{x}$ about $\mathcal{S}$?
   (b) What are the eigenvalues of each of these reflectors?
5. Update the QR decomposition with time: At a certain time t, we have available m row vectors $\boldsymbol{a}_i^T \in \mathbb{R}^n$ and their corresponding desired values b_i, for $i = 1, \ldots, m$, to form the matrix $\boldsymbol{A}_t \in \mathbb{R}^{m\times n}$ and $\boldsymbol{b}_t \in \mathbb{R}^m$. The QR decomposition $\boldsymbol{Q}_t^T \boldsymbol{A}_t = \boldsymbol{R}_t$ is available at time t to aid in the computation of the LS problem $\min_{\boldsymbol{x}} ||\boldsymbol{A}\boldsymbol{x} - \boldsymbol{b}||$. At time $t + 1$, a new $(m + 1)$th row $\boldsymbol{a}_{m+1}^T$ of $\boldsymbol{A}$ and a new $(m+1)$th element of $\boldsymbol{b}$ become available. Explain in detail how to update $\boldsymbol{Q}_t$ and $\boldsymbol{R}_t$ to get $\boldsymbol{Q}_{t+1}$ and $\boldsymbol{R}_{t+1}$. *Hint:* $\boldsymbol{A}_{t+1}$ can be decomposed as

$$\boldsymbol{A}_{t+1} = \begin{bmatrix} \boldsymbol{Q}_t & \boldsymbol{v} \\ \boldsymbol{z}^T & 1 \end{bmatrix} \begin{bmatrix} \boldsymbol{R}_t \\ \boldsymbol{a}_{m+1}^T \end{bmatrix},$$

   where $\boldsymbol{z}$ and $\boldsymbol{v}$ are vectors of zeros of appropriate length. What is the order of the FLOP count for this estimate? This process is the basis for an adaptive filtering method, used for tracking the least squares solution in nonstationary environments throughout time with a view to being as computationally efficient as possible. *Hint 2*: Consider Givens rotations.
6. Write a program to convert the Toeplitz matrix of (6.20) into a tridiagonal form.
7. Describe a QR algorithm for which the respective $\boldsymbol{R}$ is lower triangular instead of upper triangular. Describe how properties 1 and 2, stated under (6.1), are affected by this altered formulation.
8. We are given a vector $\boldsymbol{x} \in \mathbb{R}^m$. Construct an orthonormal $\boldsymbol{Q}$ such that $\boldsymbol{x}$ is rotated by θ degrees in each plane formed by the [ith, $(i + 1)$th] axis, where $i = 1, \ldots, m - 1$. Which QR method would you use to construct $\boldsymbol{Q}$?

References

1. G.H. Golub, C.F. Van Loan, *Matrix Computations*, 3rd edn. (The Johns Hopkins University, Baltimore, 1996)
2. N.K. Sinha, G.J. Lastman, *Microcomputer-Based Numerical Methods for Science and Engineering* (Oxford University Press, Oxford, 1988)
3. G. Strang, *Linear Algebra and Its Applications* (Harcourt Brace Jovanovich College Publishers, San Diego, 1988)
4. J.H. Wilkinson, *The Algebraic Eigenvalue Problem*, vol. 87 (Clarendon Press, Oxford, 1965)

Chapter 7
Linear Least Squares Estimation

Because least squares is such an important topic, it is the focus of this and the following three chapters. In Chap. 8, we discuss various LS estimation methods when the matrix $\boldsymbol{X}$ is poorly conditioned or rank deficient. Then we extend this treatment to discuss latent variable methods, which are useful for modelling poorly conditioned linear systems. Then in Chap. 9 we discuss the important concept of *regularisation*, which is an additional method for mitigating the effects of poor conditioning when modelling linear systems. The topic of the last chapter is on analysis of Toeplitz systems, which arise when solving autoregressive problems on stationary time series. This chapter describes an alternative framework for representing AR systems, which in turn lead to fast algorithms for solving AR systems and efficient methods for computing inverses, determinants, and QR decompositions.

7.1 Examples of Least Squares Analysis

In the following we present three separate examples showing diverse, practical applications of the least squares method. This section may be skipped if the reader is more interested in the mathematics of least squares analysis than in the practical examples from the field of signal processing that motivate the use of the least squares technique.

Supplementary Information The online version contains supplementary material available at (https://doi.org/10.1007/978-3-031-68915-4_7).

J. Reilly, *Fundamentals of Linear Algebra for Signal Processing*,
https://doi.org/10.1007/978-3-031-68915-4_7

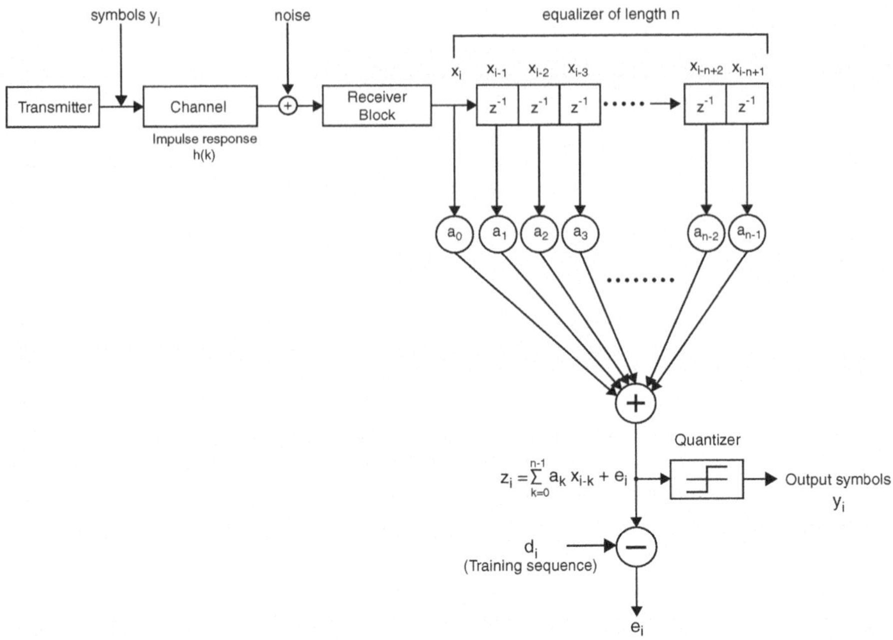

Fig. 7.1 A simplified block diagram of a digital transmission system showing the equaliser section of the receiver

7.1.1 Example 1: An Equaliser in a Communications System

In a digital communications system such as a cell phone, discrete-valued symbols $y_i, i = 1, 2, \ldots,$ are generated at the transmitter every T seconds and received at the output of the channel (or the input to the equaliser) as the quantities x_i, as shown in Fig. 7.1. The discrete-time impulse response of the channel is denoted $h_k, k = 1, \ldots, \ell$, where ℓ is the length of the channel impulse response. Ideally, this function should be a unit sample function δ_i, which is zero everywhere except for a single value of one at some specified position in time. However due to multipath propagation and other causes, in a practical system h_k often extends over several symbol periods. Since noise is always present in the receiver, the received symbols x_i may then be expressed as $x_i = h_k * y_i + e_i$, where e_i is the noise sequence and $*$ denotes the convolution operation. Therefore as a result of the convolution operation, the received symbol at time i is a weighted combination of the present and past input symbols, plus noise. The fact that the current symbol x_i contains contributions from symbols from previous time periods reduces the immunity of the receiver to noise and is referred to as *intersymbol interference* or ISI. An *equaliser* is incorporated into the communication system's receiver to alleviate the effects of ISI. A block diagram of the structure is shown in Fig. 7.1. Here, received symbols $x_i, x_{i-1}, \ldots, x_{i-n+1}$[1] at time i at the output of the receiver block are fed into the equaliser, which is an adaptive finite impulse response (FIR) filter of length n as shown. (A tutorial discussion on digital systems and digital filters is given in the Appendix of this volume.) These symbols are multiplied by their weights $a_1, a_2, \ldots, a_n$ and added together to give a set of continuous-valued output symbols z_i, as shown. In the frequency domain, the equaliser acts as a filter which attempts to invert the frequency response of the channel. From the corresponding time-domain perspective, the purpose of the equaliser is to disentangle the effect of the previous samples on the current sample and hence suppress ISI. This is accomplished by properly selecting the coefficients a_k so that the overall frequency response of the channel plus equaliser in the frequency domain is as flat as possible. The corresponding overall impulse response of the channel in combination with the equaliser therefore approximates a unit sample function, with the result that ISI is suppressed as far as possible. Thus ideally, the symbols z_i at the output of the equaliser are equal to the corresponding transmitted symbols y_i, plus noise. For more details on this topic, there are several good references on equalisers and digital communications systems at large, e.g. [9].

During the initialisation procedure of the equaliser, the transmitter sends a sequence of symbols, referred to as a *training sequence*, whose values are known at the receiver, that we denote as d_i, $i = 1, \ldots, m$, where m is the length of the training sequence. The idea of the equaliser is to generate the set of weights a_k, $k = 1, \ldots, n$, such that the output z_i is as close as possible to d_i (and hence approximately y_i). Since the equaliser must "fill up" before producing a valid output symbol z_i, the first $n - 1$ values of z_i must be discarded, leaving the relevant range of values for i as $i = n, \ldots, m$. The length ℓ of the sequence $z_i = m - n + 1$. We define the error signal e_i as the difference between z_i and d_i. Then we have

$$\begin{aligned} d_i &= z_i + e_i \\ d_i &= \sum_{k=1}^{n} a_k x(i-k) + e_i, \end{aligned} \tag{7.1}$$

where in the last line we have made use of the fact that the sequence z_i is the convolution of the input sequence x_i with the sequence a_k, as is evident from Fig. 7.1. A quantised version of the sequence z_i is the desired output of the receiver as shown.

If we observe (7.1) over ℓ sample periods, we obtain a new equation in the form of (7.1) for every value of the index $i = n, \ldots, m$, where it is assumed that $m \gg n$. We can combine these resulting m equations into a single matrix equation:

$$\underset{\ell\times 1}{\boldsymbol{d}} = \underset{\ell\times n}{\boldsymbol{X}}\ \underset{n\times 1}{\boldsymbol{a}} + \underset{\ell\times 1}{\boldsymbol{e}}, \tag{7.2}$$

[1] Here and in the sequel, for simplicity of notation, we use the subscript notation to imply the quantity $x(iT)$.

where $\boldsymbol{d} = (d_n, d_2, \ldots, d_m)^T$, similarly for $\boldsymbol{e}$. The matrices $\boldsymbol{X}$ and $\boldsymbol{a}$ are given, respectively, as

$$\boldsymbol{X} = \begin{bmatrix} x_n & x_{n-1} & \ldots & x_1 \\ x_{n+1} & x_n & \ldots & x_2 \\ \vdots & & & \vdots \\ x_m & \ldots & \ldots & x_{m-n+1} \end{bmatrix}, \quad \boldsymbol{a} = \begin{bmatrix} a_1 \\ \vdots \\ a_n \end{bmatrix}.$$

A reasonable and tractable method of choosing $\boldsymbol{a}$ is to find that value of $\boldsymbol{a}$ which minimises the 2-norm-squared difference $||\boldsymbol{e}||_2^2$ between the equaliser outputs $\boldsymbol{z} = [z_n, z_{n+1}, \ldots, z_m]^T = \boldsymbol{Xa}$ and $\boldsymbol{d} = [d_n, d_{n+1}, \ldots, d_m]^T$. Thus, we choose the optimum value $\boldsymbol{a}_0$ to satisfy

$$\boldsymbol{a}_0 = \arg\min_{\boldsymbol{a}} ||\boldsymbol{e}||_2^2 = \arg\min_{\boldsymbol{a}} ||\boldsymbol{Xa} - \boldsymbol{d}||_2^2. \tag{7.3}$$

The fact that we determine $\boldsymbol{a}_0$ by minimising $||\boldsymbol{e}||_2^2$ (*squared* 2-norm) is the origin of the term "least squares". The method of determining $\boldsymbol{a}$ to satisfy (7.3) is discussed later. Even though this example pertains specifically to an adaptive equaliser, the mathematical descriptions for other types of systems, e.g. adaptive antenna arrays, adaptive echo cancellers, adaptive filters, etc., are all identical. Basically, the mathematical framework of this section applies virtually to any type of adaptive system. In Sect. 7.10, we provide an introductory treatment on the subject of adaptive filters, which is a computationally effective method of updating the weights $\boldsymbol{a}$ in highly nonstationary environments.

Once the equaliser has been initialised, it is reasonable to say that the output sequence $\hat{y}_i$ is a close approximation to the transmitted sequence y_i. However, in mobile environments such as a cell phone, the channel changes continuously as the receiver moves about, and hence the equaliser weights must adapt to the changing environment. An adaptive filter algorithm such as the LMS method as discussed in Sect. 7.10 can be used for this purpose, as long as the filter adaptation is fast enough compared to the rate of change of the channel.

7.1.2 Example 2: Autoregressive Modelling [11]

An autoregressive (AR) process is a (stationary) random process which is the output of an all-pole discrete-time filter when excited by either white noise or a pulse train. The reason for this terminology is made apparent later. The Appendix of this volume contains a discussion on the basics of linear system theory and digital signal processing, which is required knowledge for proper understanding of this example. The Appendix is provided for the benefit of those readers who do not already have a background in this subject.

AR modelling is used extensively in signal processing applications, since in many cases it is a useful means of signal compression. It also serves as a parsi-

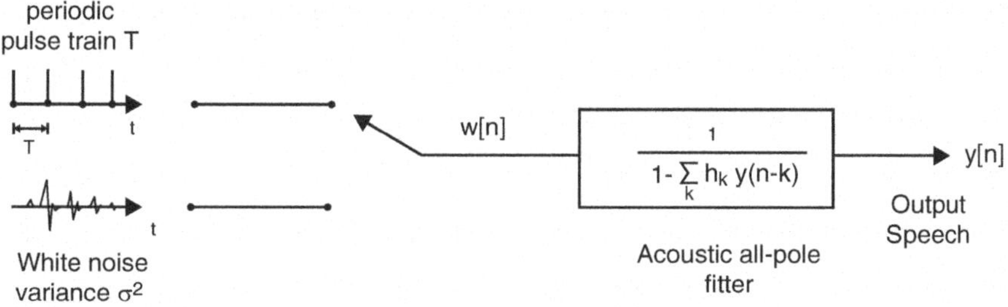

Fig. 7.2 The linear prediction model of human speech production. The switch is in the upper position for voiced sounds and the lower for unvoiced sounds. The input signal $w(n)$ is fed into the all-pole filter shown, to produce the output voice signal $y(n)$. The coefficients $h(k)$, and hence the frequency response of the filter, vary in accordance with the sound being produced. All parameters of the model change every ∼15–20 msec in accordance with the current phoneme

monious representation of an entire data sequence. For example, the human voice is an excellent example of a time-varying AR process, which can be represented by the linear predictive coding (LPC) model, a schematic for which is shown in Fig. 7.2. The human vocal tract acts as a time-varying, highly resonant all-pole acoustic filter. As shown in Fig. 7.2, the filter input is a pulse train generated by the vocal cords, or a white noise sequence generated by a restriction somewhere in the vocal tract, and whose output is the respective voice signal. The switch is in the upper position for "voiced" sounds where the vocal cords are engaged, such as "ah" or "ee", and in the lower position for unvoiced sounds, such as "sh", "s", etc. The switch position, the parameters h_k of the all-pole filter, and the period of the input pulse train all vary in accordance with the sound being produced. During the production of a single phoneme,[2] which represents an interval of about 20 msec, the voice may be considered approximately stationary and is therefore amenable to AR modelling. At an 8 KHz sampling rate, which is a typical value in telephone systems, there are 160 samples in this 20 msec interval, and an AR model representing this sequence typically consists of about 10–15 coefficients. Therefore the sequence of 160 samples can be compressed into this range of 10–15 parameters by AR modelling, thus achieving a significant degree of compression. The AR model must be updated in roughly 20 msec intervals to track the variation in phoneme production of the voice signal. More details on the modelling of voice signals and the LPC method are given in [4].

From fundamental linear system theory, an all-pole discrete-time filter has a transfer function $H(z)$ given by the expression

$$H(z) = \frac{1}{\prod_{i=1}^{n}(1 - z_i z^{-1})} \equiv \frac{1}{1 - \sum_{i=1}^{n} h_i z^{-i}}, \tag{7.4}$$

where z_i are the poles of the filter, and h_i are the coefficients of the corresponding polynomial in z.

[2] A phoneme is a single element of speech, e.g. "ah" in the English language.

Let $W(z)$ and $Y(z)$ denote the z-transforms of the input and output sequences, respectively. Then

$$H(z) = \frac{Y(z)}{W(z)} = \frac{1}{1 - \sum h_i z^{-i}}$$

or

$$Y(z)\left[1 - \sum_{i=1}^{n} h_i z^{-i}\right] = W(z). \tag{7.5}$$

We note the expression on the left is a product of z-transforms, so the corresponding time-domain expression involves the convolution of the sequence $[1, -h_1, -h_2, \ldots, -h_n]$ with $[y_1, y_2, \ldots]$. The equivalent of Eq. (7.5) in the time domain is therefore

$$y_i - \sum_{k=1}^{n} h_k y_{i-k} = w_i$$

or

$$y_i = \sum_{k=1}^{n} h_k y_{i-k} + w_i, \tag{7.6}$$

where the variance of the random sequence w_i is σ^2. From (7.6), we see that the output of an all-pole filter when driven by white noise may be given the interpretation that the present value of the output is a linear combination of past outputs weighted by the coefficients h_k of the transfer function $H(z)$, plus a random disturbance w_i. It happens that the closer the poles of the filter are to the unit circle, the more resonant is the filter, and the more predictable is the present output from its past values.

Repeating (7.6) for $i = n+1, \ldots, m+n$, we have

$$\boldsymbol{y} = \begin{bmatrix} y_n & y_{n-1} & \cdots & y_1 \\ y_{n+1} & y_n & \cdots & y_2 \\ \vdots & & & \vdots \\ y_{m+n-1} & y_{m+n-2} & \cdots & y_m \end{bmatrix} \boldsymbol{h} + \boldsymbol{w}$$

$$= \boldsymbol{Y}\boldsymbol{h} + \boldsymbol{w}, \tag{7.7}$$

where $\boldsymbol{y} = [y_{n+1}, \ldots, y_{n+m}]^T \in \mathbb{R}^m$ is the vector of predicted values, $\boldsymbol{w}$ is the vector of prediction errors, $\boldsymbol{h} = [h_1, h_2, \ldots, h_n]^T$ is the vector of prediction coefficients derived from the transfer function of (7.4), and the ith row of $\boldsymbol{Y} \in \mathbb{R}^{m \times n}$ contains the n past values of y required for the prediction of y_i according to (7.6).

The mathematical model corresponding to (7.2) and (7.7) is referred to as a *regression model.* In (7.7), the variables y are "regressed" onto themselves and hence the name "autoregressive".

Equation (7.7) is of the same form as (7.2). So again, it makes sense to choose the h's in (7.7) so that the predicting term $\boldsymbol{Yh}$ is as close as possible to the true values $\boldsymbol{y}$ in the 2-norm sense. Hence, as before, we choose the optimal $\boldsymbol{h}_o$ as the solution to

$$\begin{aligned} \boldsymbol{h}_o &= \arg\min_{\boldsymbol{h}} ||\boldsymbol{Yh} - \boldsymbol{y}||_2^2 \\ &= \arg\min_{\boldsymbol{h}} (\boldsymbol{Yh} - \boldsymbol{y})^T (\boldsymbol{Yh} - \boldsymbol{y}). \end{aligned} \tag{7.8}$$

Notice that if the parameters $\boldsymbol{h}$ and the variance σ^2 are known, the autoregressive process is completely characterised.

7.1.3 *Example 3: Hurricane Prediction Using Machine Learning*

A further example of the use of least squares is in the machine learning context. Suppose we have a training set of m observations each consisting of n meteorological features or variables such as water temperatures, air temperatures, atmospheric pressures, cloud conditions, currents, wind velocities, etc., all measured simultaneously, pertaining to various tropical areas over a range of years and seasons. We also have information whether the respective observation led to a hurricane developing or not (let us say $+1$ for developed and -1 for not). Here, the set of conditions for the ith observation, $i = 1, \ldots, m$, forms a row of a matrix $\boldsymbol{X}$, and y_i is the corresponding response [± 1]. We can formulate this situation into a linear mathematical model for this problem as follows:

$$\boldsymbol{y} = \boldsymbol{X\beta} + \boldsymbol{e}, \tag{7.9}$$

where $\boldsymbol{e}$ is the error between the linear prediction model $\boldsymbol{X\beta}$ and the observations $\boldsymbol{y}$. We can expand each of the matrix/vector quantities for clarity as follows:

$$\boldsymbol{y} = \begin{bmatrix} \pm 1 \\ \pm 1 \\ \vdots \\ \pm 1 \end{bmatrix} = \begin{bmatrix} x_{11} & x_{12} & \ldots & x_{1n} \\ x_{21} & x_{22} & \ldots & x_{2n} \\ \vdots & \vdots & \ddots & \vdots \\ x_{m1} & x_{m2} & \ldots & x_{mn} \end{bmatrix} \begin{bmatrix} \beta_1 \\ \beta_2 \\ \vdots \\ \beta_n \end{bmatrix} + \boldsymbol{e},$$

where the x_{ij} elements are the jth variable of the ith observation. As in the previous examples, we wish to determine a set $\boldsymbol{\beta}$ of predictors (or weights for each of the variables), which give the best fit between the model $\boldsymbol{X\beta}$ and the observation $\boldsymbol{y}$. The predictors $\boldsymbol{\beta}_o$ may be determined as the solution to

$$\boldsymbol{\beta}_o = \arg\min_{\boldsymbol{\beta}} ||\boldsymbol{X}\boldsymbol{\beta} - \boldsymbol{y}||_2^2.$$

Once we have solved for the optimal values $\boldsymbol{\beta}_o$ of $\boldsymbol{\beta}$, we can predict whether a hurricane will occur given a new set of previously unseen meteorological data (in the form of a new row $\boldsymbol{x}_N^T$ of $\boldsymbol{X}$) by evaluating

$$\hat{y} = \boldsymbol{x}_N^T \boldsymbol{\beta}_o,$$

where "hat" denotes an estimated value. Typically the value of $\hat{y}$ will not be ± 1 as it would be in the ideal case. But in practice a good prediction could be made by declaring a hurricane will develop if $\hat{y} \geq T$, and not develop otherwise, where T is some suitably chosen threshold such as, e.g. the value zero.

The three examples presented above illustrate only a small sample of the range of applications of the least squares technique.

7.2 The Linear Least Squares Model

It is now apparent that these examples all have the same mathematical structure. We revert to the standardised notation common to the statistical literature. We define our *regression model* or *regression equation*, corresponding to (7.2), (7.7), or (7.9), as

$$\boldsymbol{y} = \boldsymbol{X}\boldsymbol{\beta} + \boldsymbol{\epsilon}, \tag{7.10}$$

where $\boldsymbol{y}$ is a sample from random process, and we wish to determine the value $\boldsymbol{\beta}_{LS}$ which solves

$$\boldsymbol{\beta}_{LS} = \arg\min_{\boldsymbol{\beta}} ||\boldsymbol{X}\boldsymbol{\beta} - \boldsymbol{y}||_2^2, \tag{7.11}$$

where $\boldsymbol{X} \in \mathbb{R}^{m\times n}, \quad m > n, \quad \boldsymbol{y} \in \mathbb{R}^m$, and $\boldsymbol{\epsilon}$ is the unknown observational error between the observation ($\boldsymbol{y}$) and the model ($\boldsymbol{X}\boldsymbol{\beta}$). For this treatment, the matrix $\boldsymbol{X}$ is assumed full rank. The extension to the rank-deficient case is covered in the following chapters. We first consider the case where the intercept in (7.10) is zero; that is, $E(y_i) = 0$ when $\boldsymbol{x}_i^T = \boldsymbol{0}$, where $\boldsymbol{x}_i^T$ is the ith row of $\boldsymbol{X}$. We then generalise to the nonzero case later in this section.

The quantities in $\boldsymbol{X}$ are referred to as *predictor variables*, *independent variables*, or simply just *variables*. The quantities in $\boldsymbol{y}$ are referred to as *observations, responses*, or *dependent variables*. The ith row $\boldsymbol{x}_i^T$ of $\boldsymbol{X}$ is the ith instance of a set of n variables, and the jth column $\boldsymbol{x}_j$ of $\boldsymbol{X}$ contains all values of the jth variable over the whole set of m observations. The element y_i of $\boldsymbol{y}$ represents the response corresponding to the variables $\boldsymbol{x}_i^T$ in the ith row of $\boldsymbol{X}$; that is, with respect to the hurricane example above, did the hurricane develop or not for the conditions indicated by $\boldsymbol{x}_i^T$? The residual vector $\boldsymbol{r}_{LS} \triangleq \boldsymbol{X}\boldsymbol{\beta}_{LS} - \boldsymbol{y}$ is the minimised error between the observations and the model $\boldsymbol{X}\boldsymbol{\beta}_{LS}$ fitted by the LS procedure. (It should be noted that the residual vector $\boldsymbol{r}_{LS}$ is generally not equal to the error $\boldsymbol{\epsilon}$ in (7.10).) The minimum value $||\boldsymbol{X}\boldsymbol{\beta}_{LS} - \boldsymbol{y}||_2^2$ in (7.11) is referred to as the *residual sum of squares* and is given the symbol ρ_{LS}^2.

It is useful to note that the quantity $||\boldsymbol{X}\boldsymbol{\beta} - \boldsymbol{y}||_2^2$ for $\boldsymbol{\beta} = \boldsymbol{\beta}_{LS}$ in (7.11) can be written as

$$||\boldsymbol{X}\boldsymbol{\beta}_{LS} - \boldsymbol{y}||_2^2 = \sum_{i=1}^{m} (y_i - \hat{y}_i)^2 = \sum_{i=1}^{m} \left(r_{LS}(i)\right)^2, \tag{7.12}$$

where $\hat{y}_i$ is the predicted value of y_i for the ith observation given by the LS model (i.e. $\hat{y}_i = \boldsymbol{x}_i^T \boldsymbol{\beta}_{LS}$), and $r_{LS}(i)$ is the LS residual for the ith observation.

The fundamental idea with LS methods is to develop a simple (linear) model which gives the closest relationship in the 2-norm sense between the set of m noisy observations $\boldsymbol{y}$ and the linear model (i.e. linear in $\boldsymbol{\beta}$) in the form of $\boldsymbol{X}\boldsymbol{\beta}$. Each observation consists of an observation y_i and a corresponding set of n variables (in the form of the ith row $\boldsymbol{x}_i^T$ of $\boldsymbol{X}$, $i = 1, \ldots, m$) whose values are known and may be controllable. Once $\boldsymbol{\beta}$ is determined through (7.11), the model description is complete.

There are at least two reasons for building such a model. The first is that the values of $\boldsymbol{\beta}$ may be useful in their own right, as is the case with autoregressive analysis. The second reason is to be able to predict the response y corresponding to a new row $\boldsymbol{x}_N^T$ of $\boldsymbol{X}$ as $\hat{y} = E(y) = \boldsymbol{x}_N^T \boldsymbol{\beta}$, as is the case with the hurricane example. The linear form of the regression model, while simple, has proven to be extremely effective in many practical situations and leads to a convenient, closed-form solution for the parameters $\boldsymbol{\beta}$.

A geometric interpretation for the LS objective function (7.11) is given in Fig. 7.3 for the one-dimensional case as determining the slope β of the line $y = x\beta$ such that the sum of the squared distances $\left(r_{LS}(i)\right)^2$, where $r_{LS}(i) = y_i - x_i\beta_{LS}$, $i = 1, \ldots, m$, between the ith observed point (shown as red circles) and the line is minimised. In multidimensions we would have a hyperplane $\boldsymbol{y} = \boldsymbol{X}\boldsymbol{\beta}$ instead of a line, but we would still be minimising the sum of squared distances between the observed points and the hyperplane. Here are a few mathematically relevant points concerning the LS problem:

- For a well-behaved LS problem, m must be greater than n, so the system (7.11) is overdetermined and hence no solution exists in the general case for which $\boldsymbol{X}\boldsymbol{\beta} = \boldsymbol{y}$ exactly.
- Of all the commonly used values of p, only the $p = 2$ norm is differentiable for all values of $\boldsymbol{\beta}$. This property leads directly to a closed-form solution for $\boldsymbol{\beta}$. For any other value of p, the optimal solution must be solved using iterative optimisation techniques.
- Note that for an orthonormal matrix $\boldsymbol{Q}$, we have (only for $p = 2$)

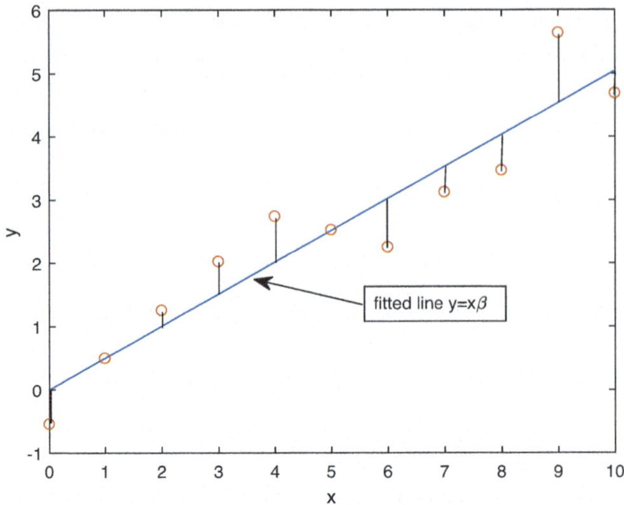

Fig. 7.3 A geometric interpretation of the LS problem for the one-dimensional case. The set of observed points is shown as the red circles, whereas the fitted line $y = x\beta_{LS}$ is shown in blue and the residuals $r_{LS}(i), i = 1, \ldots, m$, are shown as the vertical black lines. This example adopts the LS formulation which assumes the y-intercept to be zero

$$||\boldsymbol{X\beta} - \boldsymbol{y}||_2^2 = \left|\left|\boldsymbol{Q}^T\boldsymbol{X\beta} - \boldsymbol{Q}^T\boldsymbol{y}\right|\right|_2^2.$$

This fact is used to advantage later on.

- If $r = \text{rank}(\boldsymbol{X}) < n$, then there is no unique $\boldsymbol{\beta}_{LS}$ which minimises $||\boldsymbol{X\beta} - \boldsymbol{y}||_2$. Consider the set $\mathcal{S}$ of all $\boldsymbol{\beta}_{LS}$ for which $\left|\left|\boldsymbol{X\beta}_{LS} - \boldsymbol{y}\right|\right|_2$ achieves its minimum value. Then $\boldsymbol{\beta}_{LS}$ can be made unique by choosing that $\boldsymbol{\beta}_{LS} \in \mathcal{S}$ with minimum 2-norm.

7.3 The Linear Least Squares Solution

We wish to determine the parameters $\boldsymbol{\beta}$ by solving (7.11). The method we adopt here is to differentiate the quantity $||\boldsymbol{X\beta} - \boldsymbol{y}||_2^2$ with respect to $\boldsymbol{\beta}$ and set the result to zero. Thus, the remaining portion of this section is devoted to this differentiation. The result is a closed-form expression for the solution of (7.11). Other methods for solving (7.11) are given in later chapters.

The expression $||\boldsymbol{X\beta} - \boldsymbol{y}||_2^2$ can be written as

$$\begin{aligned} ||\boldsymbol{X\beta} - \boldsymbol{y}||_2^2 &= (\boldsymbol{X\beta} - \boldsymbol{y})^T(\boldsymbol{X\beta} - \boldsymbol{y}) \\ &= \boldsymbol{y}^T\boldsymbol{y} - \boldsymbol{\beta}^T\boldsymbol{X}^T\boldsymbol{y} - \boldsymbol{y}^T\boldsymbol{X\beta} + \boldsymbol{\beta}^T\boldsymbol{X}^T\boldsymbol{X\beta}. \end{aligned}$$

The solution $\boldsymbol{\beta}_{LS}$ is that value of $\boldsymbol{\beta}$ which satisfies

$$\frac{d}{d\boldsymbol{\beta}}\left[\boldsymbol{y}^T\boldsymbol{y} - \boldsymbol{\beta}^T\boldsymbol{X}^T\boldsymbol{y} - \boldsymbol{y}^T\boldsymbol{X}\boldsymbol{\beta} + \boldsymbol{\beta}^T\boldsymbol{X}^T\boldsymbol{X}\boldsymbol{\beta}\right] = \mathbf{0}. \tag{7.13}$$

Define each term in the square brackets above sequentially as $t_1(\boldsymbol{\beta}), \ldots, t_4(\boldsymbol{\beta})$, respectively. Therefore we solve

$$\frac{d}{d\boldsymbol{\beta}}\left[t_2(\boldsymbol{\beta}) + t_3(\boldsymbol{\beta}) + t_4(\boldsymbol{\beta})\right] = \mathbf{0}, \tag{7.14}$$

where we have noted the derivative $\frac{d}{d\boldsymbol{\beta}}t_1 = \mathbf{0}$, since $\boldsymbol{y}$ is independent of $\boldsymbol{\beta}$.

We see that every term of (7.14) is a scalar. To differentiate (7.14) with respect to the vector $\boldsymbol{\beta}$, we differentiate each scalar term of (7.14) with respect to each element β_k, $k = 1, \ldots, n$, of $\boldsymbol{\beta}$ and then assemble all the derivatives back into an n-length vector. We now discuss the differentiation of each term of (7.14).

Differentiation of $t_2(\mathbf{x})$ and $t_3(\mathbf{x})$ with respect to x

Let us define the quantity $\boldsymbol{c} \stackrel{\Delta}{=} \boldsymbol{X}^T\boldsymbol{y}$. This implies that the component c_k of $\boldsymbol{c}$ is $\boldsymbol{x}_k^T\boldsymbol{y}$, $k = 1, \ldots, n$, where $\boldsymbol{x}_k^T$ is the transpose of the kth column of $\boldsymbol{X}$.

Thus $t_2(\boldsymbol{\beta}) = -\boldsymbol{\beta}^T\boldsymbol{c}$. Therefore,

$$\frac{d}{d\beta_k}t_2(\boldsymbol{\beta}) = \frac{d}{d\beta_k}(-\boldsymbol{\beta}^T\boldsymbol{c}) = -c_k = -\boldsymbol{x}_k^T\boldsymbol{y}, \qquad k = 1, \ldots, n. \tag{7.15}$$

Combining these results for $k = 1, \ldots, n$ back into a column vector, we get

$$\frac{d}{d\boldsymbol{\beta}}t_2(\boldsymbol{\beta}) = -\boldsymbol{c} = -\boldsymbol{X}^T\boldsymbol{y}. \tag{7.16}$$

Since Term 3 of (7.14) is the transpose of Term 2 and both are scalars, the terms are equal. Hence,

$$\frac{d}{d\boldsymbol{\beta}}t_3(\boldsymbol{\beta}) = -\boldsymbol{X}^T\boldsymbol{y}. \tag{7.17}$$

Differentiation of $t_4(\mathbf{x})$ with respect to x

The differentiation of the quadratic form t_4 is covered in Appendix A of Chapter 2. Using the fact that $\boldsymbol{X}^T\boldsymbol{X}$ is symmetric, the result is

$$\frac{d}{d\boldsymbol{\beta}}t_4(\boldsymbol{\beta}) = \frac{d}{d\boldsymbol{\beta}}(\boldsymbol{\beta}^T\boldsymbol{X}^T\boldsymbol{X}\boldsymbol{\beta}) = 2\boldsymbol{X}^T\boldsymbol{X}\boldsymbol{\beta}. \tag{7.18}$$

Substituting (7.16), (7.17), and (7.18) into (7.13), setting the result to zero, and simplifying, we get the important desired result:

$$\boldsymbol{X}^T\boldsymbol{X}\boldsymbol{\beta} = \boldsymbol{X}^T\boldsymbol{y}. \tag{7.19}$$

The value $\boldsymbol{\beta}_{LS}$ of $\boldsymbol{\beta}$, which solves (7.19), is the least squares solution corresponding to (7.11). Equation (7.19) is called the *normal equation*. The reason for this terminology is discussed in the next section.

From (7.19), it is clear that the solution $\boldsymbol{\beta}_{LS}$ is given as

$$\boldsymbol{\beta}_{LS} = (\boldsymbol{X}^T\boldsymbol{X})^{-1}\boldsymbol{X}^T\boldsymbol{y}. \tag{7.20}$$

Note however that solving for $\boldsymbol{\beta}_{LS}$ using (7.20) is not the preferred approach. Instead, since $\boldsymbol{X}^T\boldsymbol{X}$ is square, symmetric, and positive definite, solving (7.19) for $\boldsymbol{\beta}_{LS}$ using the Cholesky decomposition is faster and more numerically accurate.

Extension to the Nonzero Intercept Case It is straightforward to deal with the situation where the intercept is known to be nonzero. An $m \times 1$ column of ones (i.e. $\mathbf{1}$) is appended to the matrix $\boldsymbol{X}$ as follows:

$$\boldsymbol{X} \leftarrow [\boldsymbol{X} \;\; \mathbf{1}]\,. \tag{7.21}$$

Then $n \leftarrow n+1$, and the LS method proceeds exactly as before. The last element of $\boldsymbol{\beta}_{LS}$ is then the value of the intercept. As we have seen from the interlacing theorem of Sect. 5.5, adding more columns to $\boldsymbol{X}$ has the effect of increasing the condition number of $\boldsymbol{X}^T\boldsymbol{X}$, thus increasing the variance of the LS estimates, as we see later. It is therefore expected that the variances of the LS estimates in the nonzero intercept case may be degraded relative to the zero intercept case.

7.3.1 Interpretation of the Normal Equations

Equation (7.19) can be written in the form

$$\boldsymbol{X}^T(\boldsymbol{y} - \boldsymbol{X}\boldsymbol{\beta}_{LS}) = \mathbf{0} \tag{7.22}$$

or

$$\boldsymbol{X}^T\boldsymbol{r}_{LS} = \mathbf{0}, \tag{7.23}$$

where

$$\boldsymbol{r}_{LS} \triangleq \boldsymbol{y} - \boldsymbol{X}\boldsymbol{\beta}_{LS}$$

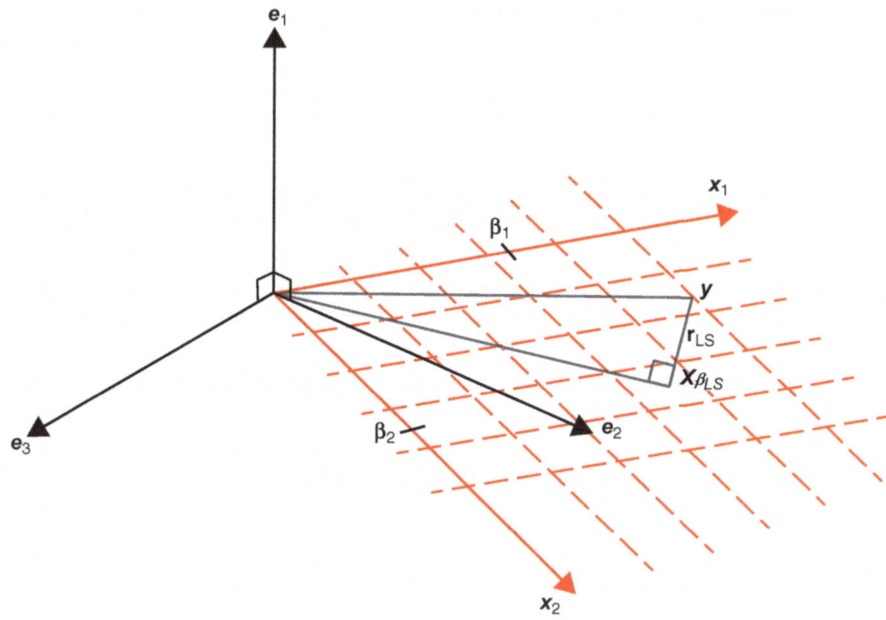

Fig. 7.4 A geometric interpretation of the LS problem for the 3×2 case. The red cross-hatched region represents a portion of $R(\boldsymbol{X})$. According to (7.22), the point $\boldsymbol{X}\boldsymbol{\beta}_{LS}$ is at the foot of a perpendicular dropped from $\boldsymbol{y}$ into $R(\boldsymbol{X})$

is the least squares residual vector between $\boldsymbol{X}\boldsymbol{\beta}_{LS}$ and $\boldsymbol{y}$. Thus, $\boldsymbol{r}_{LS}$ must be orthogonal to $R(\boldsymbol{X})$ for the LS solution, $\boldsymbol{\beta}_{LS}$, hence the name "normal equations". This fact gives an important interpretation to least squares estimation, which we now illustrate for the 3×2 case, for which (7.10) may be expressed as

$$\boldsymbol{y} = [\boldsymbol{x}_1, \boldsymbol{x}_2] \begin{bmatrix} \beta_1 \\ \beta_2 \end{bmatrix} + \boldsymbol{\epsilon},$$

where in this example all vectors are of length 3. The above vector relation is illustrated in Fig. 7.4, where we see from (7.23) that the point $\boldsymbol{X}\boldsymbol{\beta}_{LS}$ is at the foot of a perpendicular dropped from $\boldsymbol{y}$ into $R(\boldsymbol{X})$, as shown in the figure. The solution $\boldsymbol{\beta}_{LS}$ consists of the coefficients of the linear combination of $\boldsymbol{x}_1$ and $\boldsymbol{x}_2$ which equal the "foot vector", $\boldsymbol{X}\boldsymbol{\beta}_{LS}$, which is also the predicted value of $\boldsymbol{y}$.

This interpretation may be augmented as follows. From (7.19), we see that $\boldsymbol{\beta}_{LS}$ is given by

$$\boldsymbol{\beta}_{LS} = \left(\boldsymbol{X}^T\boldsymbol{X}\right)^{-1}\boldsymbol{X}^T\boldsymbol{y}. \tag{7.24}$$

Hence, the point $X\beta_{LS}$ which is in $R(X)$ is given by

$$X\beta_{LS} = X\left(X^T X\right)^{-1} X^T y \equiv Py, \tag{7.25}$$

where P is the projector onto $R(X)$. Thus, we see from another point of view that the least squares solution is the result of projecting y (the observation) onto $R(X)$.

It is seen from (7.10) that in the noise-free case, the vector y is equal to the vector $X\beta_{LS}$. The fact that $X\beta_{LS}$ should be at the foot of a perpendicular from y into $R(X)$ makes intuitive sense, because a perpendicular is the shortest 2-norm distance from y into $R(X)$. This, after all, is the objective of the LS problem as expressed by Eq. (7.11).

There is a further point we wish to address in the interpretation of the normal equations. Substituting (7.25) into (7.3.1), we have

$$\begin{aligned} r_{LS} &= y - X(X^T X)^{-1} X^T y \\ &= (I - P)y \\ &= P_\perp y. \end{aligned} \tag{7.26}$$

Thus, r_{LS} is the projection of y onto $R(X)_\perp$, as expected from Fig. 7.4.

We can now determine the value ρ_{LS}^2, which is the squared 2-norm of the LS residual:

$$\rho_{LS}^2 \triangleq ||r_{LS}||_2^2 = ||P_\perp y||_2^2 . \tag{7.27}$$

The fact that r_{LS} is orthogonal to $R(X)$ is of fundamental importance. In fact, it is easy to show that choosing β so that $r_{LS} \perp R(X)$ is a sufficient condition for the least squares solution. Often in analysis, β_{LS} is determined this way, instead of through the normal equations. This concept is referred to as the *principle of orthogonality* [13].

We repeat the normal equations from (7.19) for convenience:

$$X^T X\beta = X^T y. \tag{7.28}$$

From our discussions on covariance matrices in Sect. 2.3, we realise the matrix $X^T X$ on the left above is m times the $n \times n$ covariance matrix estimate $\hat{R}$ of the variables in X. Similarly, the vector $X^T y \in \mathbb{R}^n$ on the right, which we designate as $\hat{p}$, is m times the vector of covariance estimates between y and each of the n variables in X. Thus, with this interpretation, the normal equations can be written in the form

$$\hat{R}\beta_{LS} = \hat{p}, \tag{7.29}$$

where β_{LS} is the solution to (7.28). In this form the normal equations are referred to as the *Wiener–Hopf* equations.

LS Estimation as a Means of Noise Suppression It is clear from the regression equation $\boldsymbol{y} = \boldsymbol{X}\boldsymbol{\beta} + \boldsymbol{\epsilon}$ that $\boldsymbol{\epsilon}$ and hence $\boldsymbol{y}$ contain m dimensions of noise. That is, in the case where the elements ϵ_i are independent, identically distributed (*iid*), then the total variance of $\boldsymbol{\epsilon}$ is $E||\boldsymbol{\epsilon}||_2^2 = m\sigma^2$, where σ^2 is the variance of a single element of $\boldsymbol{\epsilon}$. However, $R(\boldsymbol{X})$ contains only $n < m$ dimensions. As we have seen, the normal equations implicitly project $\boldsymbol{\epsilon}$ into $R(\boldsymbol{X})$, giving an effective noise vector equal to $\boldsymbol{P\epsilon}$ with n dimensions (degrees of freedom), where $\boldsymbol{P}$ is the projector onto $R(\boldsymbol{X})$. Thus the total effective noise variance of the term $\boldsymbol{P\epsilon}$, which contaminates the LS estimate, is only $n\sigma^2$. Therefore the $\boldsymbol{\beta}_{LS}$ produced by the normal equations has $m - n$ dimensions of noise suppressed, resulting in an improved estimate. It follows that the smaller the n is for a given m, the better the estimate $\boldsymbol{\beta}_{LS}$ will be. It is this noise suppression property of the LS estimator that leads to its desirable properties, as discussed next.

7.4 Properties of the LS Estimate

As we have seen above, since $\boldsymbol{\epsilon}$ is a random noise vector, $\boldsymbol{y}$ is also a random variable, and therefore from (7.24) $\boldsymbol{\beta}_{LS}$ is a random variable, since it is a function of $\boldsymbol{y}$ and hence contains error. We would like to quantify this error. Measures for quantifying the error of an estimate of a vector random variable in the presence noise are *bias*, *variance*, and *covariance*. The bias $\boldsymbol{b}$ of some estimated parameter vector, which we denote as $\boldsymbol{\theta}$, is defined as $E(\hat{\boldsymbol{\theta}} - \boldsymbol{\theta}_o)$, where the expectation is taken over all possible values of the parameter estimate $\hat{\boldsymbol{\theta}}$, and $\boldsymbol{\theta}_o$ is the true value of the parameter. An example of a biased estimator is a person's estimate of how long it will take to complete a particular task—people often underestimate the required time, so in this case the expected value of the estimate is less than the true value, and the bias is negative.

The covariance matrix of the parameter estimate is discussed at some length in Chapter 2 and is repeated here as

$$\mathrm{cov}(\hat{\boldsymbol{\theta}}) = E\left[\left(\hat{\boldsymbol{\theta}} - E(\hat{\boldsymbol{\theta}})\right)\left(\hat{\boldsymbol{\theta}} - E(\hat{\boldsymbol{\theta}})\right)^T\right]. \tag{7.30}$$

An additional means to quantify the error of an estimated parameter $\hat{\boldsymbol{\theta}}$ is the *mean-squared error* $\mathrm{mse}(\hat{\boldsymbol{\theta}})$. This quantity is a vector of the same size as $\boldsymbol{\theta}$. One means of defining *mse* is given as

$$\mathrm{mse}(\hat{\boldsymbol{\theta}}) = E\left(\hat{\boldsymbol{\theta}} - \boldsymbol{\theta}_o\right)^{\cdot 2}, \tag{7.31}$$

where the notation $(\cdot)^{\cdot 2}$ refers to element-by-element squaring of each element of the respective vector. If the bias is zero, then the mean-squared error and the variance are equivalent. We can simplify the expression for mse by noting that

$$\boldsymbol{\theta}_o = E(\hat{\boldsymbol{\theta}}) - \boldsymbol{b}. \tag{7.32}$$

By substituting (7.32) into (7.31), we have

$$\begin{aligned}\text{mse}(\hat{\boldsymbol{\theta}}) &= E\left(\hat{\boldsymbol{\theta}} - \boldsymbol{\theta}_o\right)^{\cdot 2} \\ &= E\left(\hat{\boldsymbol{\theta}} - E(\hat{\boldsymbol{\theta}}) + \boldsymbol{b}\right)^{\cdot 2} \\ &= E\left(\boldsymbol{e} + \boldsymbol{b}\right)^{\cdot 2},\end{aligned}$$

where $\boldsymbol{e} \triangleq \hat{\boldsymbol{\theta}} - E(\hat{\boldsymbol{\theta}})$. From the above we have

$$\begin{aligned}\text{mse}(\hat{\boldsymbol{\theta}}) &= E(\boldsymbol{e})^{\cdot 2} + 2\boldsymbol{b}E(\boldsymbol{e}) + \boldsymbol{b}^{\cdot 2} \\ &= \boldsymbol{\sigma}^{\cdot 2} + \boldsymbol{b}^{\cdot 2}.\end{aligned} \tag{7.33}$$

This is the desired result. In the above we used the fact that $\boldsymbol{b}$ is a constant and $E(\boldsymbol{e}) = \boldsymbol{0}$. Here $\boldsymbol{\sigma}^{\cdot 2}$ is the vector of variances corresponding to each element of $\boldsymbol{e}$, which from (7.33) is also the vector of variances of the elements of $\hat{\boldsymbol{\theta}}$. This shows that the total error (mean-squared error) of each element of an estimated vector quantity is the sum of its variance + its bias^2. Thus to completely characterise the error structure of $\hat{\boldsymbol{\theta}}$, we need to only evaluate the bias and variance/covariance. Another common way to express *mse* is to take the sum over all the elements of the *mse* vector in (7.33).

To develop expressions for the bias and covariance of the specific parameter estimate $\boldsymbol{\beta}$, we invoke two assumptions:

A1 $\boldsymbol{\epsilon}$ is a zero-mean random vector with identically distributed, uncorrelated elements; that is, $E(\boldsymbol{\epsilon}\boldsymbol{\epsilon}^T) = \sigma^2\boldsymbol{I}$.

A2 $\boldsymbol{X}$ is a constant matrix, which is known with negligible error. That is, there is no *uncertainty* in $\boldsymbol{X}$.

7.4.1 $\boldsymbol{\beta}_{LS}$ Is an Unbiased Estimate of $\boldsymbol{\beta}_o$, the True Value

To show this, we have from (7.24)

$$\boldsymbol{\beta}_{LS} = \left(\boldsymbol{X}^T\boldsymbol{X}\right)^{-1}\boldsymbol{X}^T\boldsymbol{y}. \tag{7.34}$$

We realise that the observed data $\boldsymbol{y}$ are generated from the *true* values $\boldsymbol{\beta}_o$ of $\boldsymbol{\beta}$. Hence substituting the regression equation for $\boldsymbol{y}$, we have

$$\begin{aligned}\boldsymbol{\beta}_{LS} &= \left(\boldsymbol{X}^T\boldsymbol{X}\right)^{-1}\boldsymbol{X}^T(\boldsymbol{X}\boldsymbol{\beta}_o + \boldsymbol{\epsilon}) \\ &= \boldsymbol{\beta}_o + \left(\boldsymbol{X}^T\boldsymbol{X}\right)^{-1}\boldsymbol{X}^T\boldsymbol{\epsilon}. \end{aligned} \tag{7.35}$$

Therefore, $E(\boldsymbol{\beta}_{LS})$ is given as

$$\begin{aligned} E(\boldsymbol{\beta}_{LS}) &= \boldsymbol{\beta}_o + (\boldsymbol{X}^T\boldsymbol{X})^{-1}\boldsymbol{X}^T E(\boldsymbol{\epsilon}) \\ &= \boldsymbol{\beta}_o, \end{aligned} \tag{7.36}$$

where in the first line immediately above we note that $\boldsymbol{X}$ is constant from Assumption A2 and so the expectation applies only to the variable $\boldsymbol{\epsilon}$, and in the last line we note that $\boldsymbol{\epsilon}$ is zero mean from A1. Therefore under the stated assumptions, the expectation of $\boldsymbol{\beta}$ is its true value, and $\boldsymbol{\beta}_{LS}$ is *unbiased.*

7.4.2 *Covariance Matrix of* $\boldsymbol{\beta}_{LS}$

From (7.35) and (7.36), $\boldsymbol{\beta}_{LS} - E(\boldsymbol{\beta}_{LS}) = \left(\boldsymbol{X}^T\boldsymbol{X}\right)^{-1}\boldsymbol{X}^T\boldsymbol{\epsilon}$. Substituting these values into (7.30), we have

$$\text{cov}(\boldsymbol{\beta}_{LS}) = E\left[\left(\boldsymbol{X}^T\boldsymbol{X}\right)^{-1}\boldsymbol{X}^T\left(\boldsymbol{\epsilon}\boldsymbol{\epsilon}^T\right)\boldsymbol{X}\left(\boldsymbol{X}^T\boldsymbol{X}\right)^{-1}\right]. \tag{7.37}$$

From Assumption A2, we can move the expectation operator inside. Therefore,

$$\begin{aligned} \text{cov}(\boldsymbol{\beta}_{LS}) &= \left[\left(\boldsymbol{X}^T\boldsymbol{X}\right)^{-1}\boldsymbol{X}^T \underbrace{E\left(\boldsymbol{\epsilon}\boldsymbol{\epsilon}^T\right)}_{\sigma^2\boldsymbol{I}} \boldsymbol{X}\left(\boldsymbol{X}^T\boldsymbol{X}\right)^{-1}\right] \\ &= \left(\boldsymbol{X}^T\boldsymbol{X}\right)^{-1}\boldsymbol{X}^T(\sigma^2\boldsymbol{I})\boldsymbol{X}\left(\boldsymbol{X}^T\boldsymbol{X}\right)^{-1} \\ &= \sigma^2\left(\boldsymbol{X}^T\boldsymbol{X}\right)^{-1}\boldsymbol{X}^T\boldsymbol{X}\left(\boldsymbol{X}^T\boldsymbol{X}\right)^{-1} \\ &= \sigma^2\left(\boldsymbol{X}^T\boldsymbol{X}\right)^{-1}, \end{aligned} \tag{7.38}$$

where we have used the result that $\text{cov}(\boldsymbol{\epsilon}) = \sigma^2\boldsymbol{I}$ from A1. Thus the LS covariance matrix of $\boldsymbol{\beta}_{LS}$ is proportional to the inverse covariance matrix of the predictor variables.

It is desirable for the variances of the estimates $\boldsymbol{\beta}_{LS}$ to be as small as possible. How small does (7.38) say they are? We see that if σ^2 is large, then the variances (which are the diagonal elements of $\text{cov}(\boldsymbol{\beta}_{LS})$ are also large. This makes sense because if the variances of the elements of $\boldsymbol{\epsilon}$ are large, then the variances of the elements of $\boldsymbol{\beta}_{LS}$ could also be expected to be large. But more importantly, (7.38) also says that if $\boldsymbol{X}^T\boldsymbol{X}$ is "big" in some norm sense, then $\text{cov}(\boldsymbol{x}_{LS})$ is "small", which is desirable. We see later that a sufficient condition for $\text{cov}(\boldsymbol{x}_{LS})$ being small is that all eigenvalues of $\boldsymbol{X}^T\boldsymbol{X}$ are large. Further, if a large spread exists between the largest and smallest eigenvalues of $\boldsymbol{X}^T\boldsymbol{X}$ (i.e. the condition number of $\boldsymbol{X}^T\boldsymbol{X}$ as discussed in Chap. 5 becomes large), then the variances of $\boldsymbol{\beta}_{LS}$ will become large. Thus it is necessary for good performance of the LS estimator that the condition number of $\boldsymbol{X}^T\boldsymbol{X}$ must be small. We can also infer that if $\boldsymbol{X}$ is rank deficient, then $\boldsymbol{X}^T\boldsymbol{X}$ is rank deficient, and the variances of each component of $\boldsymbol{\beta}$ approach infinity, which implies the results are meaningless, unless proper measures as discussed in the next chapter are implemented.

We show an example of LS variance for the one-dimensional case in Fig. 7.5, specifically for the mean-centred condition, as a partial interpretation of behaviour of LS estimation. Here, the normal Eqs. (7.19) devolve into the form $\beta_{LS} = \frac{\boldsymbol{x}^T\boldsymbol{y}}{\boldsymbol{x}^T\boldsymbol{x}}$, and the variance expression (7.38) for $\text{var}(\beta_{LS})$ becomes $\sigma^2/(\boldsymbol{x}^T\boldsymbol{x})$. (Here $\boldsymbol{x}$ is denoted in lowercase since it is a vector, and β in unbolded format, since it is a scalar.) For part (a) of the figure, we see the range of $\boldsymbol{x}$-values is relatively compressed, in which case $\boldsymbol{x}^T\boldsymbol{x}$ is small, whereupon $\text{var}(\beta_{LS})$ is large. This fact is evident from the figure, in that the slope estimates β_{LS} will vary considerably over different sample sets of the observations (y_i, x_i), having the same noise variance and the same spread Δx. On the other hand, we see from part (b) that, due to the larger spread Δx in this case, the slope estimate is more stable over different samples of observations with the same noise variance, and hence the variance of $\boldsymbol{\beta}$ is smaller.

7.4.3 Variance of a Predicted Value of y

One of the main applications of LS analysis is to predict a scalar value $\hat{y}$ of $\boldsymbol{y}$ that corresponds to a new observation set $\boldsymbol{x}_N^T$ of variables in the form of a new row of $\boldsymbol{X}$. This procedure requires the availability of a training set of data in the form of a matrix $\boldsymbol{X}$ and the corresponding set of values $\boldsymbol{y}$. The availability of the training set enables us to calculate $\boldsymbol{\beta}_{LS}$ through the normal Eqs. (7.19). Once $\boldsymbol{\beta}_{LS}$ is calculated, the predicted value $\hat{y}$ that corresponds to new observations $\boldsymbol{x}_N^T$ may be determined as

$$\hat{y} = \boldsymbol{x}_N^T\boldsymbol{\beta}_{LS}, \tag{7.39}$$

which may be ascertained to be the expected value of y given $\boldsymbol{\beta}_{LS}$ and the new data $\boldsymbol{x}_N^T$. The problem of predicting responses to new observations is treated at length in Chap. 8.

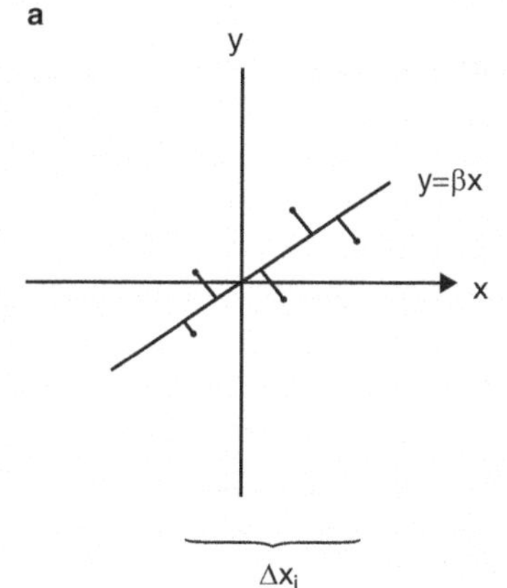

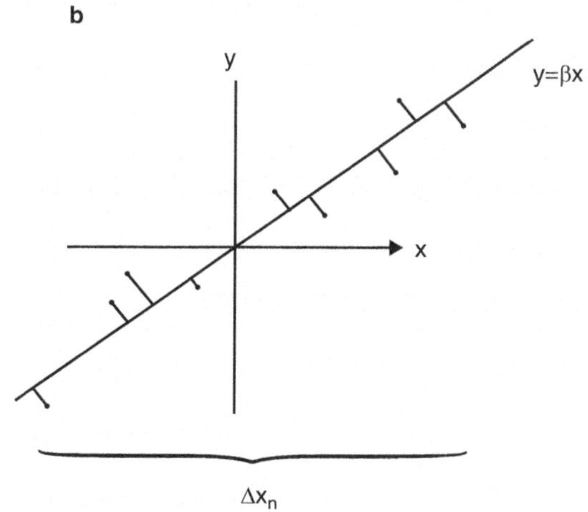

Fig. 7.5 The geometry of LS variances for the one-dimensional, mean-centred case. In the top figure, the observations (dots) are spread over a narrow range Δx_1 of x-values, whereas in the lower figure, Δx_2 is significantly larger. This implies that the respective value of $(\boldsymbol{x}^T\boldsymbol{x})^{-1}$ (which is proportional to the variance of the estimate β_{LS}) is larger for the top case than for the bottom. The y-values of the observed data vary over different sets of observed data, due to different values of the error variable. The result is that the variance of the slope estimate β_{LS} over different data sets is larger for the top case, whereas for the bottom case the variances are smaller due to the wider value of Δx

The question arises: "How good is this estimate of $\hat{y}$"? To address this issue, we evaluate the variance of the prediction $\hat{y}$. We define y_o as $y_o = \boldsymbol{x}_N^T\boldsymbol{\beta}_o$, where $\boldsymbol{\beta}_o = E(\boldsymbol{\beta}_{LS})$. The variance $\sigma_{\hat{y}}^2$ of $\hat{y}$ is calculated as follows:

$$\begin{aligned}\sigma_{\hat{y}}^2 &= E(y_o - \hat{y})(y_o - \hat{y}) \\ &= E[\boldsymbol{x}_N^T(\boldsymbol{\beta}_o - \boldsymbol{\beta}_{LS})((\boldsymbol{\beta}_o - \boldsymbol{\beta}_{LS})^T\boldsymbol{x}_N],\end{aligned}$$

where we have used (7.39) in the second line. Since $\boldsymbol{x}_N^T$ are known quantities and not random variables, the expectation operator can be moved to the inside set of brackets. The resulting expectation is given by (7.38). Therefore we can write

$$\sigma_{\hat{y}}^2 = \sigma^2\boldsymbol{x}_N^T(\boldsymbol{X}^T\boldsymbol{X})^{-1}\boldsymbol{x}_N. \tag{7.40}$$

We note that the predicted value $\hat{y}$ is also dependent on the quantity $(\boldsymbol{X}^T\boldsymbol{X})^{-1}$, and therefore as we will see, even one small eigenvalue can result in large variances in the estimate $\hat{y}$. So again in this context, we see it is necessary that $\boldsymbol{X}^T\boldsymbol{X}$ be well-conditioned for our predictions to be accurate. Various latent variable approaches for mitigating this effect are discussed in Chapter 8.

We see later in this chapter that at least one small eigenvalue of the matrix $\boldsymbol{X}^T\boldsymbol{X}$ can cause the variances of $\boldsymbol{\beta}_{LS}$ to become large. This undesirable situation can be mitigated by using the pseudo-inverse method discussed in the following chapter. However, the pseudo-inverse introduces bias into the estimate. With reference to (7.33), in many cases this bias is small, whereas the reduction in variance resulting from the use of the pseudo-inverse is considerable, resulting in a significant reduction in mean-squared error. Thus, the idea of an unbiased estimator is not always desirable, and there may be biased estimators that perform better on average than their unbiased counterparts.

7.5 Least Squares Estimation from a Probabilistic Approach

In this section, we investigate the properties of the LS estimator in a probabilistic sense, when the underlying probability density function (pdf) of the noise (and hence of $\boldsymbol{\beta}_{LS}$) is known. In order to conduct the analysis, we make the additional assumption A3:

A3: For the following properties, we further assume $\boldsymbol{\epsilon}$ is a jointly distributed *Gaussian* random variable with mean 0 and covariance $\sigma^2\boldsymbol{I}$.

Let us reconsider the LS linear regression model, which we reproduce here for convenience:

$$\boldsymbol{y} = \boldsymbol{X}\boldsymbol{\beta}_o + \boldsymbol{\epsilon}. \tag{7.41}$$

To obtain a deeper understanding of LS analysis, we investigate the probability density function (pdf) of $\boldsymbol{\beta}_{LS}$ given all the parameters that are necessary to specify the distribution—in this case these are $\boldsymbol{\beta}_o$, $\boldsymbol{X}$, and σ. It is a fundamental property of Gaussian-distributed random variables that any linear transformation of a Gaussian-distributed quantity is also Gaussian. From (7.24) we see that $\boldsymbol{\beta}_{LS}$ is a linear transformation of $\boldsymbol{y}$, which is Gaussian by hypothesis, and therefore the pdf of $\boldsymbol{\beta}_{LS}$ is also Gaussian. Recall that mean and covariance are all that are needed to completely specify a Gaussian pdf function. From (7.36), the mean of $\boldsymbol{\beta}_{LS}$ is $\boldsymbol{\beta}_o$, and the covariance in the white noise case is given from (7.38) as $\sigma^2(\boldsymbol{X}^T\boldsymbol{X})^{-1}$. Then under these conditions, $\boldsymbol{\beta}_{LS}$ has the Gaussian pdf, as discussed in Chap. 4, given by

$$p(\boldsymbol{\beta}_{LS}|\boldsymbol{\beta}_o, \boldsymbol{X}, \sigma) =$$
$$(2\pi)^{-\frac{n}{2}}|\sigma^{-2}\boldsymbol{X}^T\boldsymbol{X}|^{\frac{1}{2}} \exp\left[-\frac{1}{2\sigma^2}(\boldsymbol{\beta}_{LS} - \boldsymbol{\beta}_o)^T\boldsymbol{X}^T\boldsymbol{X}(\boldsymbol{\beta}_{LS} - \boldsymbol{\beta}_o)\right]. \tag{7.42}$$

Recall from the discussion of Sect. 4.3 that the boundary of the joint confidence region (JCR) of $\boldsymbol{\beta}_{LS}$ is defined as the locus of points ψ where the pdf has a constant value with respect to variation in $\boldsymbol{\beta}_{LS}$. These JCRs are elliptical in shape. The probability level α of an observation falling within the JCR is the integral over the interior of the ellipse. Since the variable $\boldsymbol{\beta}_{LS}$ appears only in the exponent, the set ψ is defined as the set of points $\boldsymbol{\beta}_{LS}$ such that the quadratic form in the exponent (and hence the pdf itself) is equal to a constant—that is, the JCR ψ is defined as

$$\psi = \left\{\boldsymbol{\beta}_{LS} \middle| \frac{1}{2\sigma^2}(\boldsymbol{\beta}_{LS} - \boldsymbol{\beta}_o)^T \boldsymbol{X}^T \boldsymbol{X}(\boldsymbol{\beta}_{LS} - \boldsymbol{\beta}_o) = \gamma\right\}, \tag{7.43}$$

where the value γ is determined such that the integral of the pdf over the interior of the JCR evaluates to the specified probability α.

We can simplify the quadratic form in the exponent of (7.43) as

$$-\frac{1}{2\sigma^2}\boldsymbol{z}^T \boldsymbol{\Lambda} \boldsymbol{z},$$

where $\boldsymbol{z} \triangleq \boldsymbol{V}^T(\boldsymbol{\beta}_{LS} - \boldsymbol{\beta}_o)$ and $\boldsymbol{V}\boldsymbol{\Lambda}\boldsymbol{V}^T$ is the eigendecomposition of $\boldsymbol{X}^T\boldsymbol{X}$. Thus the set defined by $\{\boldsymbol{z} | \frac{1}{2\sigma^2}\boldsymbol{z}^T \boldsymbol{\Lambda} \boldsymbol{z} = \gamma\}$ defines the boundary of the joint confidence region. From the treatment of Chap. 4, the length of the ith principal semi-axis of the associated ellipse is then proportional to $1/\sqrt{\lambda_i}$. Specifically with reference to Fig. 7.6, λ_2 is small in this case, and so the length of the corresponding $\boldsymbol{v}_2$ semi-axis is relatively large. Since the JCR ellipse is the region where observations fall with a specified probability α, the elongated nature of the ellipse implies that the variance along the $\boldsymbol{v}_2$ axis becomes large. The projections (and hence variances) along both the β_1 and β_2 axes are also seen to be large. On the other hand, we see from Fig. 7.7, where both eigenvalues are larger, that the variances along both axes have contracted considerably.

We see that the variances of both the β_1 and β_2 components of $\boldsymbol{\beta}_{LS}$ are large due to only one of the eigenvalues being small. Generalising to multiple dimensions, we see that if all components of $\boldsymbol{\beta}_{LS}$ are to have small variance, then *all* eigenvalues of $\boldsymbol{X}^T\boldsymbol{X}$ must be large. Thus, for desirable variance properties of $\boldsymbol{\beta}_{LS}$, it is necessary that the matrix $\boldsymbol{X}^T\boldsymbol{X}$ must be well- conditioned; that is, the condition number of $\boldsymbol{X}^T\boldsymbol{X}$ (as discussed in Sect. 5.4) should be as close to unity as possible and the eigenvalues be as large as possible.[3] This is the "sense" referred to earlier in which the matrix $\boldsymbol{X}^T\boldsymbol{X}$ must be "large" in order for the variances to be small.

In conclusion, we see that even one small eigenvalue (relative to the largest) has the ability to make the variances of all components of $\boldsymbol{\beta}_{LS}$ large. Thus, $\boldsymbol{X}^T\boldsymbol{X}$ must be well-conditioned. In the following chapters, we present various methods for mitigating the effect of a poorly conditioned system that destroys the desirable

[3] The ED and the SVD of a symmetric, positive definite matrix are equivalent, so it is irrelevant whether we define the condition number in terms of the singular values or the eigenvalues.

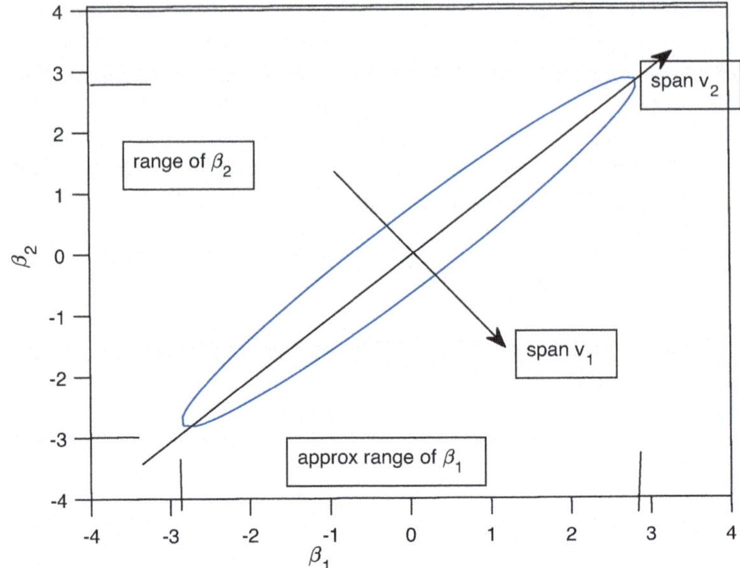

Fig. 7.6 This ellipse represents a joint confidence region at some probability level α, where the semi-axes have lengths proportional to $\frac{1}{\sqrt{\lambda_i}}$. Here λ_2 is significantly smaller than λ_1, which causes large variation along the $\boldsymbol{v}_2$-axis, which in turn causes large variation along each of the coordinate axes, leading to large variances of $\boldsymbol{\beta}$

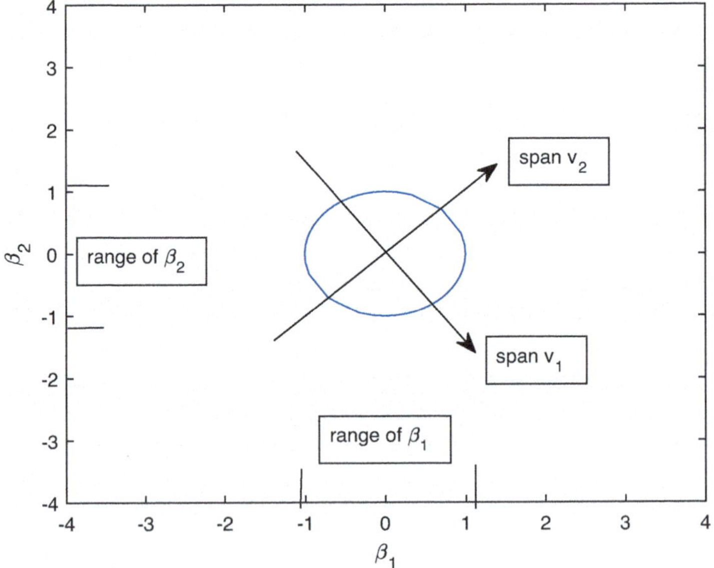

Fig. 7.7 A joint confidence region similar to that in Fig. 7.6, but where the eigenvalues are better conditioned and relatively large. In this case we see that the variation along the coordinate axes is considerably reduced

variance properties of $\boldsymbol{\beta}_{LS}$. A small eigenvalue implies that the columns of $\boldsymbol{X}$ are close to linear dependence, which implies that the correlation between the variables in $\boldsymbol{X}$ is large, suggesting that there is insufficient joint information in the variables to yield an accurate solution for $\boldsymbol{\beta}$.

*7.5.1 $\boldsymbol{\beta}_{LS}$ Is a BLUE (*The Gauss–Markov Theorem*)*

According to (7.24), we see that $\boldsymbol{\beta}_{LS}$ is a linear estimate, since it is a linear transformation of $\boldsymbol{y}$, where the transformation matrix is $(\boldsymbol{X}^T\boldsymbol{X})^{-1}\boldsymbol{X}^T$. Further, from (7.36), we see that $\boldsymbol{\beta}_{LS}$ is unbiased. We show that, amongst the class of linear unbiased estimators, $\boldsymbol{\beta}_{LS}$ has the lowest possible covariance; that is, $\boldsymbol{\beta}_{LS}$ is the *best* linear unbiased estimator (BLUE).

Theorem 2 *Consider* any *other linear* unbiased *estimate $\tilde{\boldsymbol{\beta}}$ of $\boldsymbol{\beta}$, defined by*

$$\tilde{\boldsymbol{\beta}} = \boldsymbol{B}\boldsymbol{y}, \tag{7.44}$$

where $\boldsymbol{B} \in \mathbb{R}^{n\times m}$ is an estimator or transformation matrix. Then under A1 and A2, $\boldsymbol{\beta}_{LS}$ is a BLUE.

Proof from [8]. Substituting (7.9) into (7.44), we have

$$\tilde{\boldsymbol{\beta}} = \boldsymbol{B}\boldsymbol{X}\boldsymbol{\beta}_o + \boldsymbol{B}\boldsymbol{\epsilon}, \tag{7.45}$$

where $\boldsymbol{\beta}_o$ is the true value of $\boldsymbol{\beta}$. Because $\boldsymbol{\epsilon}$ has zero mean (A1),

$$E(\tilde{\boldsymbol{\beta}}) = \boldsymbol{B}\boldsymbol{X}\boldsymbol{\beta}_o.$$

For $\tilde{\boldsymbol{\beta}}$ to be unbiased, we therefore require

$$\boldsymbol{B}\boldsymbol{X} = \boldsymbol{I}. \tag{7.46}$$

We can now write (7.45) as

$$\tilde{\boldsymbol{\beta}} = \boldsymbol{\beta}_o + \boldsymbol{B}\boldsymbol{\epsilon}.$$

The covariance matrix of $\tilde{\boldsymbol{\beta}}$ is then

$$\begin{aligned}\text{cov}(\tilde{\boldsymbol{\beta}}) &= E\left[(\tilde{\boldsymbol{\beta}} - \boldsymbol{\beta}_o)(\tilde{\boldsymbol{\beta}} - \boldsymbol{\beta}_o)^T\right] \\ &= E\left[\boldsymbol{B}\boldsymbol{\epsilon}\boldsymbol{\epsilon}^T\boldsymbol{B}^T\right] \\ &= \sigma^2\boldsymbol{B}\boldsymbol{B}^T,\end{aligned} \tag{7.47}$$

where we have used A1 in the last line.

We now consider a matrix $\boldsymbol{\Psi}$ defined as the difference of the estimator matrix $\boldsymbol{B}$ and the least squares estimator matrix $(\boldsymbol{X}^T\boldsymbol{X})^{-1}\boldsymbol{X}^T$:

$$\boldsymbol{\Psi} = \boldsymbol{B} - (\boldsymbol{X}^T\boldsymbol{X})^{-1}\boldsymbol{X}^T.$$

Now we form the matrix product $\boldsymbol{\Psi}\boldsymbol{\Psi}^T$:

$$\begin{aligned}\boldsymbol{\Psi}\boldsymbol{\Psi}^T &= \left[\boldsymbol{B} - (\boldsymbol{X}^T\boldsymbol{X})^{-1}\boldsymbol{X}^T\right]\left[\boldsymbol{B}^T - \boldsymbol{X}(\boldsymbol{X}^T\boldsymbol{X})^{-1}\right] \\ &= \boldsymbol{B}\boldsymbol{B}^T - \boldsymbol{B}\boldsymbol{X}(\boldsymbol{X}^T\boldsymbol{X})^{-1} - (\boldsymbol{X}^T\boldsymbol{X})^{-1}\boldsymbol{X}^T\boldsymbol{B}^T + (\boldsymbol{X}^T\boldsymbol{X})^{-1} \\ &= \boldsymbol{B}\boldsymbol{B}^T - (\boldsymbol{X}^T\boldsymbol{X})^{-1},\end{aligned} \tag{7.48}$$

where we have used $\boldsymbol{B}\boldsymbol{X} = \boldsymbol{I}$. Since $\boldsymbol{\Psi}\boldsymbol{\Psi}^T$ is positive definite, we have $\boldsymbol{B}\boldsymbol{B}^T \succeq (\boldsymbol{X}^T\boldsymbol{X})^{-1}$, where the symbol $\succeq$ means the matrix difference is positive semi-definite. Using the definition of positive semi-definiteness, we have

$$\boldsymbol{z}^T\left(\sigma^2\boldsymbol{B}\boldsymbol{B}^T - \sigma^2(\boldsymbol{X}^T\boldsymbol{X})^{-1}\right)\boldsymbol{z} \geq 0,$$

where in this case we have defined $\boldsymbol{z} = \boldsymbol{\beta}_{LS} - \boldsymbol{\beta}_o$ for notational convenience, and we have multiplied the expression by σ^2. We can write the above in the form

$$\boldsymbol{z}^T(\sigma^2\boldsymbol{B}\boldsymbol{B}^T)\boldsymbol{z} \geq \boldsymbol{z}^T(\sigma^2(\boldsymbol{X}^T\boldsymbol{X})^{-1})\boldsymbol{z}. \tag{7.49}$$

From (7.42), the expressions on each side of the above inequality are the exponents of the joint Gaussian pdfs for $\tilde{\boldsymbol{\beta}}$ and $\boldsymbol{\beta}_o$, respectively, and as such define the respective joint confidence regions (JCRs) of the two forms of estimator in the Gaussian case. Since (7.49) holds for any value of the quantity $\boldsymbol{z} = \boldsymbol{\beta}_{LS} - \boldsymbol{\beta}_o$ and since a smaller JCR means a more precise estimator, the JCR of the LS estimator is completely contained within the JCR of the other estimator, and therefore the latter is never better than the LS estimator.

Further, we recall the diagonal elements of the covariance matrix are the variances of the individual elements of the parameters of interest. Since the diagonal elements of the positive semi-definite matrix $\boldsymbol{\Psi}\boldsymbol{\Psi}^T$ are greater than or equal to zero,[4] by comparing diagonal elements from (7.48) we have

$$\sigma^2\left(\boldsymbol{B}\boldsymbol{B}^T\right)_{ii} \geq \sigma^2\left(\boldsymbol{X}^T\boldsymbol{X}\right)_{ii}^{-1}, \tag{7.50}$$

where the notation $(\cdot)_{ii}$ means the ith diagonal element. Because $\sigma^2\left(\boldsymbol{B}\boldsymbol{B}^T\right)$ and $\sigma^2\left(\boldsymbol{X}^T\boldsymbol{X}\right)^{-1}$ are the covariance matrices of $\tilde{\boldsymbol{\beta}}$ and $\boldsymbol{\beta}_{LS}$, respectively, from (7.50) we have that the variances of the individual elements of $\tilde{\boldsymbol{\beta}}$ are never better than those of

[4] This result holds for any positive or semi-positive matrix. Its proof is straightforward and is left as an exercise.

$\boldsymbol{\beta}_{LS}$. Thus, within the class of linear unbiased estimators and under assumptions A1 and A2, no other unbiased estimator has smaller variance than the LS estimate. □

We have seen that large variances can result if $(\boldsymbol{X}^T\boldsymbol{X})^{-1}$ is poorly conditioned. This undesirable situation can be mitigated using the pseudo-inverse method discussed in the next chapter. However, the pseudo-inverse introduces bias into the estimate. With reference to (7.33), in many cases this bias is small, whereas the reduction in variance can be large, resulting in a significant reduction in the total mean-squared error. Thus, the idea of an unbiased estimator is not always desirable, and there may be biased estimators that perform better on average than their unbiased counterparts. This possibility is discussed in Chap. 8.

7.5.2 Least Squares Estimation and the Cramer–Rao Lower Bound

In this section, we discuss the relationship between the Cramer–Rao lower bound (CRLB) and the linear least squares estimate in white and coloured noise, with respect to the estimation of the parameters $\boldsymbol{\beta}$.

For the time being we consider only the white noise case. The CRLB is a straightforward method for determining the lower bound of the covariance matrix of a parameter estimate in the presence of noise. The form of the CRLB we consider here applies only to unbiased estimators. We illustrate the CRLB for the case at hand; that is, we investigate the performance of the covariance matrix of $\boldsymbol{\beta}$ based on a set of observations $\boldsymbol{y}$, where $\boldsymbol{X}$ is known and the noise is known to be white Gaussian with a mean of zero and covariance $\sigma^2\boldsymbol{I}$. The observations $\boldsymbol{y}$ are given by the familiar regression equation

$$\boldsymbol{y} = \boldsymbol{X}\boldsymbol{\beta} + \boldsymbol{\epsilon}. \tag{7.51}$$

The pdf of relevance for the evaluation of the CRLB for the case under consideration is $p(\boldsymbol{y}|\boldsymbol{X}, \boldsymbol{\beta}, \sigma^2)$. Since $\boldsymbol{\epsilon}$ is the only random variable in (7.51), the respective pdf is a Gaussian distribution with a mean of $\boldsymbol{X}\boldsymbol{\beta}$ and the same covariance matrix as the noise. This distribution has already been discussed and is repeated here for convenience:

$$p(\boldsymbol{y}|\boldsymbol{X}, \boldsymbol{\beta}, \sigma) = (2\pi)^{-\frac{n}{2}}\sigma^{-n}\exp\left[-\frac{1}{2\sigma^2}(\boldsymbol{y} - \boldsymbol{X}\boldsymbol{\beta})^T(\boldsymbol{y} - \boldsymbol{X}\boldsymbol{\beta})\right]. \tag{7.52}$$

Then the CRLB on the covariance matrix of the parameters of interest (in this case $\boldsymbol{\beta}$) obtained from the observations $\boldsymbol{y}$ is obtained directly from the Hessian matrix[5]

[5] Recall the Hessian of a multivariate function is the matrix of second derivatives.

of the log of the conditional probability density function. The (i, j)th element $\boldsymbol{J}_{ij}$ of the Hessian matrix $\boldsymbol{J}$ in the present context is given as

$$(\boldsymbol{J})_{ij} = -E\frac{\partial^2 \ln p(\boldsymbol{y}|\boldsymbol{X}, \boldsymbol{\beta}, \sigma)}{\partial\beta_i \partial\beta_j}. \tag{7.53}$$

The matrix $\boldsymbol{J}$ defined by (7.53) is also referred to as the *Fisher information matrix*. Now consider a matrix $\boldsymbol{U}$ which is the covariance matrix of the parameter estimate $\tilde{\boldsymbol{\beta}}$, which is obtained by some other unbiased estimation process. Then $\boldsymbol{U}$ is never a better covariance than $\boldsymbol{J}^{-1}$; more precisely, we say that

$$\boldsymbol{U} \succeq \boldsymbol{J}^{-1}, \tag{7.54}$$

where the matrix inequality symbol $\succeq$ is the same as that used in the context of (7.48). Following the same reasoning as we did previously with the BLUE subsection immediately above, we can conclude from (7.54) that in the Gaussian case, no estimator has a JCR that is smaller than that corresponding to $\boldsymbol{J}^{-1}$. Further, as with the BLUE case, we have

$$u_{ii} \geq j^{ii}, \quad i = 1, \ldots, n, \tag{7.55}$$

where j^{ii} denotes the (i, i)th element of $\boldsymbol{J}^{-1}$. Thus (7.55) tells us that the individual variances of the estimates $\tilde{\beta}_i$ obtained by any estimator are never better than the corresponding diagonal term of $\boldsymbol{J}^{-1}$.

For further details on the CRLB, refer to the excellent text by Scharf [16]. Since the second derivative of a function with respect to a parameter relates to the curvature of the function, the CRLB indicates that conditional pdfs with high curvatures have the potential to yield higher precision estimates of the parameter value than those with lower values of curvature.

To evaluate $\boldsymbol{J}$, we substitute (7.52) into (7.53) and evaluate. Note that the constant terms preceding the exponent operator in (7.52) are not functions of $\boldsymbol{\beta}$ and so are not relevant with regard to the differentiation. Thus we need to consider only the exponential term of (7.52). Because of the $\ln(\cdot)$ operator, (7.53) reduces to the second derivative matrix of the quadratic form in the exponent. The expectation operator of (7.53) is redundant in our specific case because all the second derivative quantities are constant. Thus (7.53) simplifies to

$$(\boldsymbol{J})_{ij} \equiv -\frac{\partial^2}{\partial\beta_i \partial\beta_j}\left[-\frac{1}{2\sigma^2}(\boldsymbol{y} - \boldsymbol{X}\boldsymbol{\beta})^T(\boldsymbol{y} - \boldsymbol{X}\boldsymbol{\beta})\right].$$

By expanding the term in the square brackets and differentiating twice, all the terms above disappear, except for the second derivative of the term $\frac{1}{2\sigma^2}\boldsymbol{\beta}^T\boldsymbol{X}^T\boldsymbol{X}\boldsymbol{\beta}$. Using the treatment of Appendix A, Chap. 2 for differentiation of a quadratic form, it is straightforward to show that

$$J = \frac{1}{\sigma^2} \boldsymbol{X}^T \boldsymbol{X}. \tag{7.56}$$

The CRLB value $\boldsymbol{J}^{-1}$ corresponding to (7.56) is precisely the least squares covariance matrix given by (7.38) for the white noise case. Thus, for the regression model given by $\boldsymbol{y} = \boldsymbol{X\beta} + \boldsymbol{\epsilon}$ when $\boldsymbol{\epsilon}$ is Gaussian-distributed white noise, the LS estimator satisfies the CRLB and therefore has the lowest covariance possible given an unbiased estimator with observations generated according to (7.51).

7.5.3 Least Squares Estimation and the CRLB for Gaussian Coloured Noise

Here we consider the more general coloured Gaussian noise case, where the noise covariance is denoted as $\boldsymbol{\Sigma}$, i.e. $E(\boldsymbol{\epsilon\epsilon})^T = \boldsymbol{\Sigma}$, which is not necessarily diagonal. In this case, the conditional pdf of $\boldsymbol{y}$ given all the parameters corresponding to (7.52) becomes

$$p(\boldsymbol{y}|\boldsymbol{X}, \boldsymbol{\beta}, \boldsymbol{\Sigma}) = (2\pi)^{-\frac{n}{2}} |\boldsymbol{\Sigma}|^{-\frac{1}{2}} \exp\left[-\frac{1}{2}(\boldsymbol{y} - \boldsymbol{X\beta})^T \boldsymbol{\Sigma}^{-1} (\boldsymbol{y} - \boldsymbol{X\beta})\right]. \tag{7.57}$$

By substituting (7.57) into (7.53) and following the same procedures as for the white noise case, it is straightforward to show that the Fisher information matrix $\boldsymbol{J}$ for the coloured noise case when the covariance matrix $\boldsymbol{\Sigma}$ is known is given by

$$\boldsymbol{J} = \boldsymbol{X}^T \boldsymbol{\Sigma}^{-1} \boldsymbol{X}. \tag{7.58}$$

Now suppose we use the ordinary normal Eqs. (7.19) (which assumes the background noise is white), to produce the estimate $\boldsymbol{\beta}_{LS}$ when the actual noise $\boldsymbol{\epsilon}$ is coloured, with covariance matrix $\boldsymbol{\Sigma}$. Using the same analysis as in Sect. 7.4.2, except replacing the quantity $E(\boldsymbol{y}-\boldsymbol{X\beta}_o)(\boldsymbol{y}-\boldsymbol{X\beta}_o)^T$ with $\boldsymbol{\Sigma}$ instead of $\sigma^2\boldsymbol{I}$, the covariance matrix of $\boldsymbol{\beta}_{LS}$ becomes

$$\text{cov}(\boldsymbol{\beta}_{LS}) = (\boldsymbol{X}^T\boldsymbol{X})^{-1}\boldsymbol{X}^T\boldsymbol{\Sigma}\boldsymbol{X}(\boldsymbol{X}^T\boldsymbol{X})^{-1}. \tag{7.59}$$

Note that this covariance matrix is **not** equal to $\boldsymbol{J}^{-1}$ from (7.58). Therefore the variances of the elements of $\boldsymbol{\beta}_{LS}$ in this case when the noise is coloured are necessarily larger than the minimum possible given by the CRLB.[6] Therefore using the ordinary normal equations (which assume white noise), when the noise is actually not white, results in estimates with suboptimal variances.

[6] However, it may be shown that $\boldsymbol{\beta}_{LS}$ obtained in this way in coloured noise is at least unbiased.

We now show, however, that if $\boldsymbol{\Sigma}$ is known or can be estimated, we may improve the situation by pre-whitening the noise. Let $\boldsymbol{\Sigma} = \boldsymbol{G}\boldsymbol{G}^T$, where $\boldsymbol{G}$ is the Cholesky factor as discussed in Sect. 5.3.2. Then we can whiten the noise by multiplying both sides of (7.10) by $\boldsymbol{G}^{-1}$ as we have shown previously in Sect. 5.3.3. This gives

$$\boldsymbol{G}^{-1}\boldsymbol{y} = \boldsymbol{G}^{-1}\boldsymbol{X}\boldsymbol{\beta} + \boldsymbol{G}^{-1}\boldsymbol{\epsilon}. \tag{7.60}$$

Using the above as the regression model and substituting $\boldsymbol{G}^{-1}\boldsymbol{X}$ for $\boldsymbol{X}$ and $\boldsymbol{G}^{-1}\boldsymbol{y}$ for $\boldsymbol{y}$ in (7.20), we get

$$\boldsymbol{\beta}_{LS} = (\boldsymbol{X}^T\boldsymbol{\Sigma}^{-1}\boldsymbol{X})^{-1}\boldsymbol{X}^T\boldsymbol{\Sigma}^{-1}\boldsymbol{y}. \tag{7.61}$$

The covariance matrix corresponding to this estimate is found as follows. We can write

$$E(\boldsymbol{\beta}_{LS}) = (\boldsymbol{X}^T\boldsymbol{\Sigma}^{-1}\boldsymbol{X})^{-1}\boldsymbol{X}^T\boldsymbol{\Sigma}^{-1}\boldsymbol{X}\boldsymbol{\beta}_o. \tag{7.62}$$

Therefore the quantity $\boldsymbol{\beta}_{LS} - E(\boldsymbol{\beta}_{LS})$ is given from the two previous equations as $(\boldsymbol{X}^T\boldsymbol{\Sigma}^{-1}\boldsymbol{X})^{-1}\boldsymbol{X}^T\boldsymbol{\Sigma}^{-1}(\boldsymbol{y} - \boldsymbol{X}\boldsymbol{\beta}_o)$. Substituting this value into (7.30), we get

$$\begin{aligned}
\text{cov}(\boldsymbol{\beta}_{LS}) &= E(\boldsymbol{X}^T\boldsymbol{\Sigma}^{-1}\boldsymbol{X})^{-1}\boldsymbol{X}^T\boldsymbol{\Sigma}^{-1}(\boldsymbol{y} - \boldsymbol{X}\boldsymbol{\beta}_o)(\boldsymbol{y} - \boldsymbol{X}\boldsymbol{\beta}_o)^T\boldsymbol{\Sigma}^{-1}\boldsymbol{X}(\boldsymbol{X}^T\boldsymbol{\Sigma}^{-1}\boldsymbol{X})^{-1} \\
&= (\boldsymbol{X}^T\boldsymbol{\Sigma}^{-1}\boldsymbol{X})^{-1}\boldsymbol{X}^T\boldsymbol{\Sigma}^{-1}\underbrace{E(\boldsymbol{y} - \boldsymbol{X}\boldsymbol{\beta}_o)(\boldsymbol{y} - \boldsymbol{X}\boldsymbol{\beta}_o)^T}_{\boldsymbol{\Sigma}}\boldsymbol{\Sigma}^{-1}\boldsymbol{X}(\boldsymbol{X}^T\boldsymbol{\Sigma}^{-1}\boldsymbol{X})^{-1} \\
&= (\boldsymbol{X}^T\boldsymbol{\Sigma}^{-1}\boldsymbol{X})^{-1}.
\end{aligned} \tag{7.63}$$

Notice that in the coloured noise case when the noise is pre-whitened as in (7.60), the resulting matrix $\text{cov}(\boldsymbol{\beta}_{LS})$ is exactly equal to $\boldsymbol{J}^{-1}$ in (7.58), which is the corresponding form of the CRLB; that is, the equality of the bound is now satisfied.

Hence, in the presence of coloured noise with a covariance matrix that is either known or can be estimated, pre-whitening the noise before applying the linear least squares estimation procedure also results in a minimum variance unbiased estimator of $\boldsymbol{\beta}$. We have seen this is not the case when the noise is not pre-whitened. An iterative procedure for jointly estimating both $\boldsymbol{\beta}$ and $\boldsymbol{\Sigma}$ is outlined in Problem 3 at the end of this chapter.

7.6 Simulation Examples

We illustrate the previous analysis by a set of simulation scenarios, where we evaluate the variance of the estimates $\boldsymbol{\beta}_{LS}$ for the cases where $\boldsymbol{X}$ is well-conditioned vs. poorly conditioned, both in white noise. We also examine the white vs. coloured noise cases when $\boldsymbol{X}$ is well-conditioned. For each scenario, $\boldsymbol{X} \in \mathbb{R}^{m \times n}$, where $m = 50$ and $n = 2$. This value of n was chosen so that the scatter plots of $\boldsymbol{\beta}_{LS}$

could be shown on the plane of a piece of paper. For each simulation scenario, we generated observations $\boldsymbol{y}_i \in \mathbb{R}^m$ according to the standard regression equation:

$$\boldsymbol{y}_i = \boldsymbol{X}\boldsymbol{\beta}_o + \epsilon_i, \quad i = 1, \ldots, N,$$

where N is the number of observations (or iterations); in this case, $N = 1000$. The elements of $\boldsymbol{X}$ were taken as *iid* samples drawn from a Gaussian distribution with mean zero and variance one (i.e. the "randn" command in Matlab®). The $\boldsymbol{\beta}_o$ were assigned the values $[1, 1]^T$. The quantity $\boldsymbol{\epsilon}_i$ is the noise, which were Gaussian zero-mean random variables, with variance scaled so that the signal-to-noise ratio (SNR) given by the quantity $E\frac{|||\boldsymbol{X}\boldsymbol{\beta}_o||_2^2}{||\boldsymbol{\epsilon}||_2^2}$ was equal to 1 for this simulation example.

For each of the scenarios, observations $\boldsymbol{y}_i, i = 1, \ldots, N$, were generated according to the above, where the values $\boldsymbol{X}$ and $\boldsymbol{\beta}_o$ remained constant for each i, but a new, independent sample $\boldsymbol{\epsilon}_i$, which could be either white or coloured, depending on the scenario, was drawn in each iteration. For each i, the corresponding values of $\boldsymbol{\beta}$ were estimated using the appropriate form of the normal equations. These values were stored and then used to calculate the means and covariances of the estimated $\boldsymbol{\beta}$.

For the first scenario, we look at the well-conditioned vs. the more poorly conditioned case. Since each element of $\boldsymbol{X}$ in the well-conditioned case is obtained as an independent Gaussian random variable, the columns of $\boldsymbol{X}$ are nominally uncorrelated and hence close to orthogonality, with approximately equal 2-norm. This implies that the eigenvalues of $\boldsymbol{X}^T\boldsymbol{X}$ are approximately equal, and hence the respective condition number of $\boldsymbol{X}^T\boldsymbol{X}$ is close to 1. The poorly conditioned $\boldsymbol{X}$ is identical to the well-conditioned $\boldsymbol{X}$, except that the smallest singular value of the well-conditioned case is purposefully replaced with one-tenth of its value to form the $\boldsymbol{X}$ for the poorly conditioned case. The noise in both these cases is white.

From Fig. 7.8 we see that the scatter plot for the well-conditioned case (blue) is roughly circular in character, due to the eigenvalues being approximately equal. Here the semi-axis lengths of the scatterplot ellipse are proportional to $1/\sqrt{\lambda_2}$ and $1/\sqrt{\lambda_1}$, respectively, just as they are for the semi-axis lengths of the joint confidence region as discussed previously. In this simulation example, the ratio $\sqrt{\frac{\lambda_1}{\lambda_2}}$ is 1.0793, corresponding to the almost circular character of the scatterplot ellipse. The means of the estimates of $\boldsymbol{\beta}$ are $[0.99987, 0.99617]^T$, which are close enough to their true values to suggest a lack of bias. For comparison purposes, the theoretical covariance matrix of $\boldsymbol{\beta}_{LS}$, given by $\sigma^2(\boldsymbol{X}^T\boldsymbol{X})^{-1}$, and the covariance matrix obtained by simulation as $\frac{1}{N}\sum_{i=1}^{N}\boldsymbol{\beta}_{LS}\boldsymbol{\beta}_{LS}^T$, are given, respectively, by

$$\begin{bmatrix} 0.0195 & -0.0016 \\ -0.0016 & 0.0208 \end{bmatrix} \text{ and } \begin{bmatrix} 0.0205 & -0.0016 \\ -0.0015 & 0.0198 \end{bmatrix}.$$

It may be observed that these two different evaluations of the covariance matrix are very close to each other, as they should be.

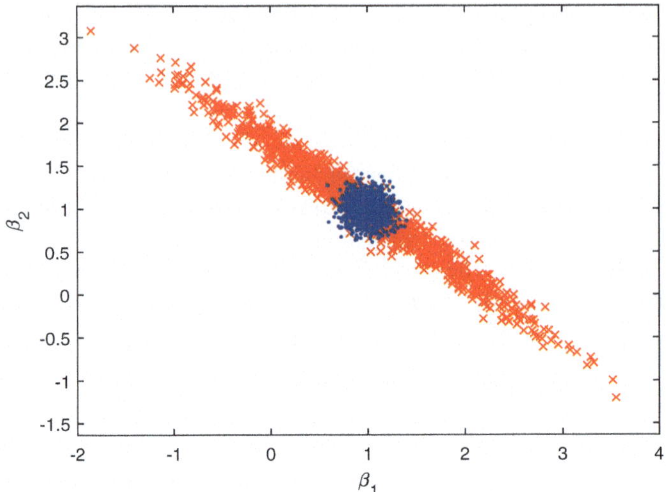

Fig. 7.8 Scatterplots of $\boldsymbol{\beta}$-estimates when $\boldsymbol{X}$ is well-conditioned (blue) and poorly conditioned (red). Notice the elongation of the scatterplot in the poorly conditioned case. The true values are at $[1, 1]^T$

We now investigate the poorly conditioned case, shown in red in Fig. 7.8. In this case the smallest singular value of $\boldsymbol{X}$ is constructed to be one-tenth of its value in the previous case, resulting in the smallest eigenvalue of $\boldsymbol{X}^T\boldsymbol{X}$ decreasing by a factor of 100. It follows that the longest semi-axis in this poorly conditioned case increases by a factor of 10. It may be verified that the major axis of the red scatterplot in Fig. 7.8 is indeed roughly ten times longer than its smaller axis. In fact, the ratio $\sqrt{\frac{\lambda_1}{\lambda_2}}$ in this case is equal to 10.7931.

The result is a significantly elongated scatterplot, resulting in larger variances along both the β_1 and β_2 axes, exactly as predicted in Sect. 7.5. The theoretical and simulation covariance matrices in this case are given, respectively, by

$$\begin{bmatrix} 0.7439 & -0.5777 \\ -0.5777 & 0.4654 \end{bmatrix} \text{ and } \begin{bmatrix} 0.7317 & -0.5655 \\ -0.5655 & 0.4533 \end{bmatrix},$$

which are significantly larger than those above for the well-conditioned case. Note that the off-diagonal covariance elements are relatively larger in this case as well. The mean values of the estimates are again close to the true values, indicating an unbiased estimator.

We now investigate the coloured noise case. The parameters for this scenario are the same as those for the well-conditioned case above, except that the noise covariance $\boldsymbol{\Sigma} \in \mathbb{R}^{m \times m}$ was implemented as a symmetric Toeplitz matrix[7] where

[7] A Toeplitz matrix is one where all elements along a given diagonal are equal. Toeplitz matrices are treated in some length in Chap. 10.

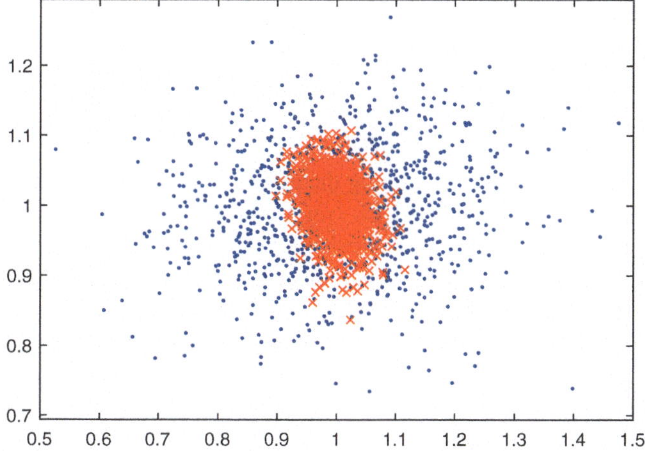

Fig. 7.9 Scatterplots of $\boldsymbol{\beta}$-estimates in coloured noise. Blue: $\boldsymbol{\beta}$ is determined using the ordinary (white noise) normal equations (7.20) when the noise is actually coloured. Red: $\boldsymbol{\beta}$ is estimated using the coloured normal equations given by (7.61)

the ith diagonal was set equal to the value $a^{i-1}, i = 1, \ldots, m$. The real constant $0 < a \leq 1$ controls the degree of colour in the noise—in this case we assigned $a = 0.9$. The noise samples $\boldsymbol{\epsilon}_i$ in this case were generated by taking $\boldsymbol{\epsilon}_i = \boldsymbol{G}\boldsymbol{n}_i$, where $\boldsymbol{G}$ is the Cholesky factor of $\boldsymbol{\Sigma}$, such that $\boldsymbol{G}\boldsymbol{G}^T = \boldsymbol{\Sigma}$, and $\boldsymbol{n}_i$ is a zero-mean, Gaussian, white noise sample with unit variance. The LS estimates $\boldsymbol{\beta}_{LS}$ were then estimated using two different methods: (*i*) the ordinary LS normal equations given by (7.19) (even though the noise is coloured) and (*ii*) the coloured version of the LS normal equations given by (7.61). The effect of the latter equation is to whiten the noise before applying the ordinary normal equations. The respective scatterplots are shown in Fig. 7.9. It is observed that whitening has the desired effect of significantly contracting the scatterplot. The theoretical covariance $(\boldsymbol{X}^T\boldsymbol{\Sigma}^{-1}\boldsymbol{X})^{-1}$ given by (7.63) and the covariance matrix determined by simulation when the coloured normal equations are used are given, respectively, by

$$\begin{bmatrix} 0.0012 & -0.0004 \\ -0.0004 & 0.0017 \end{bmatrix} \text{ and } \begin{bmatrix} 0.0013 & -0.0004 \\ -0.0004 & 0.0016 \end{bmatrix}.$$

On the other hand, the covariance is obtained from simulation when the ordinary normal equations is given as

$$\begin{bmatrix} 0.0112 & 0.0026 \\ 0.0026 & 0.0279 \end{bmatrix},$$

which is observed to be significantly higher.

7.7 Design of X

Up to now, we have assumed that $\boldsymbol{X}$ is determined before we solve the underlying LS problem. However determination of the structure of $\boldsymbol{X}$ in a practical scenario is not necessarily a straightforward procedure. There are at least two considerations we must bear in mind when formulating the matrix $\boldsymbol{X}$ with regard to an LS problem or experiment. In some but not all LS (regression) applications, we have control over which predictor variable types we include in $\boldsymbol{X}$ and also the values of these variables for each observation. It turns out as we see later that these choices can have a profound effect on the performance of the resulting regression model. We should note that none of the examples at the beginning of this chapter allow us the freedom to set the values of these X-variables (but in the hurricane example we can exercise control over which variables are included in X). However, for example, in a chemical reactor experiment we can control the concentrations, flow rates, temperatures, etc. of the input reactants (i.e. variables) and measure the concentration of the corresponding desired output chemical components (observations). We wish to choose the values of the input variables so that the resulting $\boldsymbol{X}$ produces the model with the best predictive capability (i.e. lowest possible covariances of $\boldsymbol{\beta}$). The problem of selecting the values for $\boldsymbol{X}$ is incorporated into the topic *design of experiments (DoE)*, which has a very mature history dating back to the early 1920s. For further information on DoE, refer to, e.g. [1, 2, 17]. In the following subsections, with regard to those applications that permit us to do so, we look at what variable types we should include in $\boldsymbol{X}$ as predictors and how should we determine the values of these predictors.

What Predictor Variables Should We Include in X? With regard to the hurricane example discussed earlier in the chapter, we could have a large assortment of variables to consider—these may be air temperature, wind shear, water temperature, barometric pressure, wind direction and velocity, ocean currents, tides, presence of clouds, and many other possible choices. Questions arise whether all these variables are in fact equally predictive of a hurricane event, whether they are correlated (thus leading to linear dependencies of the columns of $\boldsymbol{X}$ and thus poor conditioning), and what is the optimal number of variables to include in the model?

If we do not choose a sufficient number of variables (i.e. n is too small), then the model may not have enough degrees of freedom or flexibility to fit the observations $\boldsymbol{y}$. This condition is called *underfitting*. Choosing n too large results in *overfitting*. That is, the model, due to the large number of variables, is overly flexible may over-adapt to the available training set, but fail to predict to new data samples that are not included in the training set. Also, according to the interlacing theorem below, increasing n causes the condition number of $\boldsymbol{X}^T\boldsymbol{X}$ to increase, giving rise to a degradation of the variances of the LS estimates, as we have seen previously in this chapter.

We show that as n increases while m remains constant, cond($\boldsymbol{A}$) never improves and typically becomes worse. Because LS variances are intimately associated with condition number, increases in n tend to degrade performance, all else being equal.

It is therefore advisable to keep the value of n as small as possible. We cite a reduced version of the *Interlacing Property* [7], originally discussed in Chapter 5, repeated here for convenience:

The Interlacing Theorem Let $\boldsymbol{B}_{n+1} \in \mathbb{R}^{(n+1)\times(n+1)}$ be a symmetric matrix. Let $\boldsymbol{B}_n = \boldsymbol{B}(1:n, 1:n)$. Then

$$\lambda_{n+1}(\boldsymbol{B}_{n+1}) \leq \lambda_n(\boldsymbol{B}_n)$$

and

$$\lambda_1(\boldsymbol{B}_{n+1}) \geq \lambda_1(\boldsymbol{B}_n),$$

where $\lambda_i(\boldsymbol{B})$ indicates the ith largest eigenvalue of $\boldsymbol{B}$.

Proof [7]

Therefore, if $\boldsymbol{X}^T\boldsymbol{X}$ is of size $n \times n$, then adding a column to $\boldsymbol{X}$ adds an additional variable to the LS problem and increases $\boldsymbol{X}^T\boldsymbol{X}$ to size $(n+1) \times (n+1)$. From the Interlacing Property, we have that $\lambda_1(n+1) \geq \lambda_1(n)$ and $\lambda_{n+1}(n+1) \leq \lambda_n(n)$, where the number in round brackets indicates the size of the matrix the respective λ is associated with. Therefore the condition number (i.e. λ_1/λ_n) never improves when a column is added to $\boldsymbol{X}$. In fact, the equality only holds under special conditions, so in the general case, the condition number of $\boldsymbol{X}^T\boldsymbol{X}$ increases by adding a column, and hence the variances of the LS estimates will degrade under these circumstances. This behaviour is a manifestation of a general principle in estimation theory that as the number of parameters estimated from a given quantity of data increases, the variances of the estimates also increase. This principle applies in particular to least squares estimation.

There are two methods commonly employed for controlling the number of variables. "Variables" are also referred to as "features" in the machine learning context. These methods are *feature selection* and *feature extraction*, respectively. With variable selection in the supervised learning case (where the response $\boldsymbol{y}$ is available), variables are selected to be included in $\boldsymbol{X}$ based on their statistical dependency with the model response $\boldsymbol{y}$. For example, the minimum redundancy maximum relevance (mRMR) method [14] selects features iteratively. On the first iteration, the feature with the strongest statistical dependence on $\boldsymbol{y}$ is chosen. Then in subsequent iterations, the feature with the best combination of maximum statistical dependence with $\boldsymbol{y}$ (relevance) and minimum statistical dependence (redundancy) with the features chosen in previous iterations, is chosen. The process repeats until the number of prescribed features is selected. This process produces a set of features that are maximally predictive of the response variable and as mutually independent as possible, meaning that the columns of $\boldsymbol{X}$ are "discouraged" from being linearly dependent, thus resulting in a more favourable condition number.

The second method for controlling the number of variables is feature extraction, which is equivalent to the latent variable methods as discussed in the following chapter. Here, a prescribed number r of latent variables, each of which is some form of optimal linear combination of all the available variables, are calculated from the data. PCA regression, which we discuss in the next chapter, is an example of the feature extraction method. With this approach, the irrelevant variables would be given small weights and contribute little to the latent variables. Thus with the feature extraction method, all the variables are optimally combined into a set of r latent vectors.

The choice of m, which is the number of observations, is more straightforward than the choice of n. Generally speaking, the more observations the better, so it is desirable to choose m as large as possible. The larger the value of m, the larger the elements of $\boldsymbol{X}^T\boldsymbol{X}$ become, and consequently the smaller are the resulting variances (see (7.38)). In fact, the variances decrease as $1/m$. However, in most applications, collecting data is an expensive and time-consuming proposition, and so often we must make do with whatever quantity of data is available.

How to Choose the Values of the Predictor Variables? There are many available approaches that address this problem, if in fact the underlying environment allows us the flexibility to select the value of our predictor variables. Here we present a method that is closely related to the D-optimal design philosophy [3]. The objective with our approach is to choose the variable values of $\boldsymbol{X}$ so that the condition number (i.e. the ratio of the largest to smallest eigenvalue) of the respective $\boldsymbol{X}^T\boldsymbol{X}$ is as close to unity as possible in order to control the variances and covariances of $\boldsymbol{\beta}$ to be as small as possible. As we have seen, a poor condition number, due to relatively large off-diagonal elements of $\boldsymbol{X}^T\boldsymbol{X}$, causes the joint confidence region of the β's to become elongated, thus enlarging the variances along each of the $\boldsymbol{\beta}$ axes. On the other hand, when the condition number is small, the off-diagonal elements are also small, and the joint confidence regions become more circular (when the diagonal elements of $\boldsymbol{X}^T\boldsymbol{X}$ are equal), which is the best one can hope for.

Lemma 7.1 *For the sake of notational convenience, we define $\boldsymbol{R} \triangleq \boldsymbol{X}^T\boldsymbol{X}$. Here we show that $\lambda_1 \geq \max_k r_{kk}$ and $\lambda_n \leq \min_k r_{kk}$, where r_{ij} is the (i, j)th element of $\boldsymbol{R}$. Further, we show that the equality holds when the off-diagonal elements are all zero. Thus the condition number of $\boldsymbol{R}$ degrades in the presence of nonzero diagonal elements, a condition that occurs when the predictor variables of $\boldsymbol{X}$ are correlated.*

Proof Lemma 2.1 of Sect. 2.5 states that for any covariance matrix $\boldsymbol{R}$, the largest eigenvector $\boldsymbol{v}_1$ satisfies the expression $\arg\max_{\boldsymbol{q}} \left(\boldsymbol{q}^T\boldsymbol{R}\boldsymbol{q}\right)$, subject to the constraint $||\boldsymbol{q}||_2 = 1$. The quantity $\boldsymbol{v}_1$ is the largest eigenvector of $\boldsymbol{R}$, and $\lambda_1 = \boldsymbol{v}_1^T\boldsymbol{R}\boldsymbol{v}_1$ is the corresponding eigenvalue, which is the largest possible value of this quadratic form under a unit 2-norm constraint. Similarly, $\boldsymbol{v}_n$ satisfies $\arg\min_{\boldsymbol{q}} \left(\boldsymbol{q}^T\boldsymbol{R}\boldsymbol{q}\right)$, subject to the constraint $||\boldsymbol{q}||_2 = 1$. The quantity $\boldsymbol{v}_n$ is the smallest eigenvector of $\boldsymbol{R}$ and $\lambda_n = \boldsymbol{v}_n^T\boldsymbol{R}\boldsymbol{v}_n$ is the corresponding eigenvalue.

For any covariance matrix $\boldsymbol{R}$, we can choose an elementary vector $\boldsymbol{e}_i$ such that $\boldsymbol{e}_i^T \boldsymbol{R} \boldsymbol{e}_i = \max_i (r_{ii})$ and an $\boldsymbol{e}_j$ such that $\boldsymbol{e}_j^T \boldsymbol{R} \boldsymbol{e}_j = \min_j (r_{jj})$. Recall the elementary vector $\boldsymbol{e}_k$ is an n-length vector of zeros except for a 1 in the kth position.

From the optimality of $\boldsymbol{v}_1$ as stated by Lemma 2.1, Sect. 2.5, we have $\boldsymbol{v}_1^T \boldsymbol{R} \boldsymbol{v}_1 \geq \boldsymbol{e}_i^T \boldsymbol{R} \boldsymbol{e}_i$. It follows that

$$\lambda_1 \geq \max_i r_{ii}. \tag{7.64}$$

Following similar logic but replacing max with min, we also have

$$\lambda_n \leq \min_j r_{jj}. \tag{7.65}$$

Let us consider two versions $\boldsymbol{R}_1$ and $\boldsymbol{R}_2$ of the LS covariance matrix $\boldsymbol{X}^T \boldsymbol{X}$. The values of the variables of the $\boldsymbol{X}$ corresponding to $\boldsymbol{R}_1$ are chosen so that the columns of $\boldsymbol{X}$ are orthogonal, and so $\boldsymbol{R}_1$ is diagonal. On the other hand, the columns of the $\boldsymbol{X}$ corresponding to $\boldsymbol{R}_2$ are chosen to be not orthogonal, and so $\boldsymbol{R}_2$ is a dense matrix. The respective diagonal elements of $\boldsymbol{R}_1$ and $\boldsymbol{R}_2$ are chosen to be equal in each case. Since the r_{kk} are the eigenvalues of $\boldsymbol{R}_1$, the equalities in (7.64) and (7.65) hold, and therefore we have

$$\text{cond}(\boldsymbol{R}_1) = \frac{\max_i r_{ii}}{\min_j r_{jj}}.$$

However, from (7.64) and (7.65), respectively, we see that λ_1 of $\boldsymbol{R}_2$ is greater than or equal to the largest r_{kk} and λ_n of $\boldsymbol{R}_2$ is less than or equal to the smallest r_{kk}. Therefore

$$\text{cond}(\boldsymbol{R}_1) \leq \text{cond}(\boldsymbol{R}_2).$$

It follows that the best conditioning, and hence best performance, is obtained when $\boldsymbol{R} = \boldsymbol{X}^T \boldsymbol{X}$ is diagonal with equal diagonal elements. Further, since the covariance matrix of $\boldsymbol{\beta}$ from (7.38) depends on $\boldsymbol{R}^{-1}$, the magnitude of the diagonal elements should be as large as possible.

7.8 Algebraic Constructs Related to Whitening

We have seen that whitening the noise is necessary for LS estimates to satisfy the equality of the CRLB and thus attain the best possible covariance structure. It turns out that the requirement of whitening applies not only to LS problems but also to parameter estimation problems in general. In this section, we discuss various algebraic structures that aid us in analysing the coloured noise case.

7.8.1 Mahalanobis Distance

In the following treatment, we assume $\boldsymbol{x}$ is a zero-mean random vector or one in which the mean has been removed before processing. The Mahalanobis distance is a generalisation of the Euclidean distance. We consider the usual definition of the squared 2-norm for quantifying distances:

$$||\boldsymbol{x}||_2^2 = \boldsymbol{x}^T\boldsymbol{x}.$$

We may write this in the form $\boldsymbol{x}^T\boldsymbol{I}\boldsymbol{x}$. The squared Mahalanobis distance is given by replacing the $\boldsymbol{I}$ with a full-rank, positive-definite matrix $\boldsymbol{\Sigma}^{-1}$, to get

$$||\boldsymbol{x}||_{\boldsymbol{\Sigma}^{-1}}^2 = \boldsymbol{x}^T\boldsymbol{\Sigma}^{-1}\boldsymbol{x}. \tag{7.66}$$

We have seen in Sect. 4.2 that the set of values $\{\boldsymbol{x}|\boldsymbol{x}^T\boldsymbol{\Sigma}^{-1}\boldsymbol{x} = 1\}$ (for which the Mahalanobis distance is constant) is an ellipse. From this, we may suspect that the Mahalanobis distance varies with the direction of $\boldsymbol{x}$. To confirm this idea, we may write (7.66) in the form

$$||\boldsymbol{x}||_{\boldsymbol{\Sigma}^{-1}}^2 = \boldsymbol{x}^T\boldsymbol{V}\boldsymbol{\Lambda}^{-1}\boldsymbol{V}^T\boldsymbol{x},$$

where the eigendecomposition of $\boldsymbol{\Sigma} = \boldsymbol{V}\boldsymbol{\Lambda}\boldsymbol{V}^T$. If we define $\boldsymbol{z} \triangleq \boldsymbol{V}^T\boldsymbol{x}$, then

$$||\boldsymbol{x}||_{\boldsymbol{\Sigma}^{-1}}^2 = \boldsymbol{z}^T\boldsymbol{\Lambda}^{-1}\boldsymbol{z} = \sum_{i=1}^{n}\frac{z_i^2}{\lambda_i},$$

where $\boldsymbol{z}$ are the coefficients of $\boldsymbol{x}$ in the basis $\boldsymbol{V}$. The squared Mahalanobis distance may therefore be interpreted as measuring distances of the variable $\boldsymbol{z}$ in the basis $\boldsymbol{V}$, in units of the respective eigenvalue. In the coloured noise case the eigenvalues of $\boldsymbol{\Sigma}$ are not equal, so distances are measured differently along each eigenvector direction.

Further, let the covariance matrix of $\boldsymbol{x}$ be $\boldsymbol{\Sigma}$. If $\boldsymbol{\Sigma}^{-1} = \boldsymbol{G}^{-T}\boldsymbol{G}^{-1}$, where $\boldsymbol{G}$ is the Cholesky factor of $\boldsymbol{\Sigma}$, then $||\boldsymbol{x}||_{\boldsymbol{\Sigma}^{-1}}^2 = \boldsymbol{x}^T\boldsymbol{G}^{-T}\boldsymbol{G}^{-1}\boldsymbol{x} = \boldsymbol{z}^T\boldsymbol{z}$, where in this case $\boldsymbol{z} = \boldsymbol{G}^{-1}\boldsymbol{x}$. We have seen previously in Sect. 5.3.3 the covariance matrix of $\boldsymbol{z} = \boldsymbol{I}$. Thus we see that the Mahalanobis distance in effect subjects $\boldsymbol{x}$ to a whitening linear transformation before determining its Euclidean (i.e. 2-norm) distance. As we have seen, if $\boldsymbol{x}$ is a Gaussian-distributed random vector, then the joint confidence region corresponding to $\boldsymbol{x}$ is elliptical in shape, whereas that for $\boldsymbol{z}$ is spherical. It is for this reason that the whitening operation is also referred to as "sphering" the data.

We note that the exponent $\frac{1}{2}(\boldsymbol{x}-\boldsymbol{x}_o)^T\boldsymbol{\Sigma}^{-1}(\boldsymbol{x}-\boldsymbol{x}_o)$ of the multivariate Gaussian distribution in $\boldsymbol{x}$ is in effect a Mahalanobis distance. The quantity $(\boldsymbol{x}-\boldsymbol{x}_o)$ is a zero-mean Gaussian random variable which is subjected to the whitening transformation $\boldsymbol{G}^{-1}$ as discussed above, to give a vector of zero-mean, uncorrelated, unit-variance

Gaussian random variables. Thus the form of the Gaussian exponent transforms an arbitrary Gaussian-distributed random vector into a standard form with zero mean and covariance $\boldsymbol{I}$.

7.8.2 Generalised Eigenvalues and Eigenvectors

We have studied the ordinary eigen-problem $\boldsymbol{A}\boldsymbol{v} = \lambda\boldsymbol{v}$ at some length. We now consider the so-called *generalised* eigen-problem given by

$$\boldsymbol{A}\boldsymbol{v} = \lambda\boldsymbol{B}\boldsymbol{v}, \tag{7.67}$$

where $\boldsymbol{A}$ and $\boldsymbol{B}$ are square, positive definite, symmetric matrices, which are typically covariance matrices. By way of example, $\boldsymbol{A}$ can be the covariance matrix of a signal consisting of a desired signal component + noise and $\boldsymbol{B}$ the covariance matrix of the noise only. Another example is where $\boldsymbol{A}$ and $\boldsymbol{B}$ represent covariance matrices of two different classes of signal, between which we wish to distinguish. We consider both examples in this treatment.

Let $\boldsymbol{R}_X = \boldsymbol{X}^T\boldsymbol{X}$ correspond to the covariance matrix of a set of measurements with an additive noise component, which is uncorrelated with $\boldsymbol{X}$, whose data matrix is $\boldsymbol{Y}$, which we assume to be coloured. We substitute the covariance matrices $\boldsymbol{R}_X = \boldsymbol{X}^T\boldsymbol{X}$ and $\boldsymbol{R}_Y = \boldsymbol{Y}^T\boldsymbol{Y}$ for $\boldsymbol{A}$ and $\boldsymbol{B}$, respectively, in (7.67). We assume $\boldsymbol{R}_Y$ is full rank. Then (7.67) becomes

$$\begin{aligned}\boldsymbol{R}_X\boldsymbol{v} &= \lambda\boldsymbol{R}_Y\boldsymbol{v}\\ &= \lambda\boldsymbol{G}\boldsymbol{G}^T\boldsymbol{v},\end{aligned}$$

where the Cholesky factorisation has been applied to $\boldsymbol{R}_Y$. We define $\boldsymbol{u} = \boldsymbol{G}^T\boldsymbol{v}$ or $\boldsymbol{v} = \boldsymbol{G}^{-T}\boldsymbol{u}$, to get

$$\boldsymbol{G}^{-1}\boldsymbol{R}_X\boldsymbol{G}^{-T}\boldsymbol{u} = \lambda\boldsymbol{u}. \tag{7.68}$$

Thus the generalised eigen-problem is equivalent to an ordinary eigen-problem on a transformed matrix. As we have seen from Sect. 5.3.3, the transformation $\boldsymbol{G}^{-T}$ whitens the additive noise component $\boldsymbol{Y}$ that is present in the observations $\boldsymbol{X}$—that is the columns of $\boldsymbol{Y}\boldsymbol{G}^{-T}$ are orthonormal with the corresponding covariance matrix equal to $\boldsymbol{I}$. Therefore the effect of the generalised eigen-problem is to apply a transformation on $\boldsymbol{X}$ to whiten the noise component, thereby obtaining parameter estimates that can satisfy the equality of the Cramer–Rao lower bound, as discussed earlier in this chapter. Note that to apply this technique, either we must have samples of the noise-only signal available or $\boldsymbol{R}_Y$ must be estimated by some other means. An example of the use of the generalised eigendecomposition is the MUSIC algorithm discussed in Chap. 2 in the presence of coloured noise, where we may apply a generalised eigendecomposition instead of the ordinary eigendecomposition to obtain optimal estimates.

Common Spatial Patterns E.g. [10]. We present an additional application of the generalised eigen-problem that is relevant to the electroencephalogram (EEG) [12]. The EEG records brain activity on the scalp by placing an array of electrodes on the scalp in preset positions. The signals from the electrodes are passed through a high-gain differential amplifier and filtered. The signals on all electrodes are sampled simultaneously, typically with a sample rate of less than 1000 Hz, to produce what is referred to as a "snapshot", which is then digitised and stored in memory. A depiction is shown in Fig. 7.10.

In this example, we wish to use the EEG to discriminate between healthy brain activity and that which represents some form of pathological brain activity, arising from, e.g. coma, concussion, epilepsy, or other conditions. In this vein, we denote $\boldsymbol{x}_H(t) \in \mathbb{R}^n$, where n is the number of electrodes, as the time-varying multichannel EEG signal corresponding to a subject in the healthy state. Similarly, we denote $\boldsymbol{x}_P(t)$ as the EEG data measured from a subject in a pathological state. We denote the data matrix $\boldsymbol{X}_H \in \mathbb{R}^{m \times n}$ for the healthy case, whose rows are $\boldsymbol{x}_H^T(t_i)$, where t_i is the time corresponding to the ith snapshot, $i = 1, \ldots, m$, and m is the number of available vector samples (i.e. snapshots). Similarly $\boldsymbol{X}_P$ is the corresponding data matrix for the pathological case. These matrices serve as training data for the classification process which discriminates between the healthy and pathological states. We denote the signal $\boldsymbol{x}(t)$ to be a time-varying snapshot corresponding to an undetermined healthy or pathological condition.

We wish to discriminate between the healthy and pathological states using the EEG. In this vein, we determine a weight vector $\boldsymbol{w} \in \mathbb{R}^n$ so that the ratio given by

$$\frac{||\boldsymbol{X}_H \boldsymbol{w}||_2^2}{||\boldsymbol{X}_P \boldsymbol{w}||_2^2} \tag{7.69}$$

is maximum. Finding such a value for $\boldsymbol{w}$ gives us the scalar output waveform $z(t) \stackrel{\Delta}{=} \boldsymbol{x}^T(t)\boldsymbol{w}$ from the EEG whose variance maximally discriminates between the two conditions; that is, if $z(t)$ has sufficiently large variance, we can consider the patient healthy and otherwise pathological. Expressed mathematically, we wish to find the value of $\boldsymbol{w}$ which solves the following problem:

$$\boldsymbol{w}^* = \arg\max_{\boldsymbol{w}} \frac{\boldsymbol{w}^T \boldsymbol{R}_H \boldsymbol{w}}{\boldsymbol{w}^T \boldsymbol{R}_P \boldsymbol{w}},$$

where $\boldsymbol{R}_H = \boldsymbol{X}_H^T \boldsymbol{X}_H$ and $\boldsymbol{R}_P = \boldsymbol{X}_P^T \boldsymbol{X}_P$. We solve this optimisation problem in the usual manner by differentiating and setting the result to zero. Note that both the numerator and the denominator are scalars. Hence we can use the regular form of the quotient rule for differentiation of each element of $\boldsymbol{w}$. Differentiating and setting the result to zero, we obtain

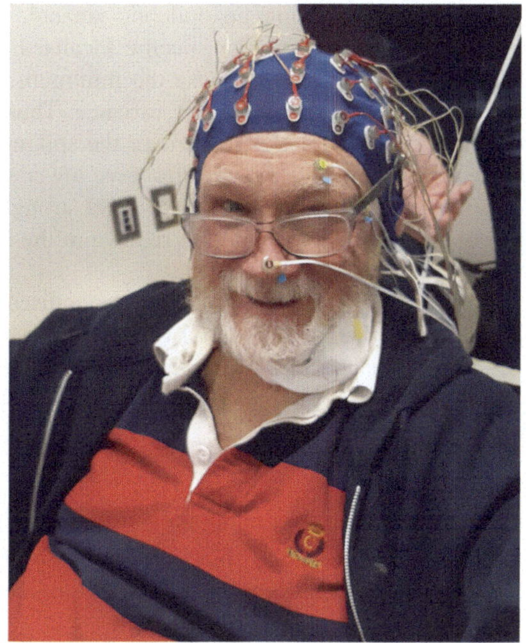

Fig. 7.10 An illustration of the EEG. The electrodes are embedded in the ovals positioned on the blue "swim hat" and make contact with the scalp. The electrodes are connected via cables to a high-gain differential amplifier, which is not visible in the above photo. The electrode on the nose serves as a ground, whereas those on the ears serve as a voltage reference for the differential amplifiers. The electrodes are coated with a conductive paste before mounting to ensure a high-conductivity connection with the scalp

$$\frac{\boldsymbol{w}^T \boldsymbol{R}_P \boldsymbol{w}\big[2\boldsymbol{R}_H \boldsymbol{w}\big] - \boldsymbol{w}^T \boldsymbol{R}_H \boldsymbol{w}\big[2\boldsymbol{R}_P \boldsymbol{w}\big]}{(\cdot)} = 0.$$

The denominator of the above is not evaluated since it is multiplied by 0 and is therefore irrelevant. The solution $\boldsymbol{w}^*$ therefore satisfies

$$\boldsymbol{R}_H \boldsymbol{w} = \lambda \boldsymbol{R}_P \boldsymbol{w}, \tag{7.70}$$

where $\lambda = \frac{\boldsymbol{w}^T \boldsymbol{R}_H \boldsymbol{w}}{\boldsymbol{w}^T \boldsymbol{R}_P \boldsymbol{w}}$ is the generalised eigenvalue. Equation (7.70) is the generalised eigen-equation, and the generalised eigenvector corresponding to the largest generalised eigenvalue is the desired solution for $\boldsymbol{w}^*$.

To apply this idea in practice, we form the scalar function of time $z(t) = \boldsymbol{x}(t)^T \boldsymbol{w}$, where $\boldsymbol{w}$ is the "largest" generalised eigenvector from a subject who we test to determine whether they are healthy or pathological. The quantity $z(t)$ is a time-varying waveform whose variance provides the maximum discrimination between the two conditions. In practice, we would set some suitably defined threshold for the variance estimate of $z(t)$, and if it exceeds the threshold over some interval of m samples, we declare the subject healthy and otherwise pathological.

Note that an equivalent formulation of this problem is to *minimise* the ratio in (7.69) rather than maximise it. In this case, the solution $\boldsymbol{w}$ is readily seen to be the generalised eigenvector corresponding to the minimum generalised eigenvalue. In this case, the function $z(t)$ becomes large in the pathological condition, rather than in the healthy condition as in the previous case.

For example, during an epileptic seizure, the EEG typically records high-intensity spike activity over specific localised regions of the scalp. In this case it is more convenient to choose the minimum solution of (7.69), so that $z(t)$ in the pathological case has higher variance. Thus we would expect the weights w_i corresponding to the electrodes over the spiking region to have larger values and those for the remaining regions to have lower values. In this manner, the spiking activity (which has high variance) is most strongly discriminated, and therefore $z(t)$ is large in the pathological case relative to the healthy case which has no spiking activity and thus low variance.

An alternate formulation of the generalised eigenvalue problem is given by approaching it using a modified version of the treatment for principal components in Chap. 2. We consider the Lagrangian function (2.47), which is at the root of the PCA approach, reproduced here for convenience:

$$L(\boldsymbol{q}) = \boldsymbol{q}^T \boldsymbol{R}_x \boldsymbol{q} + \lambda(1 - \boldsymbol{q}^T \boldsymbol{q}), \tag{7.71}$$

where we have seen the solution for $\boldsymbol{q}$ corresponds to the largest ordinary eigenvector. Thus in this case, the objective function $\boldsymbol{q}^T \boldsymbol{R}_x \boldsymbol{q}$ is maximised under a squared Euclidean norm constraint. Now (just for fun) let us consider the following modified version of the Lagrangian function:

$$L(\boldsymbol{w}) = \boldsymbol{w}^T \boldsymbol{R}_H \boldsymbol{w} + \lambda(1 - \boldsymbol{w}^T \boldsymbol{R}_P \boldsymbol{w}). \tag{7.72}$$

This Lagrangian is very similar to that above in (7.71), except now the constraint function is $\boldsymbol{w}^T \boldsymbol{R}_P \boldsymbol{w} = 1$, which is the Mahalanobis norm of $\boldsymbol{w}$ in $\boldsymbol{R}_P$. Using the same procedure as in Chap. 2, it is straightforward to show that the solution for $\boldsymbol{w}$ satisfies

$$\boldsymbol{R}_H \boldsymbol{w} = \lambda \boldsymbol{R}_P \boldsymbol{w}, \tag{7.73}$$

which is identical to (7.70). Thus, with the generalised eigen-equation, distances are implicitly measured using the Mahalanobis distance in $\boldsymbol{R}_P$, whereas the ordinary eigen-equation measures distances using the ordinary squared Euclidean distance metric. Further, with reference to (7.68) and from our interpretation of the Mahalanobis distance, we see that the generalised eigen-equation implicitly applies a whitening operation on the parameter space before solving the corresponding ordinary eigen-equation.

7.9 Solving Least Squares Using the QR Decomposition

In this section, we employ the QR decomposition to solve the LS problem for the full-rank case. We have $\boldsymbol{X} \in \mathbb{R}^{m \times n}$, $\boldsymbol{y} \in \mathbb{R}^m$, $m > n$, $\text{rank}(\boldsymbol{X}) = n$, and we wish to solve

$$\boldsymbol{\beta}_{LS} = \arg\min_{\boldsymbol{\beta}} ||\boldsymbol{X\beta} - \boldsymbol{y}||_2^2 .$$

Let the QR decomposition of $\boldsymbol{X}$ be expressed as

$$\boldsymbol{Q}^T\boldsymbol{X} = \boldsymbol{R} = \overset{n}{\begin{bmatrix} \boldsymbol{R}_1 \\ \boldsymbol{0} \end{bmatrix}} \begin{matrix} n \\ m-n, \end{matrix} \tag{7.74}$$

where $\boldsymbol{Q}$ is $m \times m$ orthonormal and $\boldsymbol{R}_1$ is full rank and upper triangular. We partition $\boldsymbol{Q}$ as

$$\boldsymbol{Q} = \underset{\quad n \quad\; m-n}{\begin{bmatrix} \boldsymbol{Q}_1 & \boldsymbol{Q}_2 \end{bmatrix}} \, m .$$

From our previous discussion and from the structure of the QR decomposition $\boldsymbol{X} = \boldsymbol{QR}$, we note that $\boldsymbol{Q}_1$ is an orthonormal basis for $R(\boldsymbol{X})$ and $\boldsymbol{Q}_2$ is an orthonormal basis for $R(\boldsymbol{X})_\perp$. We now define the quantities $\boldsymbol{c}$ and $\boldsymbol{d}$, respectively, as

$$\boldsymbol{Q}^T\boldsymbol{y} = \begin{bmatrix} \boldsymbol{Q}_1^T \\ \boldsymbol{Q}_2^T \end{bmatrix} \boldsymbol{y} = \begin{bmatrix} \boldsymbol{c} \\ \boldsymbol{d} \end{bmatrix} \begin{matrix} n \\ m-n. \end{matrix} \tag{7.75}$$

Then, we may write

$$\begin{aligned} \min_{\boldsymbol{\beta}} ||\boldsymbol{X\beta} - \boldsymbol{y}||_2^2 &= \left|\left| \boldsymbol{Q}^T\boldsymbol{X\beta} - \boldsymbol{Q}^T\boldsymbol{y} \right|\right|_2^2 \\ &= \left|\left| \begin{bmatrix} \boldsymbol{R}_1 \\ \boldsymbol{0} \end{bmatrix} \boldsymbol{\beta} - \begin{bmatrix} \boldsymbol{c} \\ \boldsymbol{d} \end{bmatrix} \right|\right|_2^2 . \end{aligned} \tag{7.76}$$

It is clear that $\boldsymbol{\beta}$ does not affect the “lower half” of the above equation. If a column vector $\boldsymbol{x}$ is partitioned as

$$\boldsymbol{x} = \begin{bmatrix} \boldsymbol{x}_1 \\ \boldsymbol{x}_2 \end{bmatrix},$$

then $||\boldsymbol{x}||_2^2 = ||\boldsymbol{x}_1||_2^2 + ||\boldsymbol{x}_2||_2^2$. Using this rule, (7.76) may be written as

$$||\boldsymbol{X\beta} - \boldsymbol{y}||_2^2 = ||\boldsymbol{R}_1\boldsymbol{\beta} - \boldsymbol{c}||_2^2 + ||\boldsymbol{d}||_2^2 .$$

Because $\boldsymbol{X}$ is assumed full rank, $\boldsymbol{R}_1$ is invertible, and the above is minimum with respect to $\boldsymbol{\beta}$ when

$$\boldsymbol{\beta}_{LS} = \boldsymbol{R}_1^{-1}\boldsymbol{c}.$$

The LS residual r_{LS} is given directly as

$$\rho_{LS} = ||\boldsymbol{d}||_2. \tag{7.77}$$

Thus the LS problem is solved. Note that this formulation using the QR decomposition requires no differentiations and so in principle is simpler than the alternative approach we used earlier in this chapter.

Further note that if a Gram–Schmidt procedure is used to compute the QR decomposition on $\boldsymbol{X}$, then there is not enough information to represent the "lower half" in (7.76). This is because this procedure only gives the partition $\boldsymbol{Q}_1$ of $\boldsymbol{Q}$, and thus $\boldsymbol{d}$ and the quantity r_{LS} cannot be computed; however, the solution $\boldsymbol{\beta}_{LS} = \boldsymbol{R}_1^{-1}\boldsymbol{c}$ is still achievable. In contrast, the Householder or Givens procedure yields a complete $m \times m$ orthonormal matrix $\boldsymbol{Q} = \begin{bmatrix} \boldsymbol{Q}_1 & \boldsymbol{Q}_2 \end{bmatrix}$, allowing a complete solution to the LS problem.

The use of the QR decomposition in solving the LS problem leads to a useful interpretation. We define

$$\boldsymbol{y} = \boldsymbol{y}_1 + \boldsymbol{y}_2,$$

where $\boldsymbol{y}_1 \in R(\boldsymbol{X})$ and $\boldsymbol{y}_2 \in R(\boldsymbol{X})_\perp$. The projectors onto these two subspaces are $\boldsymbol{Q}_1\boldsymbol{Q}_1^T$ and $\boldsymbol{Q}_2\boldsymbol{Q}_2^T$, respectively. Therefore $\boldsymbol{y}_2 = \boldsymbol{Q}_2\boldsymbol{Q}_2^T\boldsymbol{y}$. Substituting (7.75) for $\boldsymbol{Q}_2^T\boldsymbol{y}$, we have $\boldsymbol{y}_2 = \boldsymbol{Q}_2\boldsymbol{d}$. Taking 2-norms of this last expression yields $||\boldsymbol{y}_2||_2^2 = \boldsymbol{d}^T\boldsymbol{Q}_2^T\boldsymbol{Q}_2\boldsymbol{d} = ||\boldsymbol{d}||_2^2$. Thus from (7.77), ρ_{LS}^2 is the squared 2-norm of the projection of $\boldsymbol{y}$ onto $R(\boldsymbol{X})_\perp$, which is the expected result. A similar analysis shows that $||\boldsymbol{y}_1||_2^2 = ||\boldsymbol{c}||_2^2$.

The use of the QR decomposition in the case when $\boldsymbol{X}$ is rank deficient is discussed in Chap. 8.

7.10 Adaptive Filters: The LMS Algorithm

The previous LS analysis only applies when our available $\boldsymbol{X}$ and $\boldsymbol{y}$ data represent a stationary model, i.e. one in which each observation (row of $\boldsymbol{X}$ and corresponding element of $\boldsymbol{y}$) is a sample from the same environment, for which the underlying model is constant over all observations. There are however many applications of least squares problems where the underlying model exhibits significant changes with time. An example is where we use autoregressive analysis (as in Example 2 at the beginning of this chapter) when applied to a speech signal. The difficulty in this case is that speech is highly nonstationary. As we well know, speech is a sequence of phonemes (sounds) each of approximately 20 msec duration, each with its own distinct set of AR coefficients, which evolve smoothly from phoneme to another. Thus the standard form of normal equations for computing the AR coefficients is not appropriate in this case, because the normal equations assume stationarity (e.g.

constant values of $\boldsymbol{\beta}$ vs. time), and furthermore recalculating the normal equation solution at every time step is computationally prohibitive. Our objective is therefore to obtain a near-optimal set of computed AR coefficients that evolve smoothly as the underlying model changes with time, with tractable computational requirements. This is the purpose of an adaptive filter.

The field of adaptive filtering [8] is very mature, and there are a wide variety of implementations, each of which is a unique trade-off between accuracy of the solution, computational complexity, and speed of adaptation. Two such examples include the QR decomposition (QRD) adaptive filter [15] and the recursive least squares (RLS) algorithm [6]. A frequency domain approach to adaptive filtering is given in [5]. In this section we discuss the so-called least mean square (LMS) algorithm due to Widrow et al. [18] in some detail. Despite having been invented as far back as the 1960s, this algorithm remains a very popular form of adaptive filter algorithm in common use today, due to its relatively high accuracy and extremely simple computational structure compared to other algorithms.

The architectural configuration of the LMS adaptive filter structure for the autoregressive filter problem is shown in Fig. 7.11. Here the objective is to obtain a close-to-ideal estimate of the AR coefficients as they vary according to the changing input speech patterns. One of the key requirements of an adaptive filter is the provision of a so-called *desired signal* shown as $y(t)$ in the figure. The provision

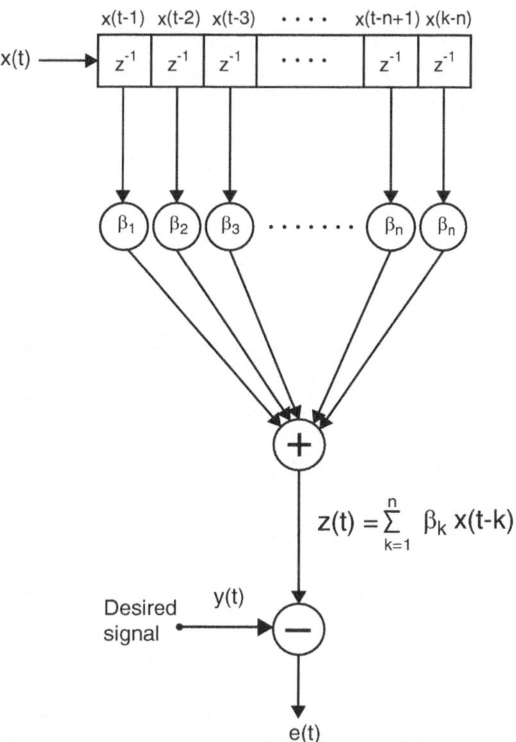

Fig. 7.11 The overall configuration of the LMS adaptive filter structure, for autoregressive analysis of speech

of the desired signal $y(t)$ in an adaptive framework is critical to its operation and sometimes, depending on the application, requires considerable ingenuity to identify such a signal. However, in the AR speech analysis and equaliser applications, the identification of $y(t)$ is straightforward. In the autoregressive speech example, the desired signal $y(t)$ is simply the present value $x(t)$ of the input signal. The objective is to adjust the weights β_i so that the tapped delay line filter output signal $z(t)$ is as close as possible in the mean-squared error sense to the desired signal $y(t)$, thus minimising the variance (power) $E[e^2(t)]$ in the error signal $e(t)$. The means to accomplish this task is discussed shortly. Once the variance of $e(t)$ is minimised, the set of coefficients β_i is then a parsimonious representation of the input speech signal $x(t)$, for as long as the signal remains stationary.

A further example of an adaptive filter application is the equaliser of Fig. 7.1, which has already been discussed to some extent. Particularly in the mobile communications environment, the channel impulse response h_k varies with time when the receiver platform is in motion in a multipath scenario. Therefore the weights a_k must adapt in accordance with the changing channel in order to maintain the functioning of the receiver. We note that the adaptive filter equaliser configuration of Fig. 7.1 is very closely related to that of the AR configuration of Fig. 7.11, with the only difference being the source of the desired signal. In the equaliser case, the desired signal $y(t)$ is a quantised version of $z(t)$, as mentioned previously, whereas in the AR analysis problem of Fig. 7.11, it is the present value of the input speech signal $x(t)$.

We now develop the LMS algorithm for updating the weight vector $\boldsymbol{\beta}(t)$ at time t. We define the vector $\boldsymbol{x}(t)$ as the set of signal samples in the tapped delay line at time t; that is, $\boldsymbol{x}(t) = [x(t-1), x(t-2), \ldots, x(t-n)]^T$. Then $z(t) = \boldsymbol{x}^T(t)\boldsymbol{\beta}(t)$. The objective function $J(\boldsymbol{\beta})$ to be considered bears a strong resemblance to that of the LS procedure we have previously discussed and is given in the adaptive filter context as

$$\begin{aligned} J(\boldsymbol{\beta}) &= E\left[e^2(t)\right] \\ &= E\left[(\boldsymbol{x}^T(t)\boldsymbol{\beta}(t) - y(t))^2\right], \end{aligned} \tag{7.78}$$

where the expectation is taken over $\boldsymbol{x}(t)$. The major difference between the ordinary LS objective function (7.11) and that of (7.78) is that the latter refers to the expectation over all possible realisations of the rows $\boldsymbol{x}^T$, whereas (7.11) in effect replaces the expectation with an arithmetic average over the m available rows of $\boldsymbol{X}$.

Recall that the negative gradient of a function with respect to its independent variables is the direction of steepest descent of the function. In this vein, the gradient $\nabla J(\boldsymbol{\beta})$ with respect to $\boldsymbol{\beta}$ can be written in the form

$$\begin{aligned} \nabla J(\boldsymbol{\beta}) &= E\left[2\left[\boldsymbol{x}^T(t)\boldsymbol{\beta}(t) - y(t)\right]\boldsymbol{x}(t)\right] \\ &= E\left[2e(t)\boldsymbol{x}(t)\right]. \end{aligned} \tag{7.79}$$

From the above, we could consider determining the weight vector $\boldsymbol{\beta}$ by assigning an arbitrary initial value to $\boldsymbol{\beta}$ and then refining the estimate of $\boldsymbol{\beta}$ in a series of steps (iterations), where at the jth step we form a new estimate according to

$$\boldsymbol{\beta}_{j+1} = \boldsymbol{\beta}_j - \lambda E\left[2e(t)\boldsymbol{x}(t)\right], \quad j = 1, 2, \ldots, \tag{7.80}$$

where the last term is the negative gradient at time t and λ is a step-size parameter to be determined. Thus in each iteration j, we take a step in the direction of steepest descent and as such $J(\boldsymbol{\beta})$ decreases in each iteration (provided the input sequence is stationary) and eventually converges to the optimal solution for $\boldsymbol{\beta}$, which corresponds to the global minimum of $J(\boldsymbol{\beta})$. In this instance convergence to a global optimum is assured since $J(\boldsymbol{\beta})$ is convex everywhere. This method of optimisation is referred to as *gradient descent.*

There is one fly in the ointment, however, and that involves the evaluation of the expectation operator in (7.80). Evaluation of the expectation requires an infinite amount of data, which is impossible to implement in practice. The LMS algorithm circumvents this difficulty in a very simple but effective manner, that is, to replace the expectation in (7.80) with the instantaneous value of the gradient, i.e. the value $2e(t)\boldsymbol{x}(t)$. The implementation of the adaptive filter therefore becomes much simpler, at the cost of what turns out to be a very noisy estimate of the true gradient. The resulting optimisation procedure is referred to as *instantaneous gradient* or *stochastic gradient*. Stochastic gradient methods are commonly used in training various forms of neural networks.

The LMS algorithm updates its weights at every sample instant. Thus from (7.80), the desired LMS update rule for the weights $\boldsymbol{\beta}$ at time t is in the direction of the negative instantaneous gradient estimate, given as

$$\boldsymbol{\beta}(t+1) = \boldsymbol{\beta}(t) - \lambda e(t)\boldsymbol{x}(t), \tag{7.81}$$

where the factor of 2 in (7.80) has been absorbed into the constant λ. Thus the weights $\boldsymbol{\beta}$ change and adapt to changing, nonstationary environments (i.e. where the parameters of the underlying model change with time) in a manner which provides an approximation to the optimal solution of the respective normal equations at each time point after convergence.

We now proceed to investigate the behaviour of the LMS algorithm with respect to its convergence properties, first assuming a stationary environment for the time being. From (7.78), we have

$$\begin{aligned} J(\boldsymbol{\beta}) &= E\left[(\boldsymbol{x}^T(t)\boldsymbol{\beta}(t) - y(t))^2\right] \\ &= E\left[\boldsymbol{x}^T(t)\boldsymbol{\beta}(t) - y(t)\right]^T\left[\boldsymbol{x}^T(t)\boldsymbol{\beta}(t) - y(t)\right] \\ &= \boldsymbol{\beta}^T(t)E\left[\boldsymbol{x}(t)\boldsymbol{x}^T(t)\right]\boldsymbol{\beta}(t) - \boldsymbol{\beta}^T(t)E\left[y(t)\boldsymbol{x}(t)\right] - E\left[y(t)\boldsymbol{x}(t)\right]^T\boldsymbol{\beta}(t) \\ &\qquad + E\left[y(t)^2\right], \end{aligned} \tag{7.82}$$

where $\boldsymbol{\beta}$ has been taken outside the expectation operator since in this case the expectation is taken over $\boldsymbol{x}(t)$, and hence $\boldsymbol{\beta}$ is considered as a constant. We note from the discussion immediately above (7.29) that the expression $E\left[\boldsymbol{x}(t)\boldsymbol{x}^T(t)\right]$ in the first term above and $E\left[y(t)\boldsymbol{x}(t)\right]$ from the second (and third) terms above are equal to $\boldsymbol{R}$ and $\boldsymbol{p}$, respectively. Substituting these values into (7.82), we get

$$J(\boldsymbol{\beta}) = \boldsymbol{\beta}^T\boldsymbol{R}\boldsymbol{\beta} - \boldsymbol{\beta}^T\boldsymbol{p} - \boldsymbol{p}^T\boldsymbol{\beta} + E\left[y^2\right],$$

where we have suppressed the dependence on time for notational convenience. From the normal equation representation of (7.29), we have $\boldsymbol{p} = \boldsymbol{R}\boldsymbol{\beta}_{LS}$, where $\boldsymbol{\beta}_{LS}$ is the LS solution corresponding to the normal equations. Substituting this into the above, we have

$$J(\boldsymbol{\beta}) = \boldsymbol{\beta}^T\boldsymbol{R}\boldsymbol{\beta} - \boldsymbol{\beta}^T\boldsymbol{R}\boldsymbol{\beta}_{LS} - \boldsymbol{\beta}_{LS}^T\boldsymbol{R}\boldsymbol{\beta} + E\left[y\right]^2.$$

Consider the following identity:

$$(\boldsymbol{\beta} - \boldsymbol{\beta}_{LS})^T\boldsymbol{R}(\boldsymbol{\beta} - \boldsymbol{\beta}_{LS}) = \\ \boldsymbol{\beta}^T\boldsymbol{R}\boldsymbol{\beta} - \boldsymbol{\beta}^T\boldsymbol{R}\boldsymbol{\beta}_{LS} - \boldsymbol{\beta}_{LS}^T\boldsymbol{R}\boldsymbol{\beta} + \boldsymbol{\beta}_{LS}^T\boldsymbol{R}\boldsymbol{\beta}_{LS}.$$

Then, comparing the two equations immediately above, we finally have

$$J(\boldsymbol{\beta}) = (\boldsymbol{\beta} - \boldsymbol{\beta}_{LS})^T\boldsymbol{R}(\boldsymbol{\beta} - \boldsymbol{\beta}_{LS}) + K, \tag{7.83}$$

where $K = E\left[y\right]^2 - \boldsymbol{\beta}_{LS}^T\boldsymbol{R}\boldsymbol{\beta}_{LS}$. Note that $\boldsymbol{\beta}_{LS}$ is a constant, independent of the variable $\boldsymbol{\beta}$, as is the term $E\left[y\right]^2$, so these two terms are lumped together into the constant K.

Thus we see that our desired final result of (7.83) is a quadratic form in the independent variable $\boldsymbol{\beta}' = \boldsymbol{\beta} - \boldsymbol{\beta}_{LS}$ (i.e. with $\boldsymbol{\beta}$ translated by $\boldsymbol{\beta}_{LS}$) and shifted along the vertical axis by the constant K.

Recall from the treatment of the quadratic form in Chap. 4 that $J(\boldsymbol{\beta}')$ has an elliptical bowl-shaped form, where contours of constant $J(\boldsymbol{\beta}')$ are ellipsoidal in shape, where the length of the ith semi-axis is proportional to $1/\sqrt{\lambda_i}$, where λ_i is the ith eigenvalue of $\hat{\boldsymbol{R}}$, which is to be defined.

We show an example of the convergence behaviour of the LMS algorithm in Figs. 7.12 and 7.13. Here, the global minimum of $J(\boldsymbol{\beta}')$ is located at $\boldsymbol{\beta}' = \boldsymbol{0}$. Figures 7.12 and 7.13 show elliptical contours of constant $J(\boldsymbol{\beta}')$ with successively increasing values of $J(\boldsymbol{\beta}')$ as the ellipses grow larger, superimposed on the trajectory of $\boldsymbol{\beta}'$ as given by the LMS algorithm. As may be seen, the algorithm is initialised with an arbitrary value for $\boldsymbol{\beta}'$, and at each time point it takes a step according to (7.81). Because the instantaneous gradient is very noisy, the step directions do not point consistently in the downhill direction. However, on average over some interval of time, it may be seen that the trajectory of the estimated point $\boldsymbol{\beta}'$ does converge towards the optimum point $\boldsymbol{0}$ as desired. It may be observed that once the global

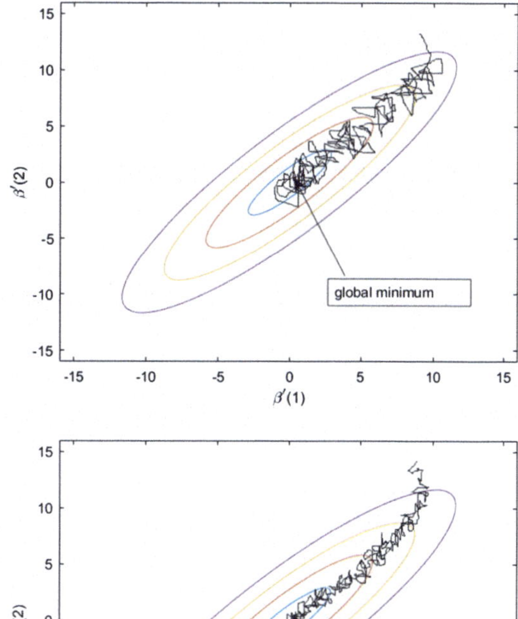

Fig. 7.12 Convergence behaviour of the LMS algorithm for a relatively large value of λ. The blue trace shows the trajectory of the estimates of $\boldsymbol{\beta}'$ vs. time

Fig. 7.13 Similar to Fig. 7.12, but for a smaller value of λ

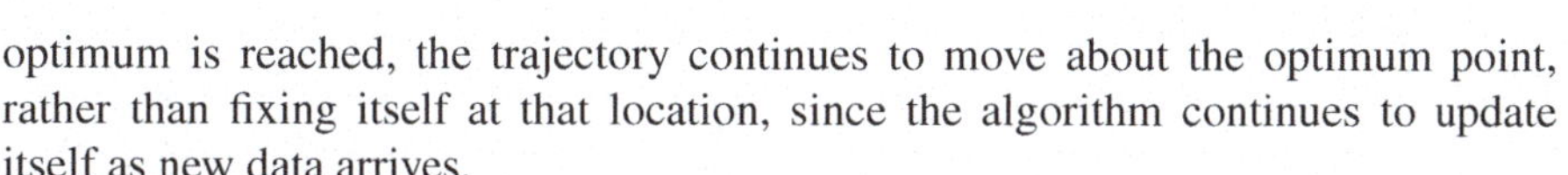

optimum is reached, the trajectory continues to move about the optimum point, rather than fixing itself at that location, since the algorithm continues to update itself as new data arrives.

The value of λ shown in Fig. 7.12 is relatively larger than that in Fig. 7.13. We see that the step sizes are larger for the large λ case, which implies that the algorithm converges faster towards the global optimum. However, the penalty for this accelerated convergence is that the steady-state error, which occurs after the algorithm has converged, is larger in the large λ case as is evident by comparing the two figures.

For the purposes of analysis, we can form the data matrix $\boldsymbol{X}$ in a manner similar to what we have done previously. In this case, the ith row $\boldsymbol{x}_i^T$ of $\boldsymbol{X}$ consists of the data in the tapped delay line filter at the ith time instant; that is, the ith row of $\boldsymbol{X}$ is $[x(i-1), x(i-2), \ldots, x(i-n)]$, for $i = 1, \ldots, m$. Then as before, the finite-sample estimate $\hat{\boldsymbol{R}}$ of $\boldsymbol{R}$ in (7.83) is given by $\hat{\boldsymbol{R}} = \frac{1}{m}\boldsymbol{X}^T\boldsymbol{X}$, where m is the number of rows in $\boldsymbol{X}$. We define the quantity $\mathcal{E}(\hat{\boldsymbol{R}}) = \sqrt{\frac{\lambda_1}{\lambda_n}}$, which is the ratio of the largest

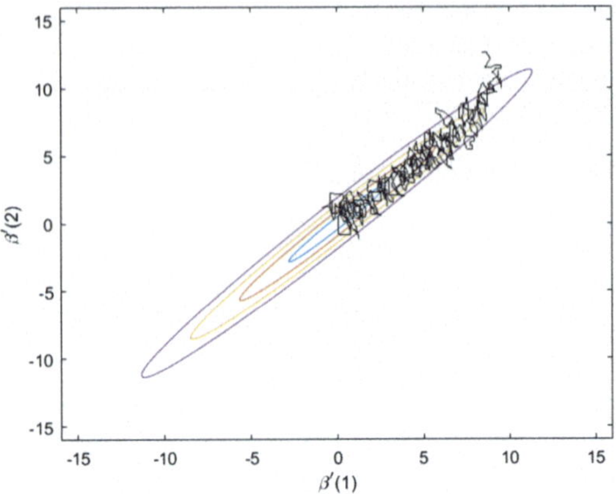

Fig. 7.14 Convergence behaviour of the LMS algorithm for the poorly conditioned case. The trajectory predominately oscillates along the minor axis direction with time, with very slow progress along the major axis direction towards the global optimum

to smallest semi-axes of the ellipsoids corresponding to the quadratic form in $\hat{\boldsymbol{R}}$), formed by $J(\boldsymbol{\beta}') = constant$. Further, it is the square root of the condition number as discussed in Chap. 5 and is also a measure of the eccentricity of the ellipsoids.

The quantity $\mathcal{E}(\hat{\boldsymbol{R}})$ has a profound effect on the behaviour of the LMS algorithm. A large $\mathcal{E}(\hat{\boldsymbol{R}})$ implies that the quadratic form ellipse is highly elongated, as demonstrated in Fig. 7.14 (illustrated for the two-dimensional case). In this context, large values of $\mathcal{E}(\hat{\boldsymbol{R}})$ arise when the input data are highly correlated, implying that the columns of the respective data matrix $\boldsymbol{X}$ are close to linear dependence. From Fig. 7.14 see that the bottom of the trough of the quadratic form bowl, which is oriented along the major axis direction, has lower curvature and a lower directional derivative than is the case along the direction of the minor axis. (In higher dimensions, there will be additional axes exhibiting intermediate values of derivative.) This characteristic implies that the gradient of $J(\boldsymbol{\beta}')$ has a large component along the minor axis and a small component along the major axis. This implies that the trajectory of $\boldsymbol{\beta}'$ mostly traverses in the minor axis direction, with only a small component along the major axis direction. Thus in the poorly conditioned case, the trajectory progresses slowly along the bottom of the trough towards the global minimum, rendering slow convergence in these cases.

It is shown in [8] that if λ is too large, the update mechanism for $\boldsymbol{\beta}$ can become oscillatory and unstable with $\boldsymbol{\beta}$ growing exponentially in magnitude with time. It is shown that this condition can be avoided by choosing $\lambda < 2/\lambda_{\max}$, with $\lambda_{\max}$ being the largest eigenvalue of $\boldsymbol{R}$.

7.10.1 The Normalised LMS Adaptive Filter

Here we consider an alternate formulation of the LMS algorithm. Rather than using the objective function of (7.78), the normalised LMS optimisation problem may be stated as minimising the quantity $||\delta\boldsymbol{\beta}(t+1)||_2^2$, where

$$\delta\boldsymbol{\beta}(t+1) = \boldsymbol{\beta}(t+1) - \boldsymbol{\beta}(t)$$

subject to the constraint

$$\boldsymbol{\beta}^T(t+1)\boldsymbol{x}(t) = y(t),$$

where $y(t)$ is the desired signal as shown in Fig. 7.11. The constraint implies that the weights $\boldsymbol{\beta}(t+1)$ at time $t+1$ are determined so that the filter output at time t using the weights $\boldsymbol{\beta}(t+1)$ with the current filter input $\boldsymbol{x}(t)$ is exactly equal to the desired signal $y(t)$, making the error at time t zero. The solution to this problem is obtained through the use of Lagrange multipliers and is given as [8]

$$\boldsymbol{\beta}(t+1) = \boldsymbol{\beta}(t) - \frac{\mu}{||\boldsymbol{x}(t)||_2^2}\boldsymbol{x}(t)e(t),$$

where μ is a step-size parameter, similar in nature to λ in the ordinary version of the algorithm. We can see that this update rule is similar to the ordinary version, except the weight update is normalised by the factor $||\boldsymbol{x}(t)||_2^2$, a feature giving the method its name. Since the term $\frac{\mu}{||\boldsymbol{x}(t)||_2^2}$ varies with time, the effective step-size parameter can also be seen to be a time-varying quantity. The normalised version is more robust to the choice of μ and has been shown to have the potential for faster convergence with both correlated and uncorrelated inputs (the poor- and well-conditioned cases, respectively). The normalised LMS algorithm is discussed in much more detail in [8], Ch. 6.

7.10.2 Computational Complexity of Adaptive Filtering Algorithms

It is readily verified that the LMS algorithm requires $\mathcal{O}(n)$ multiply/add operations per time step. This is significantly lower than other forms of adaptive filter, specifically the RLS and QRD algorithms, which each require $\mathcal{O}(n^2)$ operations per time step. An advantage of these more complex methods is that they converge faster and are more accurate in steady state. The computational simplicity of the LMS algorithm is the reason for its widespread use in many modern applications, such as the cell phone. Every cell phone requires an equaliser, and since the cell phone is a consumer product, it must be implemented in as inexpensive a manner as possible. The low computational demands of the LMS algorithm allow a simple enough implementation to satisfy this requirement, while at the same time exhibiting adequate convergence performance.

7.10.3 Simulations

In this section we illustrate the above theoretical analysis with two sets of simulations. The first is when the input data sequence is white where the corresponding value of $\mathcal{E}(\hat{\boldsymbol{R}})$ is close to 1, whereas in the second case the input data is correlated, resulting in a higher value of $\mathcal{E}(\hat{\boldsymbol{R}})$.

In the first case, the input data sequence $x(t)$ of length 1000 samples is an *iid* Gaussian process with mean zero and unit variance. The data are generated using the Matlab® *randn* command. The adaptive filter is of length $n = 5$. The data $\boldsymbol{x}(t)$ in the tapped-delay line filter of Fig. 7.11 at time t is formed by $\boldsymbol{x}(t) = [x(t-1), x(t-2), \ldots, x(t-n)]^T$. Then the filter output $z(t)$ is simulated by $z(t) = \boldsymbol{x}(t)^T \boldsymbol{\beta}_o(t)$, where the true values $\boldsymbol{\beta}_o$ in all these simulations are assigned the values $[1, 1, 1, -1, -1]^T$. The desired signal $y(t)$ at each time t is computed by a noisy version of $z(t)$ in the form of $y(t) = z(t) + \epsilon(t)$, where $\epsilon(t)$ is a Gaussian-distributed random error variable with zero mean and standard deviation equal to the value 0.1 throughout all simulations. It is clear with this arrangement that the values of $\boldsymbol{\beta}(t)$ must converge to the true values $\boldsymbol{\beta}_o$ for the error $e(t)$ to be reduced to a minimum.

The coefficients $\boldsymbol{\beta}$ are all initialised to zero. At each time step, $z(t)$ is calculated as $z(t) = \boldsymbol{x}^T(t)\boldsymbol{\beta}(t)$ as shown in Fig. 7.11 using the current values of $\boldsymbol{x}(t)$ and $\boldsymbol{\beta}(t)$. The desired signal $y(t)$ is generated as $y(t) = z(t) + \boldsymbol{\epsilon}(t)$, as previously described. Then $\boldsymbol{\beta}(t)$ is updated according to (7.81), t is incremented by one sample, and the process iterates over time as new data arrives.

A plot of the error signal $e(n)$ vs. time is shown in Fig. 7.15, for $\lambda = 0.02$ when the input data to the filter is white. As seen, the error starts off with a relatively large value, since the $\boldsymbol{\beta}$-coefficients do not have the correct value. However, we see that after a few hundred iterations $\boldsymbol{\beta}(t)$ approaches its true value and the error settles into a steady-state condition which does not decrease any further with time. This behaviour is due to the additive noise in the desired signal $y(t)$ and the fact that the $\boldsymbol{\beta}$'s continue to move about the true value $\boldsymbol{\beta}_o$ after the filter has converged, as previously explained.

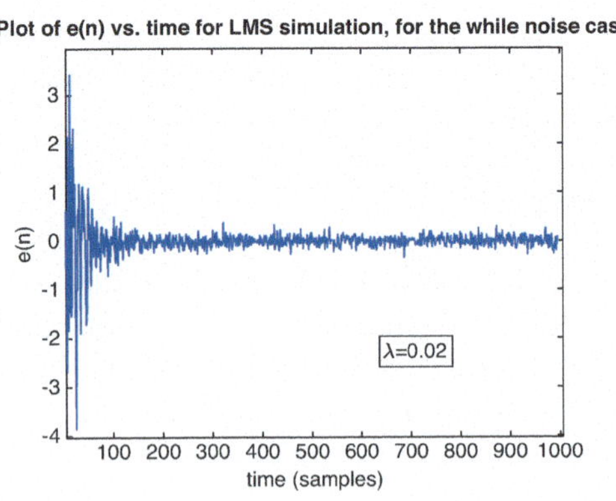

Fig. 7.15 Simulation example showing the error signal $e(n)$ vs. time when the input data is white, for $\lambda = 0.02$

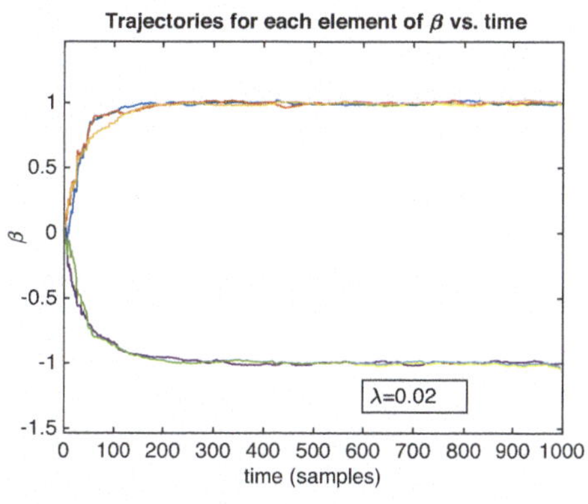

Fig. 7.16 Trajectories of $\boldsymbol{\beta}$ vs. time when the input data is white, for $\lambda = 0.02$. The true values are $[1, 1, 1, -1, -1]^T$, and the initial values are all zero. It is seen that $\boldsymbol{\beta}$ converges to its true values, as expected

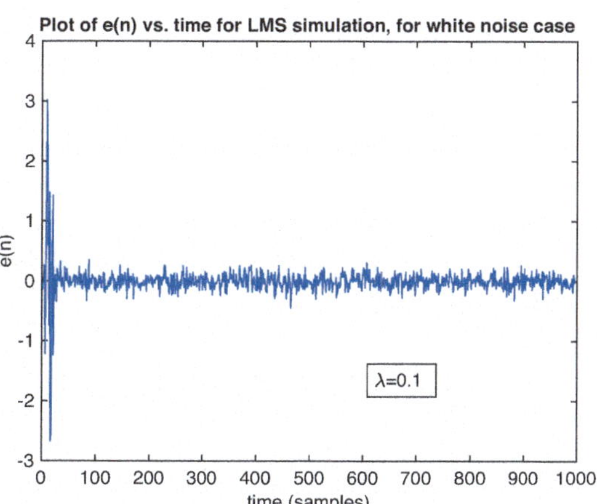

Fig. 7.17 Simulation example showing the error signal $e(n)$ vs. time when the input data is white, for $\lambda = 0.1$

The trajectories of each of the five elements of $\boldsymbol{\beta}$ vs. time are shown superimposed in Fig. 7.16, for the white data case and for $\lambda = 0.02$, which is the same value as that of Fig. 7.15. A steady-state error criterion ζ on the converged values of $\boldsymbol{\beta}$ was calculated, according to

$$\zeta = \frac{1}{|C|} \sum_{t \in C} ||\boldsymbol{\beta}(t) - \boldsymbol{\beta}_o||_2,$$

where C is the region where the filter is in the steady state condition, $|C|$ is the cardinality of C in samples, and $\boldsymbol{\beta}_o$ is the true value of $\boldsymbol{\beta}$. In this case, $C = [500, 1000]$. In words, ζ is the mean of the converged error in $\boldsymbol{\beta}$. The value of ζ for this example is $\zeta = 5.4675 \times 10^{-4}$.

For the sake of comparison, Figs. 7.17 and 7.18 show the corresponding behaviour of the adaptive filter when $\lambda = 0.1$. It is seen the convergence is much faster in this case, since inspection of these figures shows that the filter has reached steady state after about 30 samples, as opposed to about 200 samples in the

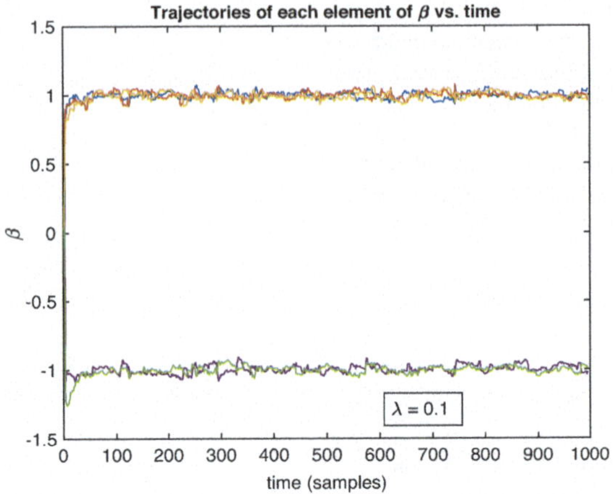

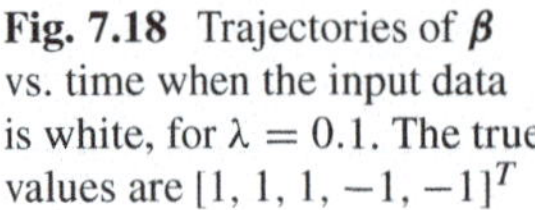
Fig. 7.18 Trajectories of $\boldsymbol{\beta}$ vs. time when the input data is white, for $\lambda = 0.1$. The true values are $[1, 1, 1, -1, -1]^T$

$\lambda = 0.02$ case. However, the value of ζ in the $\lambda = 0.1$ case is 3.4605×10^{-3}, which is seen to be significantly larger than the value 5.4675×10^{-4} in the previous case. This behaviour is an illustration of the previous discussion, that indicated for larger values of λ, convergence is faster but steady-state error is larger than in the case where λ is smaller.

It is interesting to note that for this example, the filter becomes oscillatory when $\lambda = 0.35$ or greater. In this case where the input data are white, $\hat{\boldsymbol{R}}$ is approximately diagonal with approximately equal diagonal elements, so all eigenvalues are approximately equal and the value $\mathcal{E}(\hat{\boldsymbol{R}})$ is close to 1. In all the simulations presented here for the white data case, the condition number varied between the values 1 and 1.2, consistent with our theoretical expectations.

In our second set of simulations the input data are correlated or coloured, which results in the eigenvalues of $\hat{\boldsymbol{R}}$ to become more disparate. In this case, the input data are simulated by passing white noise through a third-order Butterworth filter with a normalised cutoff frequency of 0.25 cycles/sample. This filtering operation removes the high-frequency components from the input signal, thus causing the data to vary more smoothly, thus introducing correlations amongst the data samples, resulting in a non-diagonal $\hat{\boldsymbol{R}}$ with disparate eigenvalues. All other parameters for this second simulation remain unchanged from the previously discussed white data case.

Figures 7.19 and 7.20 show $e(n)$ and trajectories of each element of $\boldsymbol{\beta}$ vs. time, respectively, when the input data is correlated, for the value $\lambda = 0.02$. Note that in these figures, the time axis has been extended to 20,000 samples, whereas in the previous white data figures the time axis extended only to 1000 samples. As may be seen from Fig. 7.20, the convergence of $\boldsymbol{\beta}$ in this correlated data case is far slower than the corresponding white data case of Fig. 7.16. As discussed previously, the directional derivative along the direction of the bottom of the trough (i.e. along the direction of the major axis of the ellipsoid) of the $J(\boldsymbol{\beta})$ objective function becomes small, rendering progress along this direction to be slow. In this simulation example, the value $\mathcal{E}(\hat{\boldsymbol{R}})$ is 10.052.

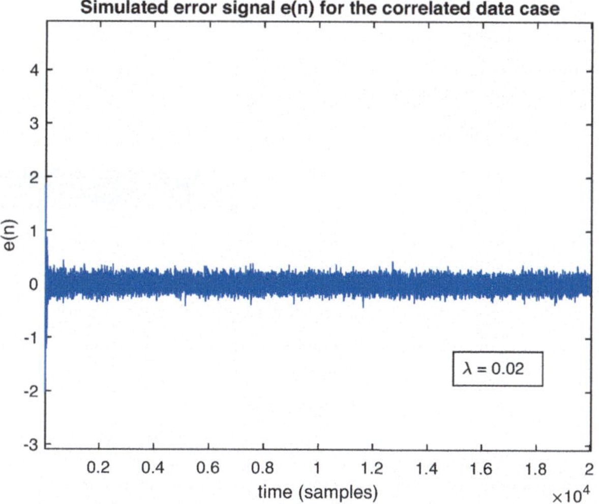

Fig. 7.19 Simulation example showing the error signal $e(n)$ vs. time when the input data is correlated, for $\lambda = 0.02$

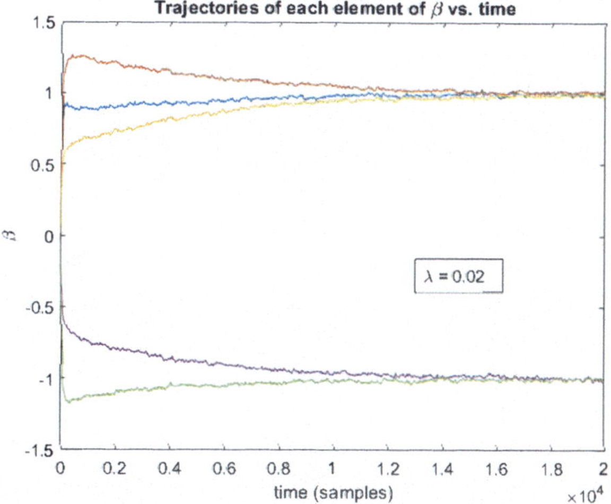

Fig. 7.20 Trajectories of $\boldsymbol{\beta}$ vs. time for correlated input data, for $\lambda = 0.02$. The true values are $[1, 1, 1, -1, -1]^T$

Note from Fig. 7.19 that the error signal $e(n)$ converges about as quickly in this case as it does in the case of Fig. 7.15. In both instances this is about 200 samples, even though the convergence behaviour in Fig. 7.19 is not evident due to the compression of the time axis. The reasoning behind this phenomenon is the basis of Problem 14.

In Figs. 7.21 and 7.22 we show the error signal $e(n)$ and trajectories of $\boldsymbol{\beta}$, respectively, for the correlated data case for the larger value of $\lambda = 0.1$. The same observations as in the white data case persist—convergence is faster and the steady-state error in $\boldsymbol{\beta}$ is larger compared to the $\lambda = 0.02$ case (see Figs. 7.19 and 7.20 for comparisons).

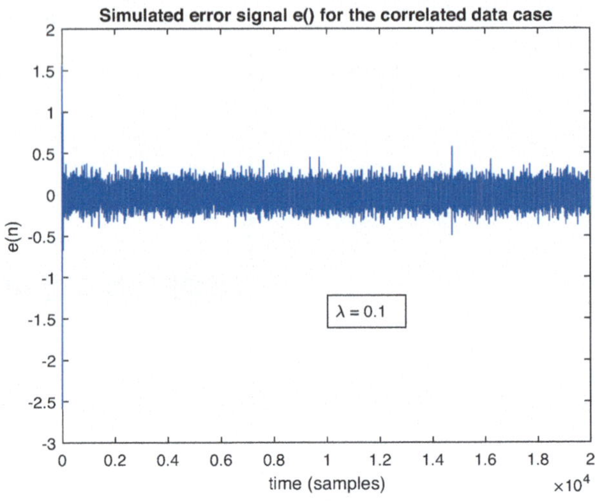

Fig. 7.21 Simulation example showing the error signal $e(n)$ vs. time when the input data is correlated, for $\lambda = 0.1$

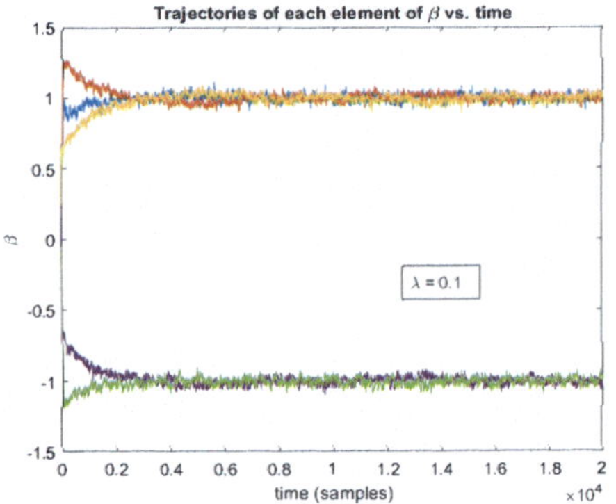

Fig. 7.22 Trajectories of $\boldsymbol{\beta}$ vs. time for correlated input data, for $\lambda = 0.1$. The true values are $[1, 1, 1, -1, -1]^T$

In conclusion, we can say that the adaptive filter is better behaved in the white data case relative to the correlated data case. Convergence of $\boldsymbol{\beta}$ to its true values is slower in the correlated data case, whereas convergence of the error signal $e(n)$ is not greatly affected by correlated input data. The effect of λ is to trade off convergence speed with accuracy in the steady state. Larger values of λ give faster convergence but larger steady-state error.

7.11 Problems

1. Here we consider an unsupervised clustering problem. In the supplementary material described in the opening pages of this chapter, you will find a file Ch7Q1.mat that contains a matrix $\boldsymbol{X}_n$, which consists of 100 independent measurements of 5 different variables. Find a subspace so that these data cluster well into two distinct regions (i.e. classes). Explain how you obtained this

subspace. Plot each data point in this two-dimensional subspace to illustrate the resulting clustering behaviour. To which class do the additional data samples $\mathbf{x}_{tst1}$ and $\mathbf{x}_{tst2}$ in the file belong? Explain your methodology carefully. *Hint:* Do not use a clustering algorithm such as k-means.

2. We have seen the LS procedure produces a value β_{LS} for which the sum of the squared lengths of the residuals $r_{LS}(i)$ is minimised. In Fig. 7.3 the LS residuals are dropped vertically from the observed point y_i into the line $y = x\beta$. On the other hand, inspection of Fig. 7.3 reveals that the residuals would have shorter lengths if they were oriented at a direction orthogonal to this line, as opposed to being vertical. Explain this apparent contradiction.
3. In the supplementary material described in the opening pages of this chapter, you will find a file Ch7Q3.mat, which contains the variables $\boldsymbol{X} \in \mathbb{R}^{m \times n}$ and $\boldsymbol{Y} \in \mathbb{R}^{m \times k}$. Here, $m = 10$, $n = 4$, and $k = 5000$. Each column $\boldsymbol{y}_i$ of $\boldsymbol{Y}$ is generated according to $\boldsymbol{y}_i = \boldsymbol{X}\boldsymbol{\beta}_o + \boldsymbol{\epsilon}_i, i = 1, \ldots, k$. For this problem, the $\boldsymbol{\epsilon}_i$ are coloured. Write a Matlab® program to estimate each $\boldsymbol{\beta}_{LS}(i), i = 1, \ldots, k$, corresponding to each observation $\boldsymbol{y}_i$ so that the estimates have the minimum possible variance. Also jointly estimate the noise covariance matrix $\boldsymbol{\Sigma}$.
Hint: This will require an iterative procedure as follows:

   (a) Initialise the iteration index j to zero and the noise covariance matrix estimate $\hat{\boldsymbol{\Sigma}}_o$ to some value (e.g. $\boldsymbol{I}$).
   (b) Using the current $\hat{\boldsymbol{\Sigma}}_j$, use the appropriate form of normal equations to calculate all k values of $\boldsymbol{\beta}_{LS}(i), i = 1, \ldots, k$.
   (c) For a more stable estimate, calculate the mean $\bar{\boldsymbol{\beta}}$ over all k LS estimates.
   (d) The noise vectors $\boldsymbol{\epsilon}_i$ can then be estimated using $\bar{\boldsymbol{\beta}}$, from which an updated $\hat{\boldsymbol{\Sigma}}_j$ can be determined.
   (e) Increment j and go to (b) until convergence.

   Also calculate the initial $\text{cov}(\boldsymbol{\beta}_{LS})$ which assumes white noise and also the final covariance estimate obtained after convergence. Comment on the differences.
4. Consider the overdetermined set of regression equations

$$\boldsymbol{y} = \boldsymbol{X}\boldsymbol{\beta} + \boldsymbol{\epsilon},$$

   where $\text{cov}(\boldsymbol{\epsilon}) = \sigma^2 \boldsymbol{I}$. Suppose, in a given experiment, we had control over the values of the matrix $\boldsymbol{X}$, so that $||\boldsymbol{x}_i||_2^2 = a, i = 1, \ldots, n$, where a is an arbitrary constant > 0, and $\boldsymbol{x}_i$ is the ith column of $\boldsymbol{X}$. Explain how to choose the values of $\boldsymbol{X}$ so that the variance of the LS estimate $\boldsymbol{\beta}_{LS}$ of $\boldsymbol{\beta}$ is minimum.
5. (a) Given $\boldsymbol{X} \in \mathbb{R}^{m \times n}$ and a set $\boldsymbol{\beta}_i \in \mathbb{R}^m$, $\boldsymbol{y}_i \in \mathbb{R}^n$, $i = 1, \ldots, k$, find a set of coefficients a_i so that

$$\left|\left|\boldsymbol{X} - \sum_{i=1}^{k} a_i \boldsymbol{\beta}_i \boldsymbol{y}_i^T\right|\right|_F^2 \tag{7.84}$$

   is minimised.
   (b) What are the sets $\boldsymbol{\beta}_i$, $\boldsymbol{y}_i$ that minimise the minimum in (7.84)?
   (c) What constraint is there on k so that the solution is unique?

6. Let $\boldsymbol{X} \in \mathbb{R}^{m \times n}$, and $\boldsymbol{x} \in \mathbb{R}^m$. Find $\boldsymbol{\beta}$ so that $||\boldsymbol{X} - \boldsymbol{x}\boldsymbol{\beta}^T||_F^2$ is minimised.
7. The supplementary material described in the opening pages of this chapter contains a link to a Matlab® data file Ch7Q7.mat, which contains a variable "speech". Here we look at evaluating the spectrum of the vocal track for a segment of this speech signal. Using a least squares approach similar to the second example in the first section of this chapter, determine the frequency response $H(z)$ of the vocal tract used to generate the speech samples 5600:6200 from the variable "speech". Assume the speech signal is stationary over this interval. Experiment with prediction orders between 8 and 12.
8. Assuming the noise is Gaussian and Assumptions A1 and A2 of Sect. 7.4 hold, calculate a 95% joint confidence region for $\boldsymbol{\beta}_{LS}$.
9. Given that we are provided a table of values for t_i and corresponding y_i in the following equation, devise a method for determining the time constant τ and the scalar value a:

$$y_i = a \exp\left(-\frac{t_i}{\tau}\right),$$

where a is a real constant.
10. With regard to the common spatial patterns method, assume $\boldsymbol{R}_H$ and $\boldsymbol{R}_P$ are given, respectively, by

$$\boldsymbol{R}_H = \begin{bmatrix} 1 & 0.5 \\ 0.5 & 1 \end{bmatrix}, \qquad \boldsymbol{R}_P = \begin{bmatrix} 1 & -0.5 \\ -0.5 & 1 \end{bmatrix}.$$

What is the optimum weight vector $\boldsymbol{w}$ in this case so that the healthy case is most prominent? What is the ratio of the variance of $y_H(t)$ to that of $y_P(t)$ for this choice of $\boldsymbol{w}$?
11. Let $P(x)$ be a polynomial in x. We are given a table of values of $P(x_i)$ vs. $x_i, i = 1, \ldots, n$. Explain a method to determine the polynomial coefficients. How would you estimate the order of the polynomial?
12. What happens with the generalised eigendecomposition as the quantity $||\boldsymbol{R}_H - \boldsymbol{R}_P||_2$ (as defined in the problem above) becomes small?
13. Consider a signal $\boldsymbol{x}(t)$ given by

$$\boldsymbol{x}(t) = \sum_{i=1}^{n} a_i \boldsymbol{g}(t - \tau_i) + \boldsymbol{\epsilon}(t),$$

where $\boldsymbol{g}(t) \in \mathbb{R}^m$ is a basic Gaussian pulse of length m samples, similar to that shown in Fig. 2.10, whose peak value is unity and delay is zero. The quantities a_i and τ_i are the respective amplitudes and delays of the pulse components comprising $\boldsymbol{x}(t)$. It is possible to estimate the τ_i using a variation of the MUSIC algorithm. However, it is also possible to accomplish this task using an LS procedure, by developing an estimator that is a function of the τ_i only. In this vein, we eliminate the a_i by treating the above as a regression equation for an assumed set of τ_i-values and substitute the LS estimate of the a_i's for the actual values. The resulting expression is a function of the τ_i only. The number of components n is assumed known.

(a) Develop this estimator and explain how to estimate the delays τ_i when the noise $\boldsymbol{\epsilon}$ is white.
(b) As above, when the noise has an arbitrary covariance $\boldsymbol{\Sigma}$.
(c) Once the τ_i have been estimated, explain how to estimate the a_i.

14. Explain why the convergence of $e(n)$ for the scenario of Fig. 7.15 is about as fast as that of Fig. 7.19, even though the convergence properties of the respective $\boldsymbol{\beta}$'s are far different in each case.

References

1. D.R. Cox, *Planning of Experiments* (Wiley, New York, 1958)
2. D.R. Cox, N. Reid, *The Theory of the design of Experiments* (CRC Press, Boca Raton, 2000)
3. P.F. de Aguiar et al., D-optimal designs. Chemometr. Intell. Laborat. Syst. **30**(2), 199–210 (1995)
4. J.R. Deller Jr, J.G. Proakis, J.H. Hansen, *Discrete Time Processing of Speech Signals* (Prentice Hall PTR, Englewood Cliffs, 1993)
5. M. Dentino, J. McCool, B. Widrow, Adaptive filtering in the frequency domain. Proc. IEEE **66**(12), 1658–1659 (1978)
6. E. Eleftheriou, D. Falconer, Tracking properties and steady-state performance of RLS adaptive filter algorithms. IEEE Trans. Acoust. Speech Signal Process. **34**(5), 1097–1110 (1986)
7. G.H. Golub, C.F. Van Loan, *Matrix Computations*, 3rd edn. (The Johns Hopkins University, Baltimore, 1996)
8. S. Haykin, *Adaptive Filter Theory*, 4th edn. (Prentice Hall, Englewood Cliffs, 2001)
9. S. Haykin, *Communication Systems* (Wiley, New York, 2008)
10. Z.J. Koles, The quantitative extraction and topographic mapping of the abnormal components in the clinical EEG. Electroencephalogr. Clin. Neurophysiol. **79**(6), 440–447 (1991)
11. S.L. Marple Jr., *Digital Spectral Analysis* (Prentice Hall, Englewood Cliffs, 1987)
12. P.L. Nunez, R. Srinivasan, et al., *Electric Fields of the Brain: The Neurophysics of EEG* (Oxford University Press, Oxford, 2006)
13. A. Papoulis, *Random Variables and Stochastic Processes* (McGraw Hill, New York, 1994)
14. H. Peng, F. Long, C. Ding, Feature selection based on mutual information criteria of max-dependency, max-relevance, and min-redundancy. IEEE Trans. Pattern Anal. Mach. Intell. **27**(8), 1226–1238 (2005)
15. I.K. Proudler, J.G. McWhirter, T.J. Shepherd, Computationally efficient QR decomposition approach to least squares adaptive filtering, in *IEE Proceedings F (Radar and Signal Processing)*, vol. 138, no. 4 (IET, 1991), pp. 341–353
16. L.L. Scharf, C. Demeure, *Statistical Signal Processing: Detection, Estimation, and Time Series Analysis* (Prentice Hall, Englewood Cliffs, 1991)
17. M. Tanco, E. Viles, L. Pozueta, Comparing different approaches for design of experiments, in *Advances in Electrical Engineering and Computational Science* (2009), pp. 611–621
18. B. Widrow et al., Stationary and nonstationary learning characteristics of the LMS adaptive filter, in *Aspects of Signal Processing: With Emphasis on Underwater Acoustics Part 1 Proceedings of the NATO Advanced Study Institute Held at Portovenere, La Spezia, 30 August–11 September 1976* (Springer, New York, 1977), pp. 355–393

Chapter 8
The Rank-Deficient Least Squares Problem

8.1 The Pseudo-Inverse

Previously in Chap. 7 we have seen that the LS problem determines the $\boldsymbol{\beta}_{LS}$ which solves the minimisation problem given by

$$\boldsymbol{\beta}_{LS} = \arg\min_{\boldsymbol{\beta}} ||\boldsymbol{X}\boldsymbol{\beta} - \boldsymbol{y}||_2^2, \tag{8.1}$$

where the observation $\boldsymbol{y}$ is generated from the regression model $\boldsymbol{y} = \boldsymbol{X}\boldsymbol{\beta}_o + \boldsymbol{\epsilon}$. The solution $\boldsymbol{\beta}_{LS}$ is the one which gives us the best fit between the linear model $\boldsymbol{X}\boldsymbol{\beta}$ and the observations $\boldsymbol{y}$ in the 2-norm sense. For the case where $\boldsymbol{X}$ is full rank, we saw that the solution $\boldsymbol{\beta}_{LS}$ which solves (8.1) is the solution to the normal equations

$$\boldsymbol{X}^T\boldsymbol{X}\boldsymbol{\beta} = \boldsymbol{X}^T\boldsymbol{y}. \tag{8.2}$$

We have seen previously in Sect. 7.5 that even one small eigenvalue of the matrix $\boldsymbol{X}^T\boldsymbol{X}$ destroys the desirable variance properties of the LS estimate and introduces the potential for all elements of $\boldsymbol{\beta}_{LS}$ to have large variance. One small eigenvalue of $\boldsymbol{X}^T\boldsymbol{X}$ implies the matrix is poorly conditioned and close to rank deficiency. The pseudo-inverse is a means of remedying this adverse situation in many circumstances and can be very effective in reducing the error in the LS solution.

Further, if the matrix $\boldsymbol{X}$ (and consequently $\boldsymbol{X}^T\boldsymbol{X}$) is exactly rank deficient (instead of being close to rank deficient), then a unique solution to the normal equations does not exist. There are an infinity of solutions that minimise (8.1) with respect to $\boldsymbol{\beta}$. However, we can generate a unique solution if, amongst the set of $\boldsymbol{\beta}$ satisfying (8.2), we choose that value of $\boldsymbol{\beta}$ which itself has minimum norm. The pseudo-inverse fulfils this goal. In this case, the pseudo-inverse solution $\boldsymbol{\beta}_{LS}$ is the result of two 2-norm minimising procedures—the first determines a set $\{\boldsymbol{\beta}\}$ which

J. Reilly, *Fundamentals of Linear Algebra for Signal Processing*,
https://doi.org/10.1007/978-3-031-68915-4_8

minimises $||X\beta - y||_2^2$ and the second determines β_{LS} as that element of $\{\beta\}$ for which $||\beta||_2$ is minimum.

We are given $X \in \mathbb{R}^{m\times n}$, $m \geq n$, and rank$(\beta) = r \leq n$. If the SVD of X is given as $U\Sigma V^T$, then we define X^+ as the *pseudo-inverse* of X, defined by

$$X^+ = V\Sigma^+U^T. \tag{8.3}$$

The matrix Σ^+ is related to Σ in the following way. If

$$\Sigma = \text{diag}(\sigma_1, \sigma_2, \ldots, \sigma_r, 0, \ldots, 0),$$

then

$$\Sigma^+ = \text{diag}(\sigma_1^{-1}, \sigma_2^{-1}, \ldots, \sigma_r^{-1}, 0, \ldots, 0), \tag{8.4}$$

where Σ and Σ^+ are padded with zeros in an appropriate manner to maintain dimensional consistency.

Theorem 8.1 *When X is rank deficient, the unique solution β_{LS} minimising (8.1) such that $||\beta||_2$ is minimum is given by*

$$\beta_{LS} = X^+y, \tag{8.5}$$

where X^+ is defined by (8.3). Further, the squared norm r_{LS}^2 of the LS residual r_{LS} is given as

$$r_{LS}^2 = \sum_{i=r+1}^{m} (u_i^T y)^2. \tag{8.6}$$

Proof For any $\beta \in \mathbb{R}^n$, we have

$$||X\beta - y||_2^2 = \left\|U^TXV(V^T\beta) - U^Ty\right\|_2^2 \tag{8.7}$$

$$= \left\|\begin{bmatrix}\Sigma_r & 0\\ 0 & 0\end{bmatrix}\begin{bmatrix}w_1\\ w_2\end{bmatrix} - \begin{bmatrix}c_1\\ c_2\end{bmatrix}\right\|_2^2, \tag{8.8}$$

where

$$w = \begin{matrix} r \\ n-r \end{matrix}\begin{bmatrix}w_1\\ w_2\end{bmatrix} = \begin{bmatrix}V_1^T\\ V_2^T\end{bmatrix}\beta = V^T\beta \tag{8.9}$$

and

$$\begin{matrix} r \\ m-r \end{matrix} \begin{bmatrix} c_1 \\ \\ c_2 \end{bmatrix} = \begin{bmatrix} U_1^T \\ \\ U_2^T \end{bmatrix} y = U^T y \tag{8.10}$$

and

$$\Sigma_r = \mathrm{diag}[\sigma_1, \ldots, \sigma_r].$$

Note that we can write the quantity $||X\beta - y||_2^2$ in the form of (8.7), since the 2-norm is invariant to the orthonormal transformation U^T, and the quantity VV^T which is inserted between X and β is identical to I. The partitioning of the matrices U and V into $U = [U_1, U_2]$ and $V = [V_1, V_2]$, where U_1 and V_1 correspond to the first r columns, respectively, was previously described in Eqs. (3.15) and (3.16) of Sect. 3.1.1.

From (8.8), we can make several immediate conclusions, as follows:

1. If X is rank deficient of rank $r < n$, then $\sigma_{r+1}, \ldots, \sigma_n$ are zero, and hence the last $n - r$ columns of Σ in (8.8) must be zero. Therefore it is apparent that the solution w is independent of w_2, and therefore w_2 is arbitrary.
2. Note that for any vector $y = [y_1^T \;\; y_2^T]^T$, $||y||_2^2 = ||y_1||_2^2 + ||y_2||_2^2$. Since the argument of the left-hand side of (8.8) is a vector, it may therefore be expressed as

$$||X\beta - y||_2^2 = ||\Sigma_r w_1 - c_1||_2^2 + ||c_2||_2^2. \tag{8.11}$$

   It is apparent therefore that (8.8) is minimised by choosing w_1 to satisfy

$$\Sigma_r w_1 = c_1. \tag{8.12}$$

   Note that this fact is immediately apparent, without any differentiations as was the case when deriving the normal equations. This is because the SVD reveals so much about the structure of the underlying problem.
3. From (8.9), we have $\beta = Vw$. Therefore,

$$||\beta||_2^2 = ||w||_2^2 = ||w_1||_2^2 + ||w_2||_2^2.$$

   Since w_2 is arbitrary, we can minimise $||\beta||_2^2$ by setting $w_2 = \mathbf{0}$.
4. From (8.12), we have $w_1 = \Sigma_r^{-1} c_1$, where the inverse exists because Σ_r consists only of the nonzero singular values. Combining our definitions for w_1 and w_2 together, we have

$$\begin{aligned} w = \begin{bmatrix} w_1 \\ w_2 \end{bmatrix} = \begin{bmatrix} \Sigma_r^{-1} c_1 \\ \mathbf{0} \end{bmatrix} &= \begin{bmatrix} \Sigma_r^{-1} & \mathbf{0} \\ \mathbf{0} & \mathbf{0} \end{bmatrix} c \\ &= \Sigma^+ c. \end{aligned} \tag{8.13}$$

Using (8.9) and (8.10), this can be written as

$$V^T \beta_{LS} = \Sigma^+ U^T y$$

or

$$\begin{aligned} \beta_{LS} &= V \Sigma^+ U^T y \\ &= X^+ y, \end{aligned} \tag{8.14}$$

which was to be shown. Furthermore, we can say from (8.11) that

$$\begin{aligned} r_{LS}^2 &= ||c_2||_2^2 \\ &= \left|\left|U_2^T y\right|\right|_2^2 . \end{aligned}$$

□

Note that X^+ is always defined even if X is singular.

8.2 Interpretation of the Pseudo-Inverse

8.2.1 Geometrical Interpretation

In the pseudo-inverse case, β_{LS} is again the solution which corresponds to projecting y onto $R(\beta)$. For the case where X is tall, we substitute (8.14) into the expression $X\beta_{LS}$ to get

$$X\beta_{LS} = XX^+ y. \tag{8.15}$$

But, for the specific case where $m > n$, we know from our previous discussion on linear least squares that

$$X\beta_{LS} = Py, \tag{8.16}$$

where P is the projector onto $R(X)$, which still exists in the rank-deficient case. Comparing (8.15) and (8.16) and noting the projector is unique, we have

$$P = XX^+. \tag{8.17}$$

Thus, the matrix XX^+ is a projector onto $R(X)$.

This may also be seen in a different way as follows: Using the definition of $\boldsymbol{X}^+$, we have

$$\begin{aligned}\boldsymbol{X}\boldsymbol{X}^+ &= \boldsymbol{U}\boldsymbol{\Sigma}\boldsymbol{V}^T\boldsymbol{V}\boldsymbol{\Sigma}^+\boldsymbol{U}^T \\ &= \boldsymbol{U}\begin{pmatrix}\boldsymbol{I}_r & \boldsymbol{0} \\ \boldsymbol{0} & \boldsymbol{0}\end{pmatrix}\boldsymbol{U}^T \\ &= \boldsymbol{U}_1\boldsymbol{U}_1^T,\end{aligned} \tag{8.18}$$

where $\boldsymbol{I}_r$ is the $r \times r$ identity and $\boldsymbol{U}_1 = [\boldsymbol{u}_1, \ldots, \boldsymbol{u}_r]$. In the above, care must be taken to ensure the zero blocks are arranged appropriately. From our discussion on projectors, we know $\boldsymbol{U}_1\boldsymbol{U}_1^T$ is also a projector onto $R(\boldsymbol{X})$ which is the same as the *column space* of $\boldsymbol{X}$.

We also note that it is just as easy to show that for the case $m < n$ where $\boldsymbol{X}$ is short, the matrix $\boldsymbol{X}^+\boldsymbol{X}$ is a projector onto the *row space* of $\boldsymbol{X}$.

8.2.2 Relationship of the Pseudo-Inverse Solution to the Normal Equations

Suppose $\boldsymbol{X} \in \mathbb{R}^{m\times n}$ $m > n$ is full rank. The normal equations give us

$$\boldsymbol{\beta}_{LS} = (\boldsymbol{X}^T\boldsymbol{X})^{-1}\boldsymbol{X}^T\boldsymbol{y},$$

but the pseudo-inverse gives

$$\boldsymbol{\beta}_{LS} = \boldsymbol{X}^+\boldsymbol{y}.$$

In the full-rank case, these two quantities must be equal. We can indeed show this is the case, as follows: We let

$$\boldsymbol{X}^T\boldsymbol{X} = \boldsymbol{V}\boldsymbol{\Sigma}^2\boldsymbol{V}^T$$

be the ED of $\boldsymbol{X}^T\boldsymbol{X}$, and we let the SVD of $\boldsymbol{X}^T$ be defined as

$$\boldsymbol{X}^T = \boldsymbol{V}\boldsymbol{\Sigma}\boldsymbol{U}^T.$$

Using these relations, we have

$$\begin{aligned}(\boldsymbol{X}^T\boldsymbol{X})^{-1}\boldsymbol{X}^T &= (\boldsymbol{V}\boldsymbol{\Sigma}^{-2}\boldsymbol{V}^T)\boldsymbol{V}\boldsymbol{\Sigma}\boldsymbol{U}^T \\ &= \boldsymbol{V}\boldsymbol{\Sigma}^{-1}\boldsymbol{U}^T \\ &= \boldsymbol{X}^+\end{aligned} \tag{8.19}$$

as desired, where the last line follows from (8.14). Thus, for the full-rank case for $m > n$, $\boldsymbol{X}^+ = (\boldsymbol{X}^T\boldsymbol{X})^{-1}\boldsymbol{X}^T$. In a similar way, we can also show that $\boldsymbol{X}^+ = \boldsymbol{X}^T(\boldsymbol{X}\boldsymbol{X}^T)^{-1}$ for the full-rank case when $\boldsymbol{X}$ is short, i.e. when $m < n$.

8.2.3 The Pseudo-Inverse as a Generalised Linear System Solver

It is generally understood that there is no solution to an overdetermined system of equations (except when $\boldsymbol{y} \in R(\boldsymbol{\beta})$). Likewise, the solution is not unique when the system is underdetermined, or $\boldsymbol{\beta}$ is rank deficient. However, if we are willing to accept the least squares solution in the overdetermined case and accept the definition of a unique solution as that for which $||\boldsymbol{\beta}||_2$ is minimum, then $\boldsymbol{\beta} = \boldsymbol{X}^+\boldsymbol{y}$ solves the system $\boldsymbol{X}\boldsymbol{\beta} = \boldsymbol{y}$ under any conditions.

8.3 The Pseudo-Inverse (PCA) Approach for Solving Poorly Conditioned Problems

In this section, we investigate the pseudo-inverse solution as a means of addressing the LS problem when $\boldsymbol{X}$ is poorly conditioned, rather than being completely rank deficient as we have discussed previously. Even though it is not directly apparent, there is a direct link between the pseudo-inverse and PCA in solving least squares problems [4]. We have studied PCA in some depth in Chap. 2. In this chapter we first introduce the pseudo-inverse as a means of addressing the poorly conditioned LS problem. Later, we apply PCA for the same purpose, and then we show that these two methods are in fact identical, even though their derivations are quite different. Because of this fact, in the following we refer to the pseudo-inverse solution to the LS problem as the PCA solution.

We have seen previously in Chap. 7 that the covariance matrix $\mathrm{cov}(\boldsymbol{\beta}_{LS})$ of the estimates $\boldsymbol{\beta}_{LS}$ obtained by the ordinary normal equations in the white noise case is given by the expression

$$\mathrm{cov}(\boldsymbol{\beta}_{LS}) = \sigma^2(\boldsymbol{X}^T\boldsymbol{X})^{-1}.$$

If $\boldsymbol{X}^T\boldsymbol{X}$ is poorly conditioned, then it has at least one small eigenvalue and is therefore close to rank deficiency. We have seen in this case that the variances of $\boldsymbol{\beta}_{LS}$ become large. In this section, we show that use of the pseudo-inverse instead of the ordinary normal equations is very effective in restoring the variances to reasonable values, even though $\boldsymbol{X}$ may not be completely rank deficient. This is accomplished at the cost, as we will soon see, of a biased solution.

We denote the principal component LS solution as $\boldsymbol{\beta}_{PC}$. In solving for $\boldsymbol{\beta}_{PC}$, we replace the ordinary normal equation solution with the pseudo-inverse solution, even though the smaller singular values are not exactly zero. That is, if $\boldsymbol{\Sigma}$ in the SVD of $\boldsymbol{X}$ is given as $\boldsymbol{\Sigma} = \text{diag}[\sigma_1, \sigma_2, \ldots, \sigma_r, \sigma_{r+1} \ldots, \sigma_n]$, then $\boldsymbol{\beta}_{PC}$ in this case is given as

$$\boldsymbol{\beta}_{PC} = \boldsymbol{V}\boldsymbol{\Sigma}_r^+\boldsymbol{U}^T\boldsymbol{y}, \tag{8.20}$$

where $\boldsymbol{\Sigma}_r^+$ is given as

$$\boldsymbol{\Sigma}_r^+ = \begin{bmatrix} \frac{1}{\sigma_1} & & & & & & \\ & \frac{1}{\sigma_2} & & & & & \\ & & \ddots & & & & \\ & & & \frac{1}{\sigma_r} & & & \\ & & & & 0 & & \\ & & & & & \ddots & \\ & & & & & & 0 \end{bmatrix},$$

where $r \leq n$ and $\sigma_{r+1} \ldots, \sigma_n$ are assumed small enough to cause trouble and are therefore truncated. In practice, the value of r is usually determined empirically by trial-and-error methods, cross-validation, or through the use of some form of prior knowledge. Further discussion on this topic is given in the PCA section of Chap. 2.

The only difficulty with this principal component approach is that it introduces a bias in $\boldsymbol{\beta}_{PC}$, whereas we have seen previously that the ordinary normal equation $\boldsymbol{\beta}_{LS}$ is unbiased. To see this biasedness, we let the singular value decomposition of $\boldsymbol{X}$ be expressed as $\boldsymbol{X} = \boldsymbol{U}\boldsymbol{\Sigma}\boldsymbol{V}^T$ and write

$$\begin{aligned} \boldsymbol{\beta}_{PC} &= \boldsymbol{X}^+\boldsymbol{y} \\ &= \boldsymbol{V}\boldsymbol{\Sigma}_r^+\boldsymbol{U}^T(\boldsymbol{X}\boldsymbol{\beta}_o + \boldsymbol{\epsilon}). \end{aligned} \tag{8.21}$$

Thus, under assumptions A1 and A2 of Chap. 7, the expected value of $\boldsymbol{\beta}_{PC}$ may be expressed as

$$E(\boldsymbol{\beta}_{PC}) = \boldsymbol{V}\boldsymbol{\Sigma}_r^+\boldsymbol{U}^T(\boldsymbol{X}\boldsymbol{\beta}_o) \tag{8.22}$$

$$\begin{aligned} &= \boldsymbol{V}\boldsymbol{\Sigma}_r^+\boldsymbol{U}^T(\boldsymbol{U}\boldsymbol{\Sigma}\boldsymbol{V}^T\boldsymbol{\beta}_o) \\ &= \boldsymbol{V}\boldsymbol{\Sigma}_r^+\boldsymbol{\Sigma}\boldsymbol{V}^T\boldsymbol{\beta}_o \\ &= \boldsymbol{V}\begin{bmatrix} \boldsymbol{I}_r & \boldsymbol{0} \\ \boldsymbol{0} & \boldsymbol{0} \end{bmatrix}\boldsymbol{V}^T\boldsymbol{\beta}_o \end{aligned}$$

$$= \boldsymbol{V}_1\boldsymbol{V}_1^T\boldsymbol{\beta}_o = \boldsymbol{P}\boldsymbol{\beta}_o \tag{8.23}$$

$$\neq \boldsymbol{\beta}_o,$$

where $V_1 = V(:, 1:r)$ is an orthonormal basis for the row space of X and P is the projector onto this space. Hence the estimate $\boldsymbol{\beta}_{PC}$ obtained from the pseudo-inverse is in general biased. However, we see that if $\boldsymbol{\beta}_o \in R(V_1)$, then $\boldsymbol{\beta}_{PC}$ is unbiased. But if $\boldsymbol{\beta}_o$ contains significant components in $R(V_r)^{\perp}$, then $\boldsymbol{\beta}_{PC}$ could be strongly biased. Note however that in the practical scenario, if indeed $\boldsymbol{\beta}_o \in R(V_1)^{\perp}$, then from (8.23) $E(\boldsymbol{\beta}_{PC}) = \mathbf{0}$, and hence no meaningful prediction of $\boldsymbol{y}$ is possible and the respective choice of X is inappropriate for the LS model in this case. Therefore if the LS model is to be effective, $\boldsymbol{\beta}_o$ must be close to $R(V_1)$, and we may consider the estimate $\boldsymbol{\beta}_{PC}$ to be approximately unbiased in this case.

We now look at the covariance matrix of $\boldsymbol{\beta}_{PC}$. Similar to the treatment of Sect. 7.4.2, we have

$$\mathrm{cov}(\boldsymbol{\beta}_{PC}) = E(\boldsymbol{\beta}_{PC} - E(\boldsymbol{\beta}_{PC}))(\boldsymbol{\beta}_{PC} - E(\boldsymbol{\beta}_{PC}))^T. \tag{8.24}$$

Substituting the first line of (8.22) for $E(\boldsymbol{\beta}_{PC})$, using (8.21) for $\boldsymbol{\beta}_{PC}$, simplifying, and assuming that $E(\boldsymbol{\epsilon}\boldsymbol{\epsilon}^T) = \sigma^2 \boldsymbol{I}$, we get

$$\begin{aligned}\mathrm{cov}(\boldsymbol{\beta}_{PC}) &= E(\boldsymbol{V}\boldsymbol{\Sigma}_r^+ \boldsymbol{U}^T \boldsymbol{\epsilon}\boldsymbol{\epsilon}^T \boldsymbol{U}\boldsymbol{\Sigma}_r^+ \boldsymbol{V}^T \\ &= \sigma^2 \boldsymbol{V}\boldsymbol{\Sigma}_r^+ \boldsymbol{U}^T \boldsymbol{I} \boldsymbol{U}\boldsymbol{\Sigma}_r^+ \boldsymbol{V}^T \\ &= \sigma^2 \boldsymbol{V}(\boldsymbol{\Sigma}_r^+)^2 \boldsymbol{V}^T. \end{aligned} \tag{8.25}$$

This expression for covariance is similar to that for $\boldsymbol{\beta}_{LS}$ from (7.38), except that it excludes the inverses of the smallest singular values and the corresponding directions which have large variation. Thus, the elements of $\mathrm{cov}(\boldsymbol{\beta}_{PC})$ can be significantly smaller than those for $\boldsymbol{\beta}_{LS}$, as desired.

Thus, we see that principal component analysis (PCA) is a trade-off between reduced variance on the one hand and increased bias on the other. The objective of any estimation problem is to reduce the *overall*, or mean–squared error, which is a combination of both bias and variance, to a minimum. In fact, it has been shown in Chap. 2 that the total mean-squared error $E(||\hat{\boldsymbol{\beta}} - \boldsymbol{\beta}_o||_2^2)$ of an estimate $\hat{\boldsymbol{\beta}}$ of a quantity whose true value is $\boldsymbol{\beta}_o$ is given by (7.33), reproduced here for convenience:

$$\mathrm{mse}(\hat{\boldsymbol{\beta}}) = \boldsymbol{\sigma}^{\cdot 2} + \boldsymbol{b}^{\cdot 2},$$

where the notation $(\cdot)^{\cdot 2}$ refers to element-by-element squaring of each element of the respective vector, and where $\boldsymbol{b} = E(\hat{\boldsymbol{\beta}} - \boldsymbol{\beta}_o)$ is the vector of biases and $\boldsymbol{\sigma}^2$ is the vector of variances of the estimate. If X is poorly enough conditioned, then the improvement in the variance of $\boldsymbol{\beta}_{PC}$ over that of $\boldsymbol{\beta}_{LS}$ is large, whereas the bias introduced can be small, so the overall effect of PCA can be positive. However, as X becomes better conditioned, then the improvement in variance is reduced and the bias may still exist, and the technique becomes less favourable.

The choice of the parameter r controls the trade-off between bias and variance. The smaller the value of r, the fewer the number of components in X^+; hence, the lower the variance and the higher the bias. The choice of r in the present context is closely related to the choice of r in the PCA context, as discussed in Chap. 2.

Simulation Example We show by simulation how the pseudo-inverse solution $\boldsymbol{\beta}_{PC}$ can improve the variances of the estimates. We consider two different scenarios with regard to the true value $\boldsymbol{\beta}_o$. For both cases, a 5×3 matrix $\boldsymbol{X}$ is given as

$$\boldsymbol{X} = \begin{bmatrix} 2.0000 & 1.0000 & 1.4738 \\ 4.0000 & 1.0000 & 2.4913 \\ 6.0000 & 1.0000 & 3.5069 \\ 8.0000 & 1.0000 & 4.5716 \\ 10.0000 & 1.0000 & 5.5554 \end{bmatrix}.$$

Note that the third column of $\boldsymbol{X}$ is almost equal to the arithmetic average of the first two columns, which will make the matrix poorly conditioned with one small singular value. The singular values of $\boldsymbol{X}$ are 17.1830, 1.0040, and 0.0142. As such, we choose the value $r = 2$. In this case, the matrix $\boldsymbol{V}_r$ in (8.23) consists of the first two columns of $\boldsymbol{V}$ from the SVD of $\boldsymbol{X}$. In each simulation scenario, $N = 500$ observations $\boldsymbol{y}$ were generated using the regression equation $\boldsymbol{y} = \boldsymbol{X}\boldsymbol{\beta}_o + \boldsymbol{\epsilon}$, where in each observation, $\boldsymbol{\epsilon}$ is an independently distributed Gaussian random vector with zero mean and covariance $\sigma^2\boldsymbol{I}$ with $\sigma = 0.1$. The corresponding value $\boldsymbol{\beta}_i$ is calculated in each iteration, $i = 1, \ldots, 500$, for both simulation scenarios.

For the first scenario, we choose $\boldsymbol{\beta}_o = [1 \quad 1 \quad 1]^T$. This value is very close to $R(\boldsymbol{V}_r)$—in fact, the relative error $\frac{||\boldsymbol{\beta}_o - \boldsymbol{P}\boldsymbol{\beta}_o||_2^2}{||\boldsymbol{\beta}_o||_2^2}$ between the true value and the projected value in this case has the value 0.0189, which may be considered small. Thus according to our previous analysis, this is a favourable situation, and as such the bias in the PCA estimate $\boldsymbol{\beta}_{PC}$ should also be small.

Figure 8.1 shows the scatter plot of the first and second elements of the estimates $\boldsymbol{\beta}_{LS}$ (red) and $\boldsymbol{\beta}_{PC}$ (blue) corresponding to the ordinary normal equation and pseudo-inverse solution, respectively, obtained from the $N = 500$ observations of $\boldsymbol{y}$. In this case, we see a dramatic contraction of the scatter diagram for $\boldsymbol{\beta}_{PC}$ compared to that for $\boldsymbol{\beta}_{LS}$, indicating that the variances have drastically reduced. To illustrate the point further, the covariance matrices in each case were calculated, using

$$\text{cov}(\boldsymbol{\beta}) = \frac{1}{N}\sum_{i=1}^{N}\left(\boldsymbol{\beta}_i - E(\boldsymbol{\beta})\right)\left(\boldsymbol{\beta}_i - E(\boldsymbol{\beta})\right)^T,$$

where $N = 500$ and $\boldsymbol{\beta}$ is either the LS or PC estimate. The result for the normal equation case is

$$\text{cov}(\boldsymbol{\beta}_{LS}) = \begin{bmatrix} 2.0915 & 1.8141 & -4.0824 \\ 1.8141 & 1.5880 & -3.5445 \\ -4.0824 & -3.5445 & 7.9695 \end{bmatrix},$$

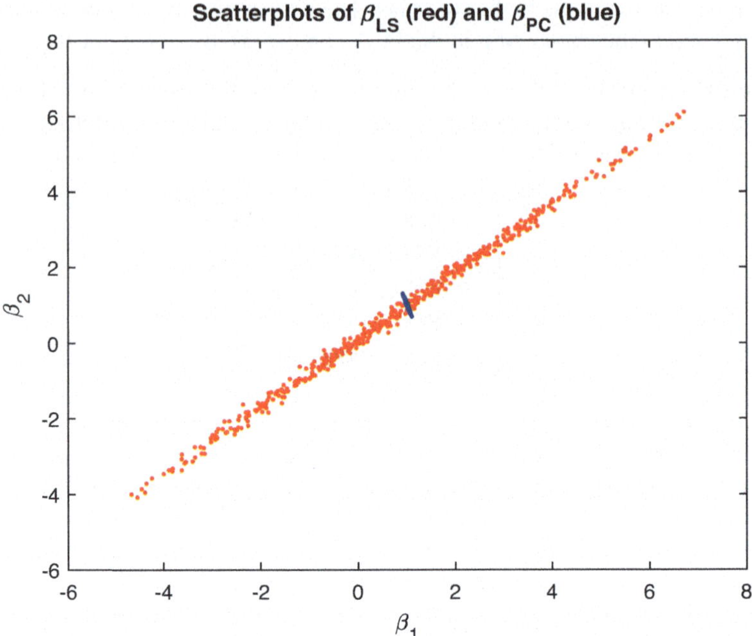

Fig. 8.1 Scatter plots for the first two dimensions β_1 and β_2 for the simulation example using both the normal equation solution $\boldsymbol{\beta}_{LS}$ (in red) and the principal component solution $\boldsymbol{\beta}_{PC}$ (in blue) when $\boldsymbol{X}$ is poorly conditioned and when $\boldsymbol{\beta}$ is very close to $R(\boldsymbol{V}_r)$. The reduction in variance for the PC case is strongly evident

whereas that for the pseudo-inverse case is

$$\operatorname{cov}(\boldsymbol{\beta}_{PC}) = \begin{bmatrix} 0.0011 & -0.0026 & -0.0012 \\ -0.0026 & 0.0093 & 0.0023 \\ -0.0012 & 0.0023 & 0.0016 \end{bmatrix}.$$

The means from the simulation for $\boldsymbol{\beta}_{LS}$ and $\boldsymbol{\beta}_{PC}$ are given, respectively, by

$$\begin{bmatrix} 1.0126 \\ 1.0102 \\ 0.9757 \end{bmatrix}, \text{ and } \begin{bmatrix} 1.0145 \\ 1.0119 \\ 0.9720 \end{bmatrix},$$

which are both close to the true values, as expected. Thus, we see that the pseudo-inverse technique has significantly improved the covariance structure in this case where $\boldsymbol{X}$ is poorly conditioned. By inspection, we also observe that the bias in the ordinary LS estimate of the means is approximately equivalent to that of the PC estimate. Thus it appears that in this example, the bias in the PC estimate may be considered negligible, especially in view of the significant reduction in variance.

Table 8.1 Simulation results showing the means of the normal equation (LS) and pseudo-inverse (PC) estimates of $\boldsymbol{\beta}$, along with the true value $\boldsymbol{\beta}_o$ shown in the first column. These results correspond to the less favourable case for the pseudo-inverse method, where $\boldsymbol{\beta}_o$ is chosen randomly

	β_o	β_{LS}	β_{PC}
$\beta(1)$	0.2329	0.2719	−0.1959
$\beta(2)$	0.6540	0.6856	0.2274
$\beta(3)$	−0.8450	−0.9219	0.0134

We now consider the second simulation example, where $\boldsymbol{\beta}_o$ is chosen randomly using the "randn" command in Matlab®. In this case $\boldsymbol{\beta}_o$ is not likely to lie entirely in the row space of $\boldsymbol{x}$, and so from (8.23) we expect the pseudo-inverse estimates to suffer some performance degradation due to bias. The simulation results confirm these theoretical predictions. Table 8.1 shows the means of the estimates $\boldsymbol{\beta}_{LS}$ and $\boldsymbol{\beta}_{PC}$ over the 500 trials, along with the true $\boldsymbol{\beta}_o$. It is seen the mean of the $\boldsymbol{\beta}_{LS}$ estimates has some degree of error, whereas the mean of $\boldsymbol{\beta}_{PC}$ is significantly biased. The covariance matrix corresponding to $\boldsymbol{\beta}_{PC}$ is still greatly contracted compared to that of $\boldsymbol{\beta}_{LS}$, as it was in the previous case.

8.4 The Rank-Deficient LS Problem Using the QR Decomposition

We have discussed the rank-deficient QR decomposition at some length in Sect. 6.5, where the final form of the decomposition may be expressed as (6.28). Given $\boldsymbol{X} \in \mathbb{R}^{m\times n}$, $m > n$, $\text{rank}(\boldsymbol{X}) = r < n$, $\boldsymbol{y} \in \mathbb{R}^n$, then

$$||\boldsymbol{X\beta} - \boldsymbol{y}||_2^2 = \left|\left|(\boldsymbol{Q}^T\boldsymbol{X\Pi})\boldsymbol{\Pi}^T\boldsymbol{\beta} - \boldsymbol{Q}^T\boldsymbol{y}\right|\right|_2^2, \tag{8.26}$$

where $\boldsymbol{\Pi}$ is a permutation matrix as discussed in Sect. 6.5. We wish to determine the $\boldsymbol{\beta}$ that solves the above minimisation problem. Recall that in the rank-deficient case there is no unique solution for (8.26). Hence, unless an extra constraint is imposed on $\boldsymbol{\beta}$, the LS solution obtained by a particular algorithm can wander throughout the infinite set of possible solutions which minimise the quantity $||\boldsymbol{X\beta} - \boldsymbol{y}||_2^2$, resulting in very large variances. As in the pseudo-inverse case, the constraint of minimum norm is a convenient one to apply in this case, in order to specify a unique solution. However, unlike the development of the pseudo-inverse solution, we will see that the direct use of the QR decomposition does not lead directly to the minimum norm solution $\boldsymbol{\beta}_{LS}$. However, it is still possible to derive an elegant solution to the rank-deficient LS problem, as discussed in [3], using only the QR decomposition procedure. We now discuss how this is achieved.

Let

$$\boldsymbol{\Pi}^T\boldsymbol{\beta} = \begin{bmatrix} \boldsymbol{w} \\ \boldsymbol{z} \end{bmatrix} \begin{matrix} r \\ n-r \end{matrix}. \tag{8.27}$$

Substituting (6.28), (8.27), and (7.75) into (8.26), we have

$$||\boldsymbol{X}\boldsymbol{\beta} - \boldsymbol{w}||_2^2 = \left|\left| \begin{bmatrix} \boldsymbol{R}_{11} & \boldsymbol{R}_{12} \\ \boldsymbol{0} & \boldsymbol{0} \end{bmatrix} \begin{bmatrix} \boldsymbol{w} \\ \boldsymbol{z} \end{bmatrix} - \begin{bmatrix} \boldsymbol{c} \\ \boldsymbol{d} \end{bmatrix} \right|\right|_2^2.$$

The minimum residual of norm $||\boldsymbol{d}||_2$ is obtained when

$$\boldsymbol{R}_{11}\boldsymbol{w} + \boldsymbol{R}_{12}\boldsymbol{z} = \boldsymbol{c}.$$

By solving for $\boldsymbol{w}$ above and substituting into (8.27), we have

$$\boldsymbol{\Pi}^T\boldsymbol{\beta} = \begin{bmatrix} \boldsymbol{R}_{11}^{-1}(\boldsymbol{c} - \boldsymbol{R}_{12}\boldsymbol{z}) \\ \boldsymbol{z} \end{bmatrix}. \tag{8.28}$$

We see from above that the vector $\boldsymbol{z}$ is arbitrary. It may seem that a reasonable approach to determine the $\boldsymbol{\beta}_{LS}$ with minimum norm is to choose $\boldsymbol{z} = 0$. However, it is shown [3] that this is not the case unless $\boldsymbol{R}_{12} = \boldsymbol{0}$.

We therefore seek a more efficient means of determining $\boldsymbol{\beta}_{LS}$ in this case. We note that the desired solution $\boldsymbol{\beta}_{LS}$ to (8.26) is that solution which minimises $||\boldsymbol{X}\boldsymbol{\beta} - \boldsymbol{y}||_2$ with respect to $\boldsymbol{\beta}$ and simultaneously minimises $||\boldsymbol{\beta}||_2$. Hence, this solution would possess the same properties as the pseudo-inverse solution, and because of uniqueness, this solution would be identical to the pseudo-inverse solution.

To address this goal, we consider the complete orthogonal decomposition (COD). Consider the rank-deficient QR form given by (6.28), reproduced here for convenience:

$$\boldsymbol{Q}^T\boldsymbol{X}\boldsymbol{\Pi} = \begin{matrix} r & n-r \end{matrix} \atop \begin{bmatrix} \boldsymbol{R}_{11} & \boldsymbol{R}_{12} \\ \boldsymbol{0} & \boldsymbol{0} \end{bmatrix} \begin{matrix} r \\ m-r. \end{matrix}$$

The idea is to eliminate $\boldsymbol{R}_{12}$. Then finding the $\boldsymbol{\beta}_{LS}$ with minimum norm is straightforward. There exists an orthonormal $\boldsymbol{Z} \in \mathbb{R}^{n\times n}$ such that

$$\begin{bmatrix} \boldsymbol{R}_{11} & \boldsymbol{R}_{12} \\ \boldsymbol{0} & \boldsymbol{0} \end{bmatrix} \boldsymbol{Z} = \begin{bmatrix} \boldsymbol{T}_{11} & \boldsymbol{0} \\ \boldsymbol{0} & \boldsymbol{0} \end{bmatrix}, \tag{8.29}$$

where T_{11} is nonsingular and upper triangular of dimension $r \times r$. Therefore,

$$Q^T X \Pi Z = \begin{bmatrix} T_{11} & \mathbf{0} \\ \mathbf{0} & \mathbf{0} \end{bmatrix}. \tag{8.30}$$

Equation (8.30) is called the *complete orthogonal decomposition* of the matrix A.

The fact that an orthonormal matrix Z can exist may be understood by taking the transpose of both sides of (8.29). Then (8.29) becomes an ordinary QR decomposition on $\begin{bmatrix} R_{11}^T \\ R_{12}^T \end{bmatrix}$, with the exception that the result is T_{11}^T, which is lower triangular instead of upper triangular, as expected. However, it is easy to modify the ordinary QR decomposition procedure to yield a lower instead of an upper triangular matrix.

Now solving the LS problem is straightforward:

$$\begin{aligned} ||X\beta - y||_2^2 &= \left|\left|(Q^T X \Pi Z)(Z^T \Pi^T \beta) - Q^T y\right|\right|_2^2 \\ &= \left|\left|\begin{pmatrix} T_{11} & \mathbf{0} \\ \mathbf{0} & \mathbf{0} \end{pmatrix}\begin{pmatrix} p \\ q \end{pmatrix} - \begin{pmatrix} c \\ d \end{pmatrix}\right|\right|_2^2, \end{aligned} \tag{8.31}$$

where

$$Z^T \Pi^T \beta = \begin{pmatrix} p \\ q \end{pmatrix} \begin{matrix} r \\ n-r \end{matrix},$$

and c, d are defined in (7.75) as before. Clearly, q is arbitrary, and d is independent of both p and q. We can write (8.31) in the form

$$||X\beta - y||_2^2 = ||Tp - c||_2^2 + ||d||_2^2,$$

which is minimum when $p = T^{-1}c$. We also have

$$\beta_{LS} = \Pi Z \begin{pmatrix} p \\ q \end{pmatrix}, \tag{8.32}$$

which clearly has minimum norm when $q = \mathbf{0}$.

The β_{LS} calculated in this way is identical to the pseudo-inverse solution. However, the computational cost with the COD is significantly less. The COD requires only two QR decompositions; the SVD is computed using an iterative procedure involving one QR decomposition per iteration.

8.5 Latent Variable Methods

We now discuss the so-called latent variable (LV) methods [1, 2, 5–7], which are prevalent in the statistical community, for solving poorly conditioned LS problems. Here our objective for model building is to identify the linear model that gives the best fit to the observed data as we have done before and predict response values $\hat{\boldsymbol{y}}^T$ corresponding to new values $\boldsymbol{x}_N^T$ in the form of a new row of $\boldsymbol{X}$. We investigate three types of latent variable methods, which are PCA (revisited), partial least squares (PLS), and canonical correlation analysis (CCA).

As before, we are given a set of independent input variables $\boldsymbol{X} \in \mathbb{R}^{m \times n}$. However in this section we generalise the LS problem to include a set of k multiple response values denoted by $\boldsymbol{Y} \in \mathbb{R}^{m \times k}$, which is in contrast to previous use where we had only one response $\boldsymbol{y}$ of observations. The respective regression equation is now expressed as

$$\boldsymbol{Y} = \boldsymbol{X}\boldsymbol{\beta} + \boldsymbol{E}, \tag{8.33}$$

where $\boldsymbol{\beta} \in \mathbb{R}^{n \times k}$ is the LS solution and $\boldsymbol{E}$ is the error matrix, which is the same size as $\boldsymbol{Y}$. Equation (8.33) is again referred to as the regression equation, where we adopt the terminology that $\boldsymbol{Y}$ is *regressed* onto $\boldsymbol{X}$.

Latent variable methods are founded on the same idea as the pseudo-inverse, i.e. that $\boldsymbol{X}$ (and often $\boldsymbol{Y}$) are close to rank deficiency with an effective rank of r, where often we have $r \ll \min(n, k)$. As with the pseudo-inverse case, we solve the LS problem by eliminating the components of $\boldsymbol{X}$ and $\boldsymbol{Y}$ associated with the smaller singular values that give us trouble, thus alleviating the conditioning problem which results in large variances of the parameter estimates, as we have seen previously. "Latent" in this context implies "unseen", and the latent variables in this context are actually basis vectors which are used to represent $\boldsymbol{X}$ and/or $\boldsymbol{Y}$. In the PCA case, the latent variables are the principal eigenvectors of $\boldsymbol{X}^T\boldsymbol{X}$; however, for the *partial least squares* (PLS) and *canonical correlation analysis* (CCA) methods which we discuss later, the latent variables are functions of both $\boldsymbol{X}$ and $\boldsymbol{Y}$. Since r is small relative to n or k, the latent variable basis is referred to as being "incomplete"; that is, $\boldsymbol{X}$ or $\boldsymbol{Y}$ cannot be represented in the LV basis without error. However, by careful choice of the LV basis, this error can be controlled, while at the same time the error in $\boldsymbol{\beta}$ or in the prediction of $\boldsymbol{Y}$ values corresponding to new values of $\boldsymbol{X}$ can be considerably reduced.

One of the objectives of this section is to *predict* or estimate the unknown response $\hat{\boldsymbol{y}}^T \in \mathbb{R}^k$ corresponding to a new, previously unseen set of input variables $\boldsymbol{x}_N^T \in \mathbb{R}^n$ in the form of a new row of $\boldsymbol{X}$. The predicted value of the corresponding vector response $\hat{\boldsymbol{y}}$ is given as $\hat{\boldsymbol{y}} = \boldsymbol{x}_N^T\boldsymbol{\beta}$, which is the expected value of $\boldsymbol{y}$. It is straightforward to show that the variance $\sigma_{\hat{y}}^2$ of each element $\hat{y}_i$ of $\hat{\boldsymbol{y}}^T$ is equal to the quantity σ_y^2 in the case where all elements of $\boldsymbol{E}$ from (8.33) are independent, identically distributed (*iid*). Then $\sigma_{\hat{y}}^2$ is given from (7.40) (repeated here for convenience) as

$$\sigma_y^2 = \sigma^2 x_N^T (X^T X)^{-1} x_N.$$

Notice that this form and the expression for the variance of $\boldsymbol{\beta}_{LS}$ in (7.38) both involve the term $(X^T X)^{-1}$. Thus, as X becomes poorly conditioned, both the above forms of variance degrade. The objective of LV methods is to mitigate this effect by choosing an appropriate r-dimensional subspace which eliminates as many noise components as possible.

In most applications of latent variable methods, it is common practice to normalise the data before the modelling process begins. In many cases, the variables involved may have significantly different scales, or a difference in units, e.g. some variables may be measured in microvolts and others in degrees Celsius. Also the variables may have significantly different means. To alleviate these disparities, each variable (column) is typically converted to their corresponding *z-score*, where all values in the jth column are subjected to the transformation

$$x_{ij} \leftarrow \frac{x_{ij}-\mu_j}{\sigma_j}, \quad i = 1, \ldots, m,$$

where μ_j and σ_j are the mean and standard deviation over the jth column of X. Thus each variable is transformed so that it has zero mean and unit standard deviation.

8.6 Principal Component Analysis Applied to Poorly Conditioned LS Problems

We have already developed the PCA approach in Chap. 2 from the perspective of data compression, denoising signals, and classification. Here we present PCA in the latent variable context for solving poorly conditioned LS problems.

The main idea with PCA in an LS context is to construct a rank r approximation X_r to the LS data matrix X such that X_r is as close as possible to X in the Frobenius norm sense. In this way we eliminate the small components of X that give rise to large variances in $\boldsymbol{\beta}$, as we did with the pseudo-inverse earlier in this chapter, while at the same time perturbing the original regression model as little as possible. As a matter of housekeeping, we define the SVD of X as $X = U\Sigma V^T$, where we partition U and V as $U = [U_1 \; U_2]$ and V as $[V_1 \; V_2]$, where U_1 and V_1 contain the first r columns of their respective matrices. We also define the quantity Σ_r as that diagonal matrix whose diagonal entries are $[\sigma_1, \ldots, \sigma_r]$.

We choose X_r so that it is the projection of the quantity X into an optimal subspace such that the quantity $||X - X_r||_2^2$ is minimised. We have already dealt with this problem before in the PCA context in Sect. 2.5 and specifically with respect to the Approximation Theorem given by Theorem 3.3 in Sect. 3.3. As we have seen, the idea is to choose a basis for X_r such that its Frobenius norm is maximum. The result of this treatment is that the optimal matrix X_r is given by any of the three forms below, corresponding to (3.33), (3.36), or (3.40), that have been shown to be equivalent in Sect. 3.3:

$$\begin{aligned} \boldsymbol{X}_r &= \boldsymbol{U}_1 \boldsymbol{\Sigma}_r \boldsymbol{V}_1^T && (8.34)\\ &= \boldsymbol{U}_1 \boldsymbol{U}_1^T \boldsymbol{X} && (8.35)\\ &= \boldsymbol{X} \boldsymbol{V}_1 \boldsymbol{V}_1^T. && (8.36) \end{aligned}$$

We can now determine the LS parameter $\boldsymbol{\beta}_{pca}$. Since $\boldsymbol{X}_r$ is rank deficient, we estimate the PCA version $\boldsymbol{\beta}_{pca}$ of $\boldsymbol{\beta}$ using the pseudo-inverse of $\boldsymbol{X}_r$:

$$\boldsymbol{\beta}_{pca} = \boldsymbol{X}_r^+ \boldsymbol{Y}. \tag{8.37}$$

We can also determine an estimated response $\hat{\boldsymbol{y}}^T \in \mathbb{R}^{1\times k}$ corresponding to a new set of variables $\boldsymbol{x}_N^T \in \mathbb{R}^{1\times n}$ in the form of a new row of $\boldsymbol{X}$. Given our estimate of $\boldsymbol{\beta}_{pca}$, we solve for $\hat{\boldsymbol{y}}^T$ by taking the expectation of the regression equation corresponding to the new row of $\boldsymbol{X}$:

$$\hat{\boldsymbol{y}}^T = \boldsymbol{x}_N^T \boldsymbol{\beta}_{pca}. \tag{8.38}$$

It may be seen through (8.35) or (8.36) that either $\boldsymbol{U}_1$ or $\boldsymbol{V}_1$ serve as the latent variables for this PCA version of least squares analysis. We see that (8.35) forms $\boldsymbol{X}_r$ by projecting columns of $\boldsymbol{X}$ onto the basis formed by $\boldsymbol{U}_1$, whereas (8.36) projects rows of $\boldsymbol{X}$ onto the basis formed by $\boldsymbol{V}_1$. Further, if the pseudo-inverse $\boldsymbol{X}_r^+$ is evaluated from (8.34) and substituted into (8.37), we have

$$\boldsymbol{\beta}_{pca} = \boldsymbol{V}_1 \boldsymbol{\Sigma}_r^+ \boldsymbol{U}_1^T \boldsymbol{Y}, \tag{8.39}$$

which is identical to the LS pseudo-inverse solution given by (8.14) for the case $k = 1$. Thus we have the significant result that the pseudo-inverse and PCA approaches presented in this chapter are equivalent, even though they are derived from completely different viewpoints.

We see from (8.34) that the dimension of the row and column spaces of $\boldsymbol{X}$ has been reduced to r in the formulation of $\boldsymbol{X}_r$. It is therefore reasonable to ask whether we should project $\boldsymbol{x}_N^T$ onto the r-dimensional row space defined by $\boldsymbol{V}_1$ before finding the corresponding $\hat{\boldsymbol{y}}^T$ through (8.38), in order to eliminate some noise components. The projector for the row space[1] is $\boldsymbol{V}_1 \boldsymbol{V}_1^T$. To address this question, we substitute the projected version $\boldsymbol{x}_N^T \boldsymbol{V}_1 \boldsymbol{V}_1^T$ for $\boldsymbol{x}_N^T$ in (8.38) to obtain

$$\begin{aligned} \hat{\boldsymbol{y}}^T &= \boldsymbol{x}_N^T \boldsymbol{V}_1 \boldsymbol{V}_1^T \boldsymbol{\beta}_{pca}\\ &= \boldsymbol{x}_N^T \boldsymbol{V}_1 \boldsymbol{V}_1^T \boldsymbol{V}_1 \boldsymbol{\Sigma}_r^+ \boldsymbol{U}_1^T \boldsymbol{Y} \end{aligned}$$

[1] Recall that to project a row vector we post-multiply by the projector matrix.

$$= \boldsymbol{x}_N^T \boldsymbol{V}_1 \boldsymbol{\Sigma}_r^+ \boldsymbol{U}_r^T \boldsymbol{Y}$$

$$= \boldsymbol{x}_N^T \boldsymbol{\beta}_{pca}, \tag{8.40}$$

where we have substituted (8.39) for $\boldsymbol{\beta}_{pca}$ in the second and fourth lines and used the identity $\boldsymbol{V}_1^T \boldsymbol{V}_1 = \boldsymbol{I}$. Since (8.40) is identical to (8.38), there is no need to project $\boldsymbol{x}_N^T$ onto the row space of $\boldsymbol{X}_r$ before estimating the new responses.

8.7 Partial Least Squares (PLS) and Canonical Correlation Analysis (CCA)

The PCA latent variables are determined solely from $\boldsymbol{X}$ and are independent of the $\boldsymbol{Y}$ variables and capture the directions of major variation in $\boldsymbol{X}$ only. For the PLS and CCA methods on the other hand, we form a set of latent variables, one in the $\boldsymbol{X}$-space and another in the $\boldsymbol{Y}$-space, such that their covariance is maximum in the PLS case, or in the CCA case, the correlation is maximum. By forming these latent variable spaces that are as closely aligned as possible to each other, we expect that the PLS and CCA methods might be better at predicting $\boldsymbol{Y}$ corresponding to a new set of $\boldsymbol{X}$ values.

As a preliminary, we recall the defining relationships for the SVD, as outlined in Sect. 3.1.3, which we repeat here for convenience. Consider a matrix $\boldsymbol{A} = \boldsymbol{U}\boldsymbol{\Sigma}\boldsymbol{V}^T$. From this definition of the SVD, it follows that

$$\boldsymbol{A}\boldsymbol{v}_i = \sigma_i \boldsymbol{u}_u \tag{8.41}$$

$$\boldsymbol{A}^T \boldsymbol{u}_i = \sigma_i \boldsymbol{v}_i. \tag{8.42}$$

We use these relations later in this section.

Consider random vectors $\boldsymbol{x}$ and $\boldsymbol{y} \in \mathbb{R}^m$. The sample covariance estimate r_{xy} between these vectors is given from (2.26) as

$$r_{xy} = \frac{1}{m}\boldsymbol{x}^T \boldsymbol{y}, \tag{8.43}$$

whereas the sample correlation estimate ρ_{xy} is given as

$$\rho_{xy} = \frac{\boldsymbol{x}^T \boldsymbol{y}}{||\boldsymbol{x}||_2 ||\boldsymbol{y}||_2}. \tag{8.44}$$

The inner product in both cases can be written in the form

$$\boldsymbol{x}^T \boldsymbol{y} = ||\boldsymbol{x}||_2 ||\boldsymbol{y}||_2 \cos(\theta), \tag{8.45}$$

where θ is the angle between the two vectors. Thus from (8.43), we note that the covariance depends on $||\boldsymbol{x}||_2$, $||\boldsymbol{y}||_2$, and θ. Comparing (8.45) with (8.44), we have

$$\rho_{xy} = \cos(\theta),$$

and so ρ_{xy}, unlike r_{xy}, depends only on the angle between the vectors and is independent of the norms. Thus ρ_{xy} lies in the range $-1 \leq \rho_{xy} \leq 1$ and gives an idea how closely the random variables $\mathcal{X}$ and $\mathcal{Y}$ agree with other on average.

The idea of covariances and correlations can be generalised to the multidimensional case where we have matrices $\boldsymbol{X} \in \mathbb{R}^{m\times n}$ and $\boldsymbol{Y} \in \mathbb{R}^{m\times k}$ instead of vectors $\boldsymbol{x}$ and $\boldsymbol{y}$. The covariance matrix $\boldsymbol{R}_{XY} \in \mathbb{R}^{n\times k}$ for the multidimensional case is given as

$$\boldsymbol{R}_{XY} = \boldsymbol{X}^T\boldsymbol{Y}. \tag{8.46}$$

To define a multidimensional version of correlation, we must first define matrices $\boldsymbol{G}_X$ and $\boldsymbol{G}_Y$ which are square-root factors (e.g. Cholesky factors) of the covariance matrices $\boldsymbol{X}^T\boldsymbol{X}$ and $\boldsymbol{Y}^T\boldsymbol{Y}$, respectively. Then the matrices $\tilde{\boldsymbol{X}}$ and $\tilde{\boldsymbol{Y}}$ defined, respectively, as

$$\tilde{\boldsymbol{X}} = \boldsymbol{X}\boldsymbol{G}_X^{-1} \tag{8.47}$$

and

$$\tilde{\boldsymbol{Y}} = \boldsymbol{Y}\boldsymbol{G}_Y^{-1} \tag{8.48}$$

have orthonormal columns. (The proof is left as an exercise.) The presence of the tilde indicates the respective quantity has been orthonormalised. The correlation matrix $\tilde{\boldsymbol{R}}_{XY}$ is then defined as

$$\tilde{\boldsymbol{R}}_{XY} = \tilde{\boldsymbol{X}}^T\tilde{\boldsymbol{Y}}. \tag{8.49}$$

Note that both forms $\boldsymbol{R}_{XY}$ and $\tilde{\boldsymbol{R}}_{XY}$ from (8.46) and (8.49) respectively are $n \times k$.

In loose terms, the normalising or orthonormalising factors $\boldsymbol{G}^{-1}$ in the multidimensional case play the same role as the norm expressions in the denominator of (8.44). The CCA method proceeds in the same manner as the PLS method, except the matrix $\tilde{\boldsymbol{R}}_{XY}$ is used in place of $\boldsymbol{R}_{XY}$.

With the PLS and CCA methods, we create a set of r orthogonal basis vectors for each of the $\boldsymbol{X} \in \mathbb{R}^{m\times n}$ and $\boldsymbol{Y} \in \mathbb{R}^{m\times k}$ datasets. We refer to the respective subspaces formed by these bases as $\mathcal{S}_X$ and $\mathcal{S}_Y$. Each is of dimension $r \leq \min(n, k)$. The latent variable basis vectors for $\boldsymbol{X}$ and $\boldsymbol{Y}$ are denoted $\boldsymbol{t}_i$ and $\boldsymbol{p}_i$, $i = 1, \ldots r$, respectively. For the PLS case, we choose the $\boldsymbol{t}_1 \in \mathcal{S}_X$ and $\boldsymbol{p}_1 \in \mathcal{S}_Y$ so that their covariance, i.e. the quantity $\boldsymbol{t}_1^T\boldsymbol{p}_1$ is maximum. Then $\boldsymbol{t}_2$ and $\boldsymbol{p}_2$ are chosen, so they too have maximum covariance, under the constraint they are each orthogonal to their counterparts of the first set. The remaining basis vectors are found in a similar manner. The CCA case is similar, except we choose to maximise correlations instead of covariances. By choosing the latent variables in this manner, we provide the best possible fit between the $\boldsymbol{X}$ and $\boldsymbol{Y}$ subspaces, and therefore new values $\boldsymbol{x}_N$ of $\boldsymbol{X}$ are more likely to lead to "good" predictions of the corresponding $\boldsymbol{Y}$-values.

For the time being, we consider only the covariance or PLS case—the CCA case is addressed later. To determine $\boldsymbol{t} \in \mathcal{S}_X$ and $\boldsymbol{p} \in \mathcal{S}_Y$ with maximum covariance, we identify unit-norm vectors $\boldsymbol{s} \in \mathbb{R}^n$ and $\boldsymbol{q} \in \mathbb{R}^k$, such that the covariance between $\boldsymbol{t} = \boldsymbol{X}\boldsymbol{s}$ and $\boldsymbol{p} = \boldsymbol{Y}\boldsymbol{q}$ is maximum. Posing the problem in this manner guarantees the solutions $\boldsymbol{t}^*$ and $\boldsymbol{p}^*$ belong to their respective subspaces. This problem may be expressed in the form of the following constrained optimisation problem:

$$[\boldsymbol{s}^*, \boldsymbol{q}^*] = \arg\max_{\boldsymbol{s},\boldsymbol{q}} \boldsymbol{s}^T \boldsymbol{X}^T \boldsymbol{Y} \boldsymbol{q} \equiv \arg\max_{\boldsymbol{s},\boldsymbol{q}} \boldsymbol{s}^T \boldsymbol{R}_{xy} \boldsymbol{q}, \tag{8.50}$$

subject to

$$||\boldsymbol{q}||_2 = 1, \quad ||\boldsymbol{s}||_2 = 1.$$

The Lagrangian corresponding to this problem is given by

$$\boldsymbol{s}^T \boldsymbol{R}_{xy} \boldsymbol{q} + \gamma_s \left[1 - (\boldsymbol{s}^T \boldsymbol{s})^{\frac{1}{2}}\right] + \gamma_q \left[1 - (\boldsymbol{q}^T \boldsymbol{q})^{\frac{1}{2}}\right]. \tag{8.51}$$

We differentiate (8.51) with respect to $\boldsymbol{s}$ and $\boldsymbol{q}$. With regard to the first term, using a procedure similar to that outlined in Sect. 2.7, it is straightforward to show that

$$\begin{aligned} \frac{d}{d\boldsymbol{q}} \boldsymbol{s}^T \boldsymbol{R}_{xy} \boldsymbol{q} &= \boldsymbol{s}^T \boldsymbol{R}_{xy} \\ \frac{d}{d\boldsymbol{s}} \boldsymbol{s}^T \boldsymbol{R}_{xy} \boldsymbol{q} &= \boldsymbol{R}_{xy} \boldsymbol{q}. \end{aligned}$$

Differentiation of the second term of (8.51) with respect to $\boldsymbol{s}$ is straightforward using the chain rule. It is readily verified that

$$\frac{d}{d\boldsymbol{s}} \gamma_s \left[1 - (\boldsymbol{s}^T \boldsymbol{s})^{\frac{1}{2}}\right] = -\gamma_s \boldsymbol{s},$$

where $\boldsymbol{s}$ is normalised so that $||\boldsymbol{s}||_2^2 = 1$. A corresponding result holds for the last term:

$$\frac{d}{d\boldsymbol{q}} \gamma_q \left[1 - (\boldsymbol{q}^T \boldsymbol{q})^{\frac{1}{2}}\right] = -\gamma_q \boldsymbol{q}^T,$$

where again $||\boldsymbol{q}||_2^2 = 1$. Assembling these derivative terms with respect to $\boldsymbol{s}$ and $\boldsymbol{q}$ individually and setting the result to zero for each case, the solutions $\boldsymbol{s}$ and $\boldsymbol{q}$ must jointly satisfy the following:

$$\boldsymbol{R}_{xy}\boldsymbol{q} = \gamma_s \boldsymbol{s} \tag{8.52}$$

$$\left(\boldsymbol{R}_{xy}\right)^T \boldsymbol{s} = \gamma_q \boldsymbol{q}, \tag{8.53}$$

where we have transposed both sides of the second line above. Comparing (8.52) and (8.53) to (8.41) and (8.42), the latter of which are the defining relations for the SVD, we see that the stationary points of (8.50) are, respectively, the right and left singular vectors of $\boldsymbol{R}_{XY}$. Let the SVD of $\boldsymbol{R}_{XY}$ be expressed as $\boldsymbol{R}_{XY} = \boldsymbol{U}\boldsymbol{\Sigma}\boldsymbol{V}^T$. Therefore the optimal set of r solutions for $\boldsymbol{s}$ solving (8.50) is given as $[\boldsymbol{u}_1, \ldots, \boldsymbol{u}_r] \stackrel{\Delta}{=} \boldsymbol{U}_1$. The corresponding r solutions for $\boldsymbol{q}$ are given as $[\boldsymbol{v}_1, \ldots, \boldsymbol{v}_r] \stackrel{\Delta}{=} \boldsymbol{V}_1$, and the corresponding γ's are the respective singular values σ_i. Note that the required orthogonality property of the solutions follows directly from the orthonormality of $\boldsymbol{U}$ and $\boldsymbol{V}$.

Since we have specified $\boldsymbol{t}_i = \boldsymbol{X}\boldsymbol{s}_i$ and $\boldsymbol{p}_i = \boldsymbol{Y}\boldsymbol{q}_i, i = 1, \ldots, r$, where the solution for the $\boldsymbol{s}_i$ and $\boldsymbol{q}_i$ is $\boldsymbol{U}_1$ and $\boldsymbol{V}_1$, respectively, the corresponding vector sets $\boldsymbol{T}_1$ and $\boldsymbol{P}_1$ in $\mathcal{S}_X$ and $\mathcal{S}_Y$, respectively, with maximum covariance are therefore given as

$$\boldsymbol{T}_1 = [\boldsymbol{t}_1 \ldots, \boldsymbol{t}_r] = \boldsymbol{X}\boldsymbol{U}_1, \tag{8.54}$$

and

$$\boldsymbol{P}_1 = [\boldsymbol{p}_1 \ldots, \boldsymbol{p}_r] = \boldsymbol{Y}\boldsymbol{V}_1. \tag{8.55}$$

Note that $\boldsymbol{T}_1$ and $\boldsymbol{P}_1$ are both $m \times r$. The $\boldsymbol{T}_1$ and $\boldsymbol{P}_1$ are the desired latent variable bases for $\mathcal{S}_X$ and $\mathcal{S}_Y$, respectively.

For the PLS case, we project both $\boldsymbol{X}$ and $\boldsymbol{Y}$ onto their respective subspaces $\mathcal{S}_X$ and $\mathcal{S}_Y$ to form the desired reduced-rank regression matrices, so that the covariances of the projected matrices are maximised. A reasonable way to perform these projection operations is to form the projectors $\boldsymbol{P}_X$ and $\boldsymbol{P}_Y$, whose dimensions are both $m \times m$, onto the subspaces $\mathcal{S}_X$ and $\mathcal{S}_Y$ whose bases are $\boldsymbol{T}_1$ and $\boldsymbol{P}_1$. In this case, $\boldsymbol{X}_r = \boldsymbol{P}_X\boldsymbol{X}$ and $\boldsymbol{Y}_r = \boldsymbol{P}_Y\boldsymbol{Y}$. These operations project *columns* of $\boldsymbol{X}$ and $\boldsymbol{Y}$ onto their respective subspaces. However, these projectors are both $m \times m$, and because usually $m \gg n$ this approach for projection may not be the most efficient. A preferred method is to project *rows* of $\boldsymbol{X}$ and $\boldsymbol{Y}$ onto the subspaces formed by the bases $\boldsymbol{U}_1$ and $\boldsymbol{V}_1$, respectively. The corresponding projectors are $\boldsymbol{U}_1\boldsymbol{U}_1^T$ and $\boldsymbol{V}_1\boldsymbol{V}_1^T$. We prove in the Appendix of this chapter that these two projection methods are identical. The desired reduced-rank PLS regression matrices $\boldsymbol{X}_r$ and $\boldsymbol{Y}_r$ can therefore be expressed as

$$\boldsymbol{X}_r = \boldsymbol{X}\boldsymbol{U}_1\boldsymbol{U}_1^T \qquad \text{and} \tag{8.56}$$

$$\boldsymbol{Y}_r = \boldsymbol{Y}\boldsymbol{V}_1\boldsymbol{V}_1^T. \tag{8.57}$$

We have shown that the projections of the columns of $\boldsymbol{X}$ and $\boldsymbol{Y}$ into the subspaces $\mathcal{S}_X$ and $\mathcal{S}_Y$ yield matrices $\boldsymbol{X}_r$ and $\boldsymbol{Y}_r$ whose columns have maximum covariance. Since the corresponding matrices $\boldsymbol{X}_r$ and $\boldsymbol{Y}$ r yielded by (8.56) and (8.57) are identical to the column projection versions, these latter matrices also have maximum covariance.

Once we have available the rank r approximations $\boldsymbol{X}_r$ and $\boldsymbol{Y}_r$ given by (8.56) and (8.57), we can proceed with a PLS latent variable least squares implementation. Because $\boldsymbol{X}_r$ is rank deficient, we use the pseudo-inverse formulation to find the PLS version $\boldsymbol{\beta}_{pls}$ of $\boldsymbol{\beta}$, by analogy to the PCA approach (8.37) as

$$\boldsymbol{\beta}_{pls} = \boldsymbol{X}_r^{+}\boldsymbol{Y}_r. \tag{8.58}$$

The determination of $\boldsymbol{\beta}_{pls}$ completes the training process for the PLS method. Then, given new data in the form of a previously unseen row $\boldsymbol{x}_N^T$ of $\boldsymbol{X}$, we can use the PLS model to predict the corresponding $\hat{\boldsymbol{y}}^T$, which in the PLS case is given as

$$\hat{\boldsymbol{y}}_{pls}^T = \boldsymbol{x}_N^T \boldsymbol{\beta}_{pls}. \tag{8.59}$$

It is interesting to evaluate the covariance values corresponding to the optimal solution of (8.50). To do so we evaluate the quantity $\boldsymbol{t}_i^T \boldsymbol{p}_i$ as follows:

$$\begin{aligned} \boldsymbol{t}_i^T \boldsymbol{p}_i &= \boldsymbol{u}_i^T \boldsymbol{X}^T \boldsymbol{Y} \boldsymbol{v}_i \\ &= \boldsymbol{u}_i^T \boldsymbol{R}_{XY} \boldsymbol{v}_i \\ &= \boldsymbol{u}_i^T \boldsymbol{U} \boldsymbol{\Sigma} \boldsymbol{V}^T \boldsymbol{v}_i \\ &= \sigma_i. \end{aligned} \tag{8.60}$$

It is seen that the r maximum covariance values between $\boldsymbol{X}$ and $\boldsymbol{Y}$ are the largest r singular values of $\boldsymbol{R}_{XY}$, which are the σ_i. The directions in $\mathcal{S}_X$ and $\mathcal{S}_Y$ which result in this largest covariance are given by $\boldsymbol{t}_i = \boldsymbol{X}\boldsymbol{u}_i$ and $\boldsymbol{p}_i = \boldsymbol{Y}\boldsymbol{v}_i$, respectively.

The PLS process may now be summarised. Given a training set of corresponding $\boldsymbol{X}$ and $\boldsymbol{Y}$ values, we compute an estimate $\boldsymbol{\beta}_{pls}$ of $\boldsymbol{\beta}$ and predicted values $\hat{\boldsymbol{y}}_{pls}^T$ of $\boldsymbol{Y}$ corresponding to previously unseen values $\boldsymbol{x}_N^T$ of $\boldsymbol{X}$ in the following manner:

- Evaluate $\boldsymbol{R}_{XY}$ according to (8.46).
- Calculate the SVD of $\boldsymbol{R}_{XY}$ to give the values $\boldsymbol{U}$ and $\boldsymbol{V}$.
- Calculate the rank r versions $\boldsymbol{X}_r$ and $\boldsymbol{Y}_r$ of $\boldsymbol{X}$ and $\boldsymbol{Y}$, respectively, from (8.56) and (8.57).
- Complete the PLS model by calculating $\boldsymbol{\beta}_{pls}$ from (8.58). Predictions of $\boldsymbol{y}$-values corresponding to a new, previously unseen row $\boldsymbol{x}_N^T$ of $\boldsymbol{X}$ are obtained using (8.59).

Development for the CCA Case This is similar to that of the PLS case, except that we use $\tilde{\boldsymbol{R}}_{XY} = \tilde{\boldsymbol{X}}^T\tilde{\boldsymbol{Y}}$, as defined by (8.49), in place of $\boldsymbol{R}_{XY}$ to determine the latent variables $\tilde{U}_1$ and $\tilde{V}_1$. (Recall the presence of the tilde ($\tilde{}$) indicates the respective quantity has been whitened.) Further, because the variables $\tilde{X}$ and $\tilde{Y}$ have been transformed (whitened) by (8.47) and (8.48), respectively, the $\boldsymbol{\beta}$ calculated directly as $\boldsymbol{\beta} = \tilde{X}^+\tilde{Y}$ corresponding to (8.58) predicts the whitened version $\tilde{Y}$ from the whitened version $\tilde{X}$ and therefore does not reflect the true relationship between X and Y in the original X and Y spaces. So we must take appropriate steps to correct this effect in order to estimate a relevant version of $\boldsymbol{\beta}$. The CCA process is very similar to the PLS method, with the exception that the relevant variables require whitening. The method is described as follows:

- From the available training set consisting of X and corresponding Y values, calculate $R_X = X^TX$ and $R_Y = Y^TY$, respectively. Then calculate the respective inverse square-root factors G_X^{-1} and G_Y^{-1} (e.g. in the form of inverse Cholesky factors).
- Form $\tilde{X}$ and $\tilde{Y}$, respectively, as $\tilde{X} = XG_X^{-1}$ and $\tilde{Y} = YG_Y^{-1}$.
- Choose an appropriate value for r and form $\tilde{R}_{XY} = \tilde{X}^T\tilde{Y}$, and compute its SVD $\tilde{R}_{XY} = \tilde{U}\tilde{\Sigma}\tilde{V}^T$. Then extract the latent variables $\tilde{U}_1$ and $\tilde{V}_1$ as the first r columns of each respective matrix.
- Calculate the projected matrices $\tilde{X}_r$ and $\tilde{Y}_r$ corresponding to (8.56) and (8.57) as

$$\tilde{X}_r = X\tilde{U}_1\tilde{U}_1^T \quad \text{and}$$
$$\tilde{Y}_r = Y\tilde{V}_1\tilde{V}_1^T.$$

- The value $\boldsymbol{\beta}_1$ which relates $\tilde{X}$ and $\tilde{Y}$ in the whitened, latent space is given as

$$\boldsymbol{\beta}_1 = \tilde{X}_r^+\tilde{Y}_r.$$

However, we require a $\boldsymbol{\beta}$ which relates X and Y in the original space. We proceed in this direction by taking the following steps:

  - The respective regression equation in the whitened space is given as $XG_X^{-1}\boldsymbol{\beta}_1 = YG_Y^{-1}$ + error. Therefore a value $\boldsymbol{\beta}_2$ which relates the original X and Y is given as $\boldsymbol{\beta}_2 = G_X^{-1}\boldsymbol{\beta}_1G_Y$. Then we have $Y = X\boldsymbol{\beta}_2$ + error.
  - Since the original X may be considered rank deficient, G_X^{-1} may contain significant components in the nullspace of X, resulting in significant error. Therefore to remove these components, we form the desired estimate $\boldsymbol{\beta}_{cca}$ of $\boldsymbol{\beta}$ by projecting $\boldsymbol{\beta}_2$ into the row space of the original data matrix X. If $X = U_x\Sigma_xV_x$, then we define V_{x1} as the first r columns of V_x. The desired estimate of $\boldsymbol{\beta}$ is then given as

$$\boldsymbol{\beta}_{cca} = V_{x1}V_{x1}^T\boldsymbol{\beta}_2.$$

- Finally, we can predict values $\hat{\boldsymbol{y}}^T_{cca}$ of $\boldsymbol{y}$ by substituting $\boldsymbol{\beta}_{cca}$ above for $\boldsymbol{\beta}_{pls}$ in (8.59).

Because PLS is related to the covariance between $\boldsymbol{X}$ and $\boldsymbol{Y}$, the PLS latent variables are formed from a combination of the magnitude of the major variation in both $\boldsymbol{X}$ and $\boldsymbol{Y}$, as well as the angles between the latent vectors in the two respective subspaces. Because CCA is derived from the correlation between the variables, the CCA latent variables are determined solely from the angles between the subspaces and are independent of the magnitudes of major variation. This is a direct consequence of the fact the columns of both $\tilde{\boldsymbol{X}}$ and $\tilde{\boldsymbol{Y}}$ are orthonormal. In the CCA case only, it can be shown [3] that the $\sigma_i, i = 1, \ldots, r$, are the cosines of the r angles specifying the relative orientations between $\mathcal{S}_X$ and $\mathcal{S}_Y$. In this vein, it may be shown (Problem 3) that $0 \leq \sigma_i \leq 1$.

The presentation here for identifying the PLS and CCA latent variables is quite different from the usual treatment in the literature. Most methods, e.g. [1, 5], use the *nonlinear iterative partial least squares* (NIPALS) algorithm for extracting the latent variables. However, the method presented here using the SVD on the matrix $\boldsymbol{R}_{XY}$ as in (8.50) yields identical LVs and affords a simpler presentation.

A last topic for this section is to introduce the terminology "loadings and scores", with respect to latent variables that are in common use in the statistical literature. The variables $\boldsymbol{X}$ and $\boldsymbol{Y}$ are typically represented in their rank-r latent variable bases as

$$\boldsymbol{X} = \boldsymbol{T}\boldsymbol{P}^T + \boldsymbol{E}_x$$
$$\boldsymbol{Y} = \boldsymbol{U}\boldsymbol{Q}^T + \boldsymbol{E}_y.$$

The matrices $\boldsymbol{T}$ and $\boldsymbol{U}$ are bases for the column spaces of $\boldsymbol{X}$ and $\boldsymbol{Y}$, respectively, whereas $\boldsymbol{P}$ and $\boldsymbol{Q}$ are the corresponding row space bases. The matrices $\boldsymbol{T}$ and $\boldsymbol{U}$ are referred to as "scores", whereas $\boldsymbol{P}$ and $\boldsymbol{Q}$ are referred to as "loadings".

8.8 Simulation Example

We present a simulation example to compare the relative performances of the three latent variable methods we have discussed, with respect to accuracy of the $\boldsymbol{\beta}$-estimates and prediction of $\boldsymbol{Y}$ values corresponding to previously unseen values of $\boldsymbol{X}$. First, we construct a matrix $\boldsymbol{X} \in \mathbb{R}^{m \times n}$ whose elements are independent, zero mean, unit variance Gaussian random variables using the "randn" command in Matlab®. In our simulations, $n = 8$ and $m = 100$.

With this present simulation scenario, the singular values of $\boldsymbol{X}$ are typically in the range 10–20. To introduce near linear dependence amongst the columns of $\boldsymbol{X}$ and corresponding poor conditioning (which is necessary to illustrate the effectiveness of LV methods), we perform an SVD on $\boldsymbol{X}$ and replace the four smallest singular values of $\boldsymbol{\Sigma}$ with the values 1×10^{-6}. A new, almost rank-deficient matrix $\boldsymbol{X}$ is then reassembled from its SVD components using the modified version of $\boldsymbol{\Sigma}$. The effective rank of $\boldsymbol{X}$ is therefore 6.

An $m \times k$ matrix $\boldsymbol{Y}$ where $k = 6$ was constructed as follows:

$$\boldsymbol{Y} = \boldsymbol{X}\boldsymbol{\beta}_o + \sigma \boldsymbol{E}, \tag{8.61}$$

where

- $\boldsymbol{\beta}_o$ is an $n \times k$ matrix of true parameter values. We consider two scenarios for $\boldsymbol{\beta}_o$. The first is where $\boldsymbol{\beta}_o$ is in the row space of $\boldsymbol{X}$, thus avoiding the bias problems as discussed previously in Sect. 8.3 associated with the pseudo-inverse. In this scenario, $\boldsymbol{\beta}_o$ was constructed as $\boldsymbol{\beta} = \boldsymbol{V}_1\boldsymbol{B}$, where $\boldsymbol{B}$ is an $r \times k$ random matrix chosen by the Matlab® "randn" command, and $k = 6$, $r = 4$ and $\boldsymbol{V}_1$, consists of the first r columns of the SVD of the version of $\boldsymbol{X}$ used for training (see below). The rank of $\boldsymbol{Y}$ from (8.61) is therefore $r = 4$. The second scenario is where $\boldsymbol{\beta}_o$ is chosen randomly, so in this situation we can expect some bias and loss of performance.
- $\boldsymbol{E} \in \mathbb{R}^{m \times k}$ is additive Gaussian noise, also with zero mean and unit variance.
- σ controls the signal-to-noise ratio (SNR) of $\boldsymbol{Y}$.

The SNR is defined as

$$\text{SNR} = \frac{||\boldsymbol{X}\boldsymbol{\beta}||_F^2}{||\sigma \boldsymbol{E}||_F^2}.$$

The simulation to determine the error between the true and estimated values of $\boldsymbol{Y}$ and $\boldsymbol{\beta}$ consists of an inner and an outer loop, as shown in Fig. 8.2. In each iteration of the inner loop, a noise sample $\boldsymbol{E}$ is generated by the "randn" command in Matlab® for a given set of values for $\boldsymbol{X}$ and $\boldsymbol{\beta}$ and used to form a new sample of $\boldsymbol{Y}$ according to (8.61). The $\boldsymbol{X}$ and $\boldsymbol{Y}$ arrays are then split into training and test sets, giving $\boldsymbol{X}_{train}$ and $\boldsymbol{X}_{test}$, likewise for $\boldsymbol{Y}$. The latent variables and the $\boldsymbol{\beta}$-values are determined in the LV analysis block for each of the three methods, using only the training set data. The estimated values $\boldsymbol{\beta}$ of $\boldsymbol{\beta}_o$ and the predicted values $\hat{\boldsymbol{Y}}$ corresponding to $\boldsymbol{X}_{test}$ are then calculated using the methods discussed in Sect. 8.6 and compared to the true test values $\boldsymbol{Y}_{test}$ to generate the normalised prediction error given by

$$\text{relative prediction error} = \frac{||\boldsymbol{Y}_{test} - \hat{\boldsymbol{Y}}||_F^2}{||\boldsymbol{Y}_{test}||_F^2}. \tag{8.62}$$

A similar relation holds for the relative error in $\boldsymbol{\beta}$, where the true value is $\boldsymbol{\beta}_o$. Then after 100 iterations of the inner loop, a new iteration of the outer loop proceeds, where new values of $\boldsymbol{X}$ and $\boldsymbol{\beta}$ are calculated and 100 iterations of the inner loop are repeated for these new values. After ten iterations of the outer loop, all the

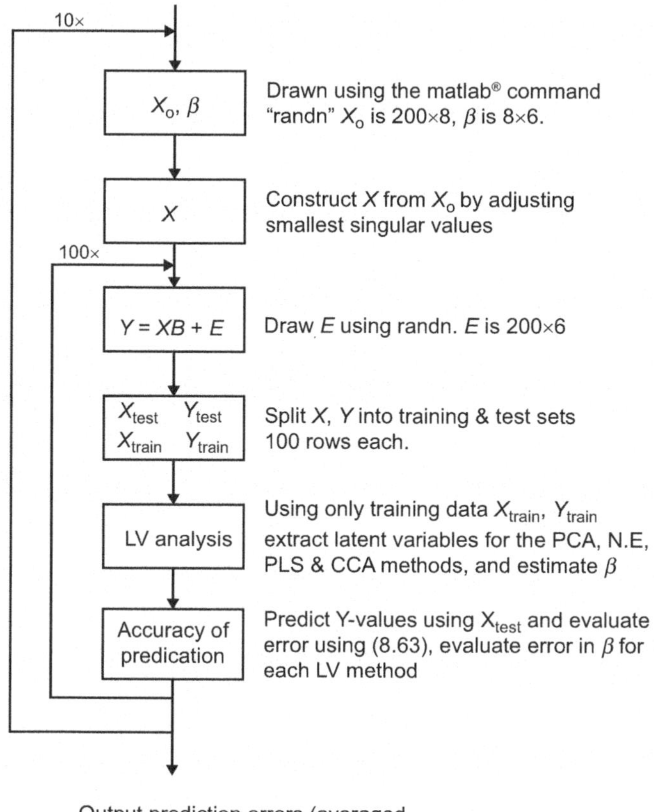

Fig. 8.2 An overview of the simulation process used to compare the normal equation, PCA, PLS, and CCA methods of LV analysis

normalised prediction errors are averaged together. In this manner, the final results reflect LV performance over many settings of $\boldsymbol{X}$ and $\boldsymbol{\beta}$ values.

We show the relative prediction errors of $\hat{y}$ vs. r (the number of latent components chosen in the construction of $\boldsymbol{X}_r$ and $\boldsymbol{Y}_r$) and the error in the estimation of $\boldsymbol{\beta}$, for SNR $= 20\,\text{dB}$ in the following figures, for each of the three LV methods discussed. Errors for $\boldsymbol{\beta}$ for the ordinary normal equations are not shown since their accuracies are very poor, due to the poor conditioning of $\boldsymbol{X}$. We show results for two different methods of generating $\boldsymbol{\beta}_o$—the first corresponds to Figs. 8.3 and 8.4, where the true $\boldsymbol{\beta}$ (i.e. $\boldsymbol{\beta}_o$) lies in $R(\boldsymbol{V}_1)$ or equivalently in the row space of $\boldsymbol{X}$. We have shown in this case from (8.23) that the estimate of $\boldsymbol{\beta}$ in the LV case is unbiased. The second case corresponds to Figs. 8.5 and 8.6. Here $\boldsymbol{\beta}$ is chosen randomly, so the $\boldsymbol{\beta}$-estimates will generally be biased and thus exhibit higher error. This effect is readily observed from the figures.

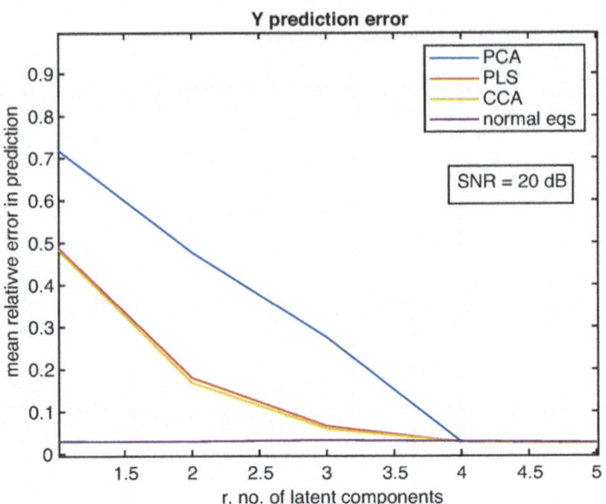

Fig. 8.3 Mean relative prediction error vs. r, for SNR = 20 dB, for the case $\boldsymbol{\beta} \in R(\boldsymbol{V}_1)$, which results in an unbiased estimate as indicated by (8.23). In the following simulations, $\boldsymbol{Y}$ is adjusted to be rank 4. Results for the PCA, PLS, CCA, and normal equation methods have been shown

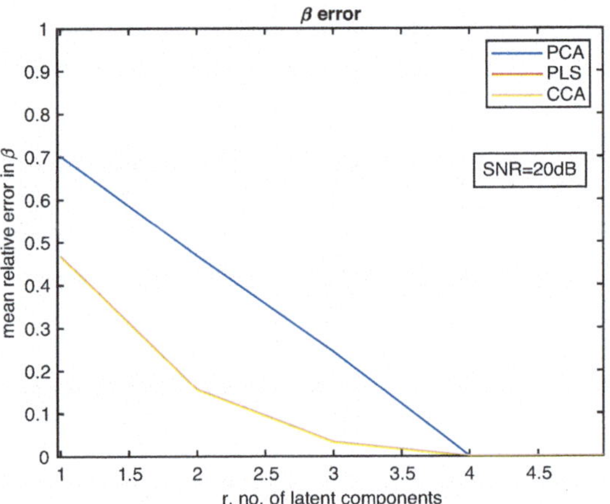

Fig. 8.4 Relative error in $\boldsymbol{\beta}$ for the various LV methods. Other parameter values are the same as in Fig. 8.3. Results for the ordinary normal equation method are not shown because the error in $\boldsymbol{\beta}$ is extremely large

The nominal rank for $\boldsymbol{Y}$ is 4, since $\boldsymbol{\beta}_o$ was constructed to be rank 4. It may be seen from Figs. 8.3 and 8.4 that the prediction error for $\boldsymbol{Y}$ drops to a plateau as r increases to the value 4 and above, whose value depends mostly on the SNR. In this case, the dimensionality of the latent variable subspaces is high enough to form an accurate model, and thus the prediction error is low. Below $r = 4$, the prediction errors rise sharply due to underfitting. In these cases, PLS uniformly performs better than PCA as expected, since PLS is inherently a more flexible model. The CCA performance is approximately comparable to that of PLS, except that performance drops off for low values of SNR (not shown). Although not shown, the performance in predicting $\boldsymbol{Y}$, and the error in estimating $\boldsymbol{\beta}$ for all three methods, has been observed to be linearly dependent on the SNR value.

The relative error in the estimation of $\boldsymbol{\beta}$ using the ordinary normal equation method has been noted to be very high—on the order of 10^{11}, whereas the prediction error in $\boldsymbol{Y}$ is roughly the same as the other methods. This phenomenon is the basis for problem 6.

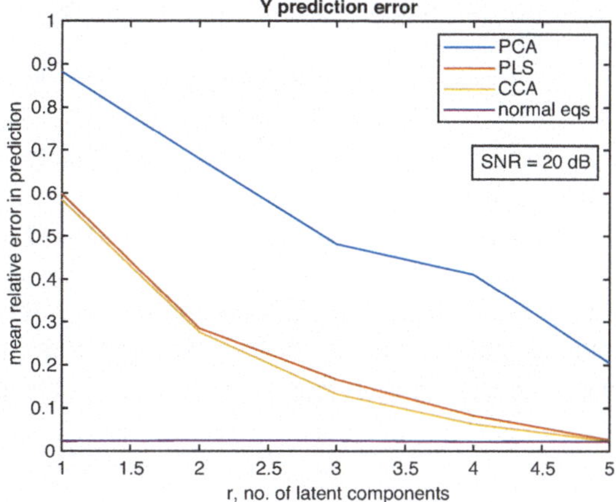

Fig. 8.5 Relative prediction error in $\boldsymbol{Y}$ for the various LV methods. Other parameter values are as previous, but in this case $\boldsymbol{\beta}$ is chosen randomly, resulting in bias and larger error for the $\boldsymbol{\beta}$ estimates

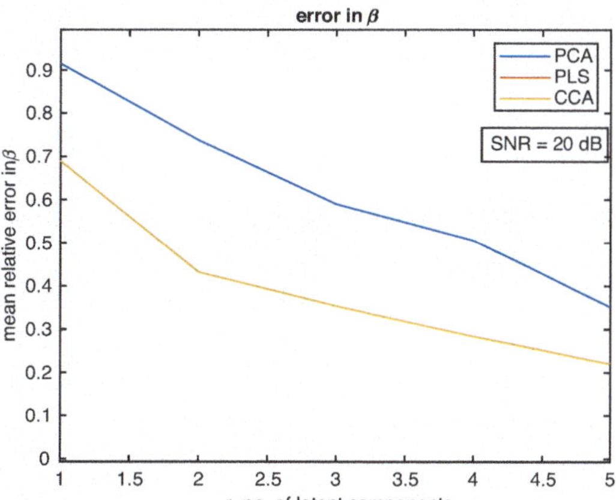

Fig. 8.6 Relative error in $\boldsymbol{\beta}$ for the various LV methods where $\boldsymbol{\beta}$ is chosen randomly

Appendix

In this appendix we show that the matrix formed by the projection of the columns of $\boldsymbol{X}$ onto the basis formed by $\boldsymbol{T}_1$ is equivalent to the projection of the rows of $\boldsymbol{X}$ onto the basis formed by $\boldsymbol{U}_1$. First, we establish notation: The SVD of $\boldsymbol{R}_{XY} = \boldsymbol{X}^T\boldsymbol{Y}$ has been given as $\boldsymbol{U}\boldsymbol{\Sigma}\boldsymbol{V}^T$. We denote the SVDs of $\boldsymbol{X}$ and $\boldsymbol{Y}$ as

$$\boldsymbol{X} = \boldsymbol{U}_X\boldsymbol{\Sigma}_X\boldsymbol{V}_X^T$$
$$\boldsymbol{Y} = \boldsymbol{U}_Y\boldsymbol{\Sigma}_Y\boldsymbol{V}_Y^T.$$

Then it is determined that the range of the matrices $\boldsymbol{U}$ and $\boldsymbol{V}_X$ is identical, likewise for $\boldsymbol{V}$ and $\boldsymbol{V}_Y$.

We now investigate the projection of columns of $\boldsymbol{X}$ onto $R(\boldsymbol{T}_1)$. From (8.54), we have $\boldsymbol{T}_1 = \boldsymbol{X}\boldsymbol{U}_1 = \boldsymbol{X}\boldsymbol{V}_{X1}$, where a subscript $(\cdot)_1$ refers to the first r columns of the respective matrix. We can write this in the form

$$\begin{aligned} \boldsymbol{T}_1 = \boldsymbol{X}\boldsymbol{V}_{X1} &= \boldsymbol{U}_X\boldsymbol{\Sigma}_X\boldsymbol{V}_X^T\boldsymbol{V}_{X1} \\ &= \boldsymbol{U}_X\boldsymbol{\Sigma}_X\begin{bmatrix}\boldsymbol{I}_r\\ \boldsymbol{0}\end{bmatrix} \\ &= \boldsymbol{U}_{X1}\boldsymbol{\Sigma}_{X1}, \end{aligned}$$

where $\boldsymbol{\Sigma}_{X1}$ consists of the upper-left, $r \times r$ block of $\boldsymbol{\Sigma}$. Thus it is clear that an orthonormal basis for $\boldsymbol{T}_1$ is $\boldsymbol{U}_{X1}$. We now form the projection of the columns of $\boldsymbol{X}$ onto $R(\boldsymbol{U}_{X1})$ which is equivalent to $R(\boldsymbol{T}_1)$. The projection is given as

$$\begin{aligned} \boldsymbol{U}_{X1}\boldsymbol{U}_{X1}^T\boldsymbol{X} &= \left[\boldsymbol{U}_{X1}\boldsymbol{U}_{X1}^T\right]\boldsymbol{U}_X\boldsymbol{\Sigma}_X\boldsymbol{V}_X^T \\ &= \boldsymbol{U}_{X1}\left[\boldsymbol{I}_r\ \boldsymbol{0}\right]\boldsymbol{\Sigma}_X\boldsymbol{V}_X^T \\ &= \boldsymbol{U}_{X1}\left[\boldsymbol{\Sigma}_{X1}\ \boldsymbol{0}\right]\begin{bmatrix}\boldsymbol{V}_{X1}^T\\ \boldsymbol{V}_{X2}^T\end{bmatrix} \\ &= \boldsymbol{U}_{X1}\boldsymbol{\Sigma}_{X1}\boldsymbol{V}_{X1}. \end{aligned} \tag{8.63}$$

We now form the projection of rows of $\boldsymbol{X}$ onto $R(\boldsymbol{V}_{X1})$. This is given as

$$\begin{aligned} \boldsymbol{X}\boldsymbol{V}_{X1}\boldsymbol{V}_{X1}^T &= \boldsymbol{U}_X\boldsymbol{\Sigma}_X\boldsymbol{V}_X^T\boldsymbol{V}_{X1}\boldsymbol{V}_{X1}^T \\ &= \boldsymbol{U}_X\boldsymbol{\Sigma}_X\begin{bmatrix}\boldsymbol{I}_r\\ \boldsymbol{0}\end{bmatrix}\boldsymbol{V}_{X1}^T \\ &= \boldsymbol{U}_X\begin{bmatrix}\boldsymbol{\Sigma}_{X1}\\ \boldsymbol{0}\end{bmatrix}\boldsymbol{V}_{X1}^T \\ &= \boldsymbol{U}_{X1}\boldsymbol{\Sigma}_{X1}\boldsymbol{V}_{X1}^T, \end{aligned} \tag{8.64}$$

which is identical to (8.63), which was to be shown.

Problems

1. Explain the effect of (independently) varying m and n with respect to the matrix $\boldsymbol{X}$ in an LV least squares problem.
2. Let the SVD of $\boldsymbol{X}$ be expressed as $\boldsymbol{X} = \boldsymbol{U}\boldsymbol{\Sigma}\boldsymbol{V}^T$. Prove the equivalence of Eqs. (8.34)–(8.36). Therefore projecting $\boldsymbol{X}$ onto the principal row space or the principal column space is equivalent. Show that the quantity $||\boldsymbol{X} - \boldsymbol{X}_r||_2^2$ is minimised using (8.34).
3. Explain the connection between $\boldsymbol{X}_r$ given by (8.34) and the rank r pseudo-inverse of $\boldsymbol{X}$.

4. Given $\boldsymbol{Z} = \boldsymbol{X}^T\boldsymbol{Y}$, find transformations $\boldsymbol{T}_x$ on $\boldsymbol{X}$ and $\boldsymbol{T}_y$ on $\boldsymbol{Y}$ (post-multiplications are implied) so that $\boldsymbol{T}_x^T\boldsymbol{Z}\boldsymbol{T}_y$ is diagonal. What are the diagonal values?
5. Prove that the canonical correlation coefficients σ_i in (8.60) satisfy $0 \leq |\sigma_i| \leq 1,\ i = 1, \ldots, r$.
6. The relative error in the estimation of $\boldsymbol{\beta}$ using the ordinary normal equations has been noted to be extremely large, whereas the relative error in the prediction of $\boldsymbol{Y}$ as shown in Fig. 8.3 using the normal equations is comparable to that of the other methods. Provide an explanation for this behaviour.
7. Explain why the $\boldsymbol{\beta}_{PC}$ results are biased towards zero in Table 8.1 where the true value $\boldsymbol{\beta}_o$ was chosen randomly.
8. We have seen that the ordinary normal equation solution yields large variances of the $\boldsymbol{\beta}$ estimates. This is because the $\boldsymbol{\beta}$ yielded by the normal equations can include an arbitrary component in $\mathcal{N}(\boldsymbol{X})$. A possible method for suppressing this effect is to project the normal equation solution for $\boldsymbol{\beta}$ onto $R(\boldsymbol{V}_1)$. Explain the feasibility of this approach, and compare it to other methods for LS estimation for poorly conditioned problems, as discussed in this chapter.

References

1. H. Abdi, Partial least square regression (PLS regression). Encyclopedia Res. Methods Soc. Sci. **6**(4), 792–795 (2003)
2. P. Geladi, B.R. Kowalski, Partial least-squares regression: a tutorial. Analytica chimica acta **185**, 1–17 (1986)
3. G.H. Golub, C.F. Van Loan, *Matrix Computations*, 3rd edn. (The Johns Hopkins University, Baltimore, 1996)
4. I.T. Jolliffe, Principal components in regression analysis, in *Principal Component Analysis* (Springer, Berlin, 1986), pp. 129–155
5. R. Rosipal, N. Krämer, Overview and recent advances in partial least squares, in *International Statistical and Optimization Perspectives Workshop" Sub-space, Latent Structure and Feature Selection* (Springer, Berlin, 2005), pp. 34–51
6. S. Wold et al., The collinearity problem in linear regression. The partial least squares (PLS) approach to generalized inverses. SIAM J. Sci. Stat. Comput. **5**(3), 735–743 (1984)
7. S. Wold, K. Esbensen, P. Geladi, Principal component analysis. Chemomet. Intell. Laborat. Syst. **2**(1–3), 37–52 (1987)

Chapter 9
Regularisation

9.1 Ridge Regression

In the case of a square, symmetric, positive definite matrix, its singular values and eigenvalues are identical. Thus, in the context of LS analysis, we may use the eigenvalues λ_i of $\boldsymbol{X}^T\boldsymbol{X}$ to determine conditioning. In this case, we have $\sigma_i^2 = \lambda_i$, where the σ_i are the singular values of $\boldsymbol{X}$. If $\boldsymbol{X}$ is poorly conditioned, then some eigenvalues of $\boldsymbol{X}^T\boldsymbol{X}$ are relatively small, and therefore, some eigenvalues of $(\boldsymbol{X}^T\boldsymbol{X})^{-1}$ become large, resulting in $(\boldsymbol{X}^T\boldsymbol{X})^{-1}$ having large elements. Therefore, in solving the ordinary normal equations $\boldsymbol{\beta}_{LS} = (\boldsymbol{X}^T\boldsymbol{X})^{-1}\boldsymbol{X}^T\boldsymbol{y}$, elements of the solution $\boldsymbol{\beta}_{LS}$ can become inappropriately large. Ridge regression imposes the prior knowledge that the solution $\boldsymbol{\beta}$ should have small norm. In this respect, a constraint on $||\boldsymbol{\beta}_{LS}||_2$ is incorporated into the LS objective function to encourage a more stable solution with a more moderate norm. In this vein, we modify the ordinary LS objective function to the form

$$\min_{\boldsymbol{\beta}} ||\boldsymbol{X}\boldsymbol{\beta} - \boldsymbol{y}||_2^2 + \lambda||\boldsymbol{\beta}||_2^2. \tag{9.1}$$

The second term of this objective function is referred to as a *penalty function*. This term encourages a solution where $||\boldsymbol{\beta}||_2^2$ is small. In effect, this modified objective function trades off fit (the first term) for a small-norm solution (imposed by the second term). The parameter $\lambda \geq 0$ controls the degree of this trade-off, where a larger value places more emphasis on the norm of the solution being small and less weight to the fit and vice versa.

A universal value for λ in general cannot be determined beforehand. Its value is typically determined by examining the effect of these trade-offs on a case-by-case basis, using trial-and-error or cross-validation techniques. A useful, more structured approach to the determination of λ is the L-curve method presented in [3].

J. Reilly, *Fundamentals of Linear Algebra for Signal Processing*,
https://doi.org/10.1007/978-3-031-68915-4_9

An alternative form for (9.1) is the following:

$$\hat{\boldsymbol{\beta}}^{rr} = \arg\min_{\boldsymbol{\beta}} ||\boldsymbol{X}\boldsymbol{\beta} - \boldsymbol{y}||_2^2$$
$$\text{subject to } ||\boldsymbol{\beta}^{rr}||_2 \leq t. \tag{9.2}$$

for some value of t, where $\hat{\boldsymbol{\beta}}^{rr}$ is the ridge regression estimate of $\boldsymbol{\beta}$. Equation (9.2) expresses (9.1) in the form of a strict upperbound on $||\boldsymbol{\beta}||_2$. It may be shown [4] there is a corresponding value of t in (9.2) for which the two solutions are identical.

There is an analytic solution to (9.1). The derivative with respect to $\boldsymbol{\beta}$ of the first term is, as before, $2\boldsymbol{X}^T\boldsymbol{X}\boldsymbol{\beta} - 2\boldsymbol{X}^T\boldsymbol{y}$. It is straightforward to verify that the derrivative of the second term is $2\lambda\boldsymbol{\beta} = 2\lambda\boldsymbol{I}\boldsymbol{\beta}$. Adding these terms and setting the result to zero, we obtain a modified set of normal equations given by

$$(\boldsymbol{X}^T\boldsymbol{X} + \lambda\boldsymbol{I})\boldsymbol{\beta} = \boldsymbol{X}^T\boldsymbol{y},$$

and therefore

$$\hat{\boldsymbol{\beta}}^{rr} = (\boldsymbol{X}^T\boldsymbol{X} + \lambda\boldsymbol{I})^{-1}\boldsymbol{X}^T\boldsymbol{y}. \tag{9.3}$$

Thus, the ridge regression method effectively adds the value λ to the diagonal elements of $\boldsymbol{X}^T\boldsymbol{X}$. Recall from the *Properties of Eigenvalues* in Chap. 2, adding a constant term λ to the diagonal elements of a matrix has the effect of adding the same value to each of its eigenvalues; i,e., each λ_i is replaced by $\lambda_i + \lambda$.[1] In the present context, the condition number $\kappa_2(\boldsymbol{X}^T\boldsymbol{X})$ that is relevant in the ordinary LS problem is given by

$$\kappa_2(\boldsymbol{X}^T\boldsymbol{X}) = \frac{|\lambda_1|}{|\lambda_n|}$$

i.e. the ratio of the largest to smallest eigenvalues values of $\boldsymbol{X}^T\boldsymbol{X}$. After regularisation, the modified condition number $\kappa_2'(\boldsymbol{X}^T\boldsymbol{X})$ corresponding to (9.3) becomes

$$\kappa_2'(\boldsymbol{X}^T\boldsymbol{X}) = \frac{|\lambda_1 + \lambda|}{|\lambda_n + \lambda|}.$$

In a poorly conditioned LS problem, $\lambda_1 \gg \lambda_n$, and so if λ is significantly greater than λ_n, $\kappa_2'(\boldsymbol{X}^T\boldsymbol{X})$ can be significantly less than $\kappa_2(\boldsymbol{X}^T\boldsymbol{X})$, without significantly perturbing the matrix $\boldsymbol{X}^T\boldsymbol{X}$ and thus significantly shifting the value of $\hat{\boldsymbol{\beta}}^{rr}$ away from its nominal value.

[1] A clarification on notation: a λ_i (with a subscript) denotes an eigenvalue, whereas λ without a subscript denotes the ridge regression regularisation parameter.

We can compare the LS solutions for the normal equation, PCA, and ridge regression methods in the following manner. Using the ordinary normal equations, the quantity $\boldsymbol{X}\boldsymbol{\beta}^n$ is given by

$$\boldsymbol{X}\boldsymbol{\beta}^n = \boldsymbol{X}(\boldsymbol{X}^T\boldsymbol{X})^{-1}\boldsymbol{X}^T\boldsymbol{y},$$

where $\boldsymbol{\beta}^n$ is the normal equation estimate of $\boldsymbol{\beta}$. When we substitute the SVD of $\boldsymbol{X} = \boldsymbol{U}\boldsymbol{\Sigma}\boldsymbol{V}^T$ into the above, we obtain the simplified form

$$\boldsymbol{X}\boldsymbol{\beta}^n = \boldsymbol{U}_n\boldsymbol{U}_n^T\boldsymbol{y},$$

where $\boldsymbol{U}_n \in \mathbb{R}^{m\times n} = [\boldsymbol{u}_1 \ldots \boldsymbol{u}_n]$. Since $\boldsymbol{X}$ is assumed tall, $m > n$, $\boldsymbol{U}_n$ is a tall matrix with orthonormal columns. Applying the outer product rule for matrix multiplication we have

$$\boldsymbol{X}\boldsymbol{\beta}^n = \sum_{i=1}^{n} \boldsymbol{u}_i\boldsymbol{u}_i^T\boldsymbol{y}. \tag{9.4}$$

It is interesting to note that the PCA solution $\boldsymbol{\beta}^{pca}$ can also be expressed in the form

$$\boldsymbol{X}\boldsymbol{\beta}^{pca} = \boldsymbol{U}_r\boldsymbol{U}_r^T\boldsymbol{y}$$

where $\boldsymbol{U}_r = [\boldsymbol{u}_1 \ldots, \boldsymbol{u}_r]$. Using the outer product rule for matrix multiplication, this can be written in the form

$$\boldsymbol{X}\boldsymbol{\beta}^{pca} = \sum_{i=1}^{r} \boldsymbol{u}_i\boldsymbol{u}_i^T\boldsymbol{y}. \tag{9.5}$$

Thus, by comparing (9.4) and (9.5), we see that the PCA solution is similar in form to the normal equation solution, but PCA applies a hard thresholding procedure to eliminate the components $[\boldsymbol{u}_{r+1} \ldots \boldsymbol{u}_n]$ which are associated with the smaller singular values of $\boldsymbol{X}$.

Now, we look at the ridge regression solution in the light of (9.4) and (9.5). From the ridge regression estimate (9.3), we get

$$\boldsymbol{X}\hat{\boldsymbol{\beta}}^{rr} = \boldsymbol{X}^T(\boldsymbol{X}^T\boldsymbol{X} + \lambda\boldsymbol{I})^{-1}\boldsymbol{X}\boldsymbol{y}.$$

Substituting the SVD for $\boldsymbol{X}$ as before and simplifying, we have

$$\boldsymbol{X}\hat{\boldsymbol{\beta}}^{rr} = \boldsymbol{U}\boldsymbol{\Sigma}(\boldsymbol{\Sigma}^2 + \lambda\boldsymbol{I})^{-1}\boldsymbol{\Sigma}\boldsymbol{U}^T\boldsymbol{y}.$$

Because the inner matrices are diagonal, we can express the above using the outer product rule for matrix multiplication as

$$X\hat{\boldsymbol{\beta}}^{rr} = \sum_{i=1}^{n} \boldsymbol{u}_i \left(\frac{\sigma_i^2}{\sigma_i^2 + \lambda} \right) \boldsymbol{u}_i^T \boldsymbol{y}. \tag{9.6}$$

Since $\lambda > 0$, the term in the round brackets above is always less than 1. For suitably chosen λ, this term is close to one for the larger singular values and small for the small singular values. By comparing (9.5) and (9.6), we see that the ridge regression approach is similar to the PCA approach, but ridge regression applies a soft instead of a hard thresholding function to suppress the effect of the components that are associated with the small singular values. It may be seen from (9.4) that the ordinary normal equation approach on the other hand applies no thresholding procedure at all.

We note that both the PCA and ridge regression methods involve a process which forces particular singular values of $\boldsymbol{X}$ to become smaller in one way or another. This same phenomenon also holds for other forms of regularisation which are not discussed here. This process of reduction of the eigevalues or singular values is the origin of the term "shrinkage", which is a term often used in the machine learning and statistical literature to describe the regularisation procedure.

9.2 Regularisation Using a Smoothness Penalty

This approach is useful if it is known that the solution $\boldsymbol{\beta}$ is smooth, i.e. changes in successive elements of $\boldsymbol{\beta}$ are small relative to $||\boldsymbol{\beta}||_2$. Consider a matrix $\boldsymbol{B}$ defined as

$$\boldsymbol{B} = \begin{bmatrix} 1 & -1 & & & \\ & 1 & -1 & & \\ & & \ddots & \ddots & \\ & & & 1 & -1 \\ & & & & 1 \end{bmatrix}$$

i.e. an identity matrix with the first upper diagonal replaced with -1. Then, the elements of $z = \boldsymbol{B}\boldsymbol{x}$ measure differences in successive elements of a vector $\boldsymbol{x}$. If the solution is to be smooth, then we want $||z||_2^2$ to be small. We can therefore modify the ordinary LS objective function to incorporate a smoothness constraint by adopting the following objective function:

$$\min_{\boldsymbol{\beta}} ||\boldsymbol{X}\boldsymbol{\beta} - \boldsymbol{y}||_2^2 + \lambda ||\boldsymbol{B}\boldsymbol{\beta}||_2^2.$$

Differentiating and setting the result to zero, the smoothness regularised solution $\hat{\boldsymbol{\beta}}^S$ satisfies

$$(\boldsymbol{X}^T\boldsymbol{X} + \lambda\boldsymbol{B}^T\boldsymbol{B})\boldsymbol{\beta} = \boldsymbol{X}^T\boldsymbol{y}.$$

The solution $\hat{\boldsymbol{\beta}}^S$ to this form of normal equations penalises a non-smooth solution. As in (9.1), the above may also be expressed in the form

$$\begin{aligned}\hat{\boldsymbol{\beta}}^S = \; & \arg\min_{\boldsymbol{\beta}} ||\boldsymbol{X}\boldsymbol{\beta} - \boldsymbol{y}||_2^2 \\ & \text{subject to } ||\boldsymbol{B}\boldsymbol{\beta}||_2 \leq t.\end{aligned}$$

Simulation Example We simulate an LS problem where the initial $\boldsymbol{X}$ is a 30×6 matrix of independent Gaussian random variables with mean zero and unit variance. In order to create a poorly conditioned $\boldsymbol{X}$, an SVD was performed on this matrix, and the smallest two singular values were both set to $0.001\times$ their original values. Then, a new $\boldsymbol{X}$ was reconstructed as $\boldsymbol{X} = \boldsymbol{U}\boldsymbol{\Sigma}'\boldsymbol{V}^T$, where $\boldsymbol{\Sigma}'$ is the modified matrix of singular values. The parameter $\boldsymbol{\beta} = [1, 1.1, 1.1, 1.1, 1.1, 1, 0.9]^T$, which can be verified by inspection to be smooth. A response vector $\boldsymbol{y}$ was simulated, such that $\boldsymbol{y} = \boldsymbol{X}\boldsymbol{\beta} + \boldsymbol{\epsilon}$, where $\boldsymbol{\epsilon}$ consists of zero-mean, independent Gaussian noise samples with a standard deviation such that the ratio of $||\boldsymbol{X}\boldsymbol{\beta}||_2$ to $||\boldsymbol{\epsilon}||_2$ is approximately 2:1. The $\boldsymbol{\beta}$ parameter was then estimated using the following methods, over 1000 iterations, each with a different value of $\boldsymbol{\epsilon}$: smoothness constraint, ridge regression, the pseudo-inverse, and the ordinary normal equations. The averaged relative error in the respective solutions is given in Table 9.1, after tuning the respective λ's for minimum error.

It is seen that imposing a smoothness penalty in this situation helps reduce the relative error in the solution by a significant margin. It also may be observed that the ordinary normal equations give a meaningless result, which arises due to the lack of regularisation, relatively high noise levels, and poor conditioning. The remaining regularised methods (ridge regression and pseudo-inverse) fare much better than the ordinary normal equation approach but, because they do not exploit the knowledge that the solution is smooth, cannot perform as well as smoothness regularisation.

Table 9.1 Relative errors in $\boldsymbol{\beta}$ for different forms of regularisation, when the solution is known to be smooth

Method	Relative error
Smoothness	0.13735
Ridge regression	0.83789
Pseudo-inverse	0.83318
Ordinary normal equations.	118.2612

9.3 Sparsity Regularisation

Regularisation by sparsity is also known as the *least absolute shrinkage and selection operator* (the LASSO) [4]. This form of regularisation imposes the prior knowledge that the solution must be sparse, i.e. have as few nonzero elements as possible. An example illustrating why sparsity is a useful form of penalty is given later in this section. The LASSO is also referred to as *basis pursuit* in some of the signal processing literature.

The least squares objective function for sparsity regularisation is given by

$$||\boldsymbol{X}\boldsymbol{\beta} - \boldsymbol{y}||_2^2 + \lambda||\boldsymbol{\beta}||_{\mathcal{S}}, \tag{9.7}$$

where $||\cdot||_{\mathcal{S}}$ denotes a norm which induces sparsity. Unlike the previous forms of regularisation, the lasso has no closed-form solution.

We now examine suitable sparsity-inducing norms for this purpose. Ideally, we would like a norm which simply counts the number of nonzero elements in the solution. Such a norm exists in the form of a modified $p = 0$ norm. However, the use of this norm results in a computationally intractable optimisation problem and so other forms of norm are more favoured. As may be seen from Fig. 9.1, the $p = 2$ norm penalty function is small for small values of its argument, and so it is ineffective at forcing small elements of the solution towards zero. On the other hand, it may be seen that the $p = 1$ norm penalty function imposes a significantly larger penalty for small values and so may be more effective at inducing sparsity. It may also be observed from the figure that the penalty imposed for $p < 1$ is larger than that for the 1-norm case for small values. The problem however is that the optimisation problem of (9.7) for $p < 1$ becomes non-convex and therefore is more difficult to compute. Therefore in practice, the lasso is implemented using the 1-norm penalty. In this case, the objective function of (9.7) is a convex quadratic program with a unique global minimum and can therefore be solved using readily available optimisation packages.

We now show a further illustration of the effect of using the 1-norm as a penalty function. In a manner similar to (9.2), it can be shown that there exists a value of t

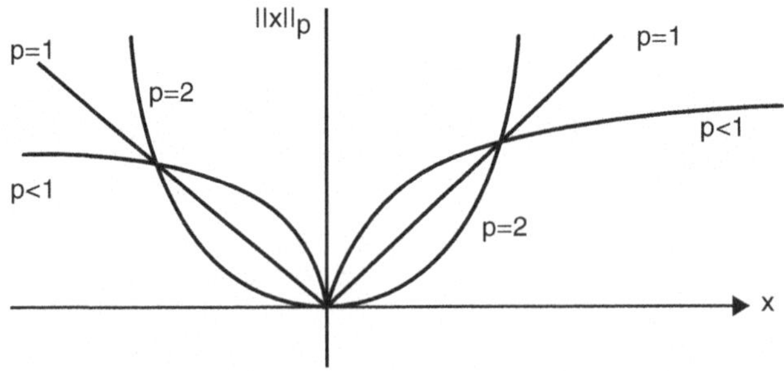

Fig. 9.1 Curves of $||\boldsymbol{x}||_p$ vs. $\boldsymbol{x}$ for various values of p for the one-dimensional case

for which the following optimisation problem has the same solution as that of (9.7), for a given value of λ:

$$\boldsymbol{\beta}^{\text{lasso}} = \arg\min_{\boldsymbol{\beta}} ||\boldsymbol{X}\boldsymbol{\beta} - \boldsymbol{y}||_2^2$$
$$\text{subject to } ||\boldsymbol{\beta}||_1 \leq t. \tag{9.8}$$

Figure 9.2 shows the elliptical contours of the joint confidence regions of the estimates $\boldsymbol{\beta}^{\text{lasso}}$ for various values of α, as discussed in Sect. 7.5. These ellipses are the contours for which $(\boldsymbol{\beta}^{\text{lasso}} - \boldsymbol{\beta}_o)^T \boldsymbol{X}^T \boldsymbol{X} (\boldsymbol{\beta}^{\text{lasso}} - \boldsymbol{\beta}_o) = k$, where the curves for various values of k are shown. The solution to (9.8) corresponds to the case where the ellipse with the lowest possible k just touches the constraint function $||\boldsymbol{\beta}||_1 \leq t$,

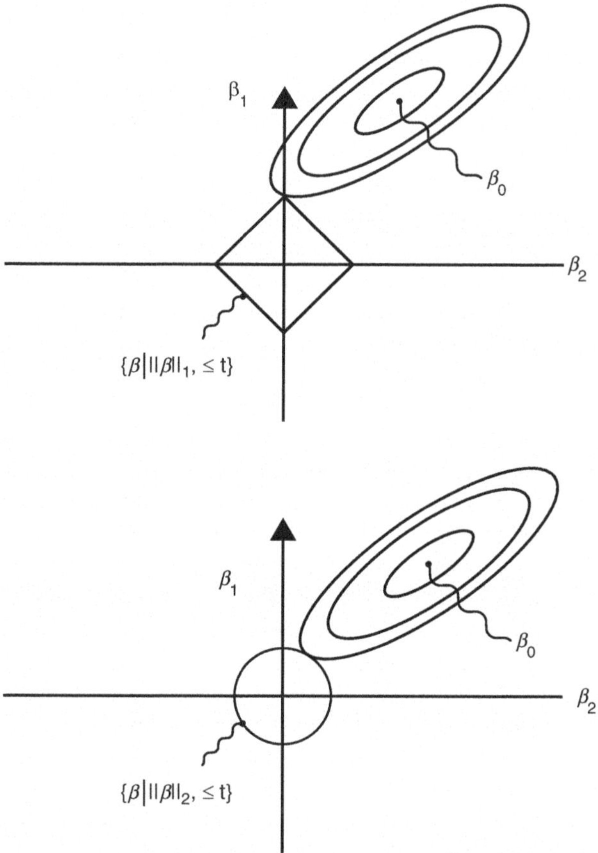

Fig. 9.2 Illustration of the effect of a 1-norm penalty in the two-dimensional case. The interior of the diamond region in the top figure is the set of points for which $||\boldsymbol{\beta}||_1 \leq t$, whereas the circular region in the lower figure corresponds to $||\boldsymbol{\beta}||_2 \leq t$. The ellipses are the contours of the LS error function

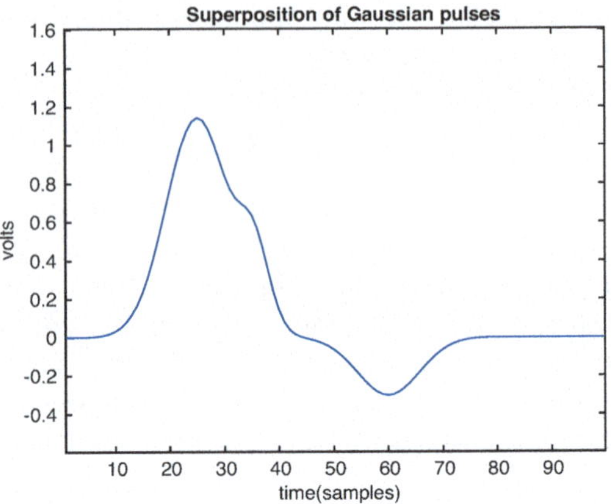

Fig. 9.3 An ERP waveform consisting of a superposition of Gaussian pulses for a lasso simulation

which is the diamond region in the top figure of Fig. 9.2. As can be seen, the ellipse touches the constraint function on the β_1 axis, where $\beta_2 = 0$, thus inducing a sparse solution in the present two-dimensional system. When this situation is extended to multiple dimensions, the "pointy" nature of the 1-norm constraint encourages a solution along one of the coordinate axes, where most of the elements of $\boldsymbol{\beta}$ are zero, again promoting sparsity in the solution. On the other hand, we see from the lower figure in Fig. 9.2 that the ellipse touches the circular constraint function $||\boldsymbol{\beta}||_2 \leq t$ at a point away from a coordinate axis, thus admitting a small value of β_2 to exist in the solution. From this example, we see that a 2-norm constraint is ineffective at encouraging a sparse solution.

We now present a simulation example. We consider a waveform which is a superposition of Gaussian pulses as shown in Fig. 9.3. This waveform loosely resembles a clean version of an event-related potential (ERP) waveform recorded from an electroencephalogram (EEG) in response to a deviant stimulus tone. For more on ERPs, see [2], whereas an excellent tutorial on the surrounding aspects of the subject is given in [1]. For this simulation, we generated 1000 such pulses, where in each waveform the pulses are subjected to timing jitter and amplitude variation, as well as additive coloured noise at an SNR of approximately 0 dB. These corrupted waveforms are characteristic of actual recorded EEG signals. We show 50 of the 1000 corrupted waveforms superimposed in Fig. 9.4.

In this simulation example, the objective is to estimate a single uncorrupted waveform, (of the type shown in Fig. 9.3) while still retaining its delay value, corresponding to the observed corrupted signals of the type shown in Fig. 9.4. We first apply the principal component method as discussed in Chap. 2, Sect. 2.6 to partially denoise the observed ERP signals $\boldsymbol{x}(t)$. We then apply a LASSO technique on the denoised signal to model each specific waveform.

In this vein, we construct a *dictionary* matrix of Gaussian pulses, as shown in Fig. 9.5, with each pulse having its own unique delay value. Here, we assume the width (standard deviation) of the pulses corresponds to those of the observed signals,

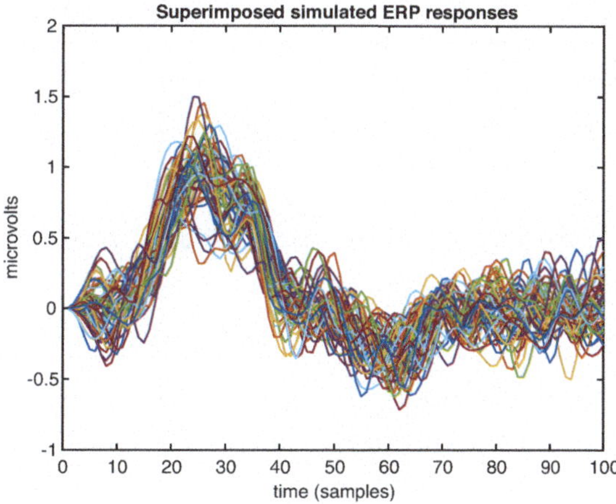

Fig. 9.4 50 superimposed corrupted waveforms simulating real ERP pulses

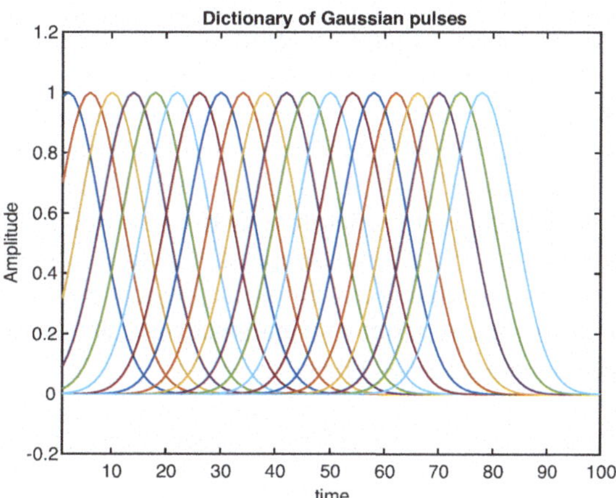

Fig. 9.5 A dictionary of Gaussian pulses, each with its own unique delay

as in Fig. 9.4. We model each observed ERP waveform $\boldsymbol{x}_i \in \mathbb{R}^m$ as a superposition of Gaussian pulses of the form

$$\boldsymbol{x}_i = \boldsymbol{D}\boldsymbol{a}_i + \boldsymbol{n}_i \tag{9.9}$$

where $\boldsymbol{D} \in \mathbb{R}^{m \times n}$ is the dictionary matrix and n is the number of entries (pulses) in $\boldsymbol{D}$. As such, every column of $\boldsymbol{D}$ represents a Gaussian pulse with its unique value of delay. Altogether there are $n = 120$ pulses (columns) in $\boldsymbol{D}$. The vector $\boldsymbol{a} \in \mathbb{R}^n$ is the vector of amplitudes associated with each pulse. The vector $\boldsymbol{n}$ is the additive noise. A naive approach for constructing the model $\boldsymbol{D}\boldsymbol{a}$ is then to solve the following LS optimisation problem:

$$\boldsymbol{a}^\star = \arg\min_{\boldsymbol{a}} ||\boldsymbol{x} - \boldsymbol{D}\boldsymbol{a}||_2^2. \tag{9.10}$$

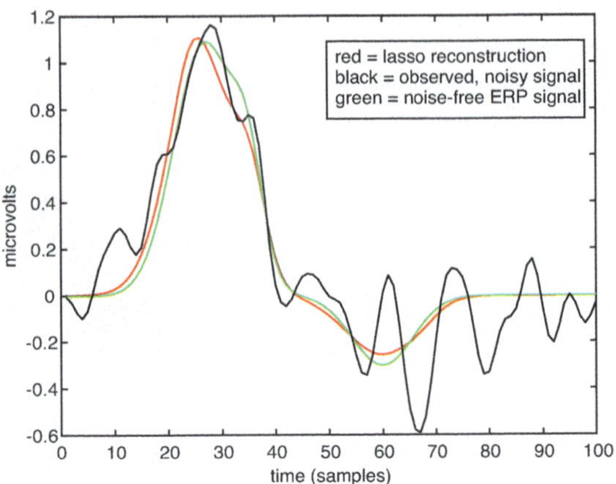

Fig. 9.6 A LASSO reconstruction of an ERP signal. Shown are the original noise-free ERP signal (green), its noisy version (black), and the LASSO reconstruction (red)

The model $\boldsymbol{D}\boldsymbol{a}^{\star}$ therefore represents the linear combination of pulses from $\boldsymbol{D}$ that best fit the observation $\boldsymbol{x}$. The problem with this naive approach is that there is no penalty on the complexity of the model (which in this case is the number of nonzero elements in $\boldsymbol{a}^{\star}$). In the presence of noise, a very large number of dictionary elements will be selected to minimise (9.10). What happens is that the approach in (9.10) will choose enough elements from $\boldsymbol{D}$ to fit the noise as well as the signal, with the result that the model will not suppress the noise. On the other hand, the *parsimonious* model is one which is as simple as possible without seriously degrading the expected fit and therefore tends to generalise well. We can encourage parsimony in (9.10) by introducing a LASSO penalty:

$$\boldsymbol{a}^{\text{lasso}} = \arg\min_{\boldsymbol{a}} ||\boldsymbol{x} - \boldsymbol{D}\boldsymbol{a}||_2^2 + \lambda||\boldsymbol{a}||_1.$$

This has the effect of significantly reducing the number of nonzero elements in $\boldsymbol{D}$, thus simplifying the model. The results of this process are shown in Fig. 9.6 for a single observation, using the value $\lambda = 0.5$. At the end of this simulation, there were a few remaining amplitudes with small value; those below a threshold value of 0.03 micrvolts were set to zero. It is seen that even in the presence of significant noise, the relative error in the lasso reconstruction is 0.0709, which in most circumstances could be considered as an acceptable level of error. It is interesting to note that of the 120 values in $\boldsymbol{a}$, only 10 values of the optimised solution exceeded the threshold value. This illustrates the effectiveness of the LASSO penalty in restricting the number of nonzero elements in the optimised solution.

Problems

1. We saw that regularisation introduces prior information into the LS model-building process. The Bayesian framework [5] (or any suitable reference on

introductory probability and statistics) is a probabilistic means to accomplish the same goal. In the LS context, this involves the creation of a *posterior* distribution $p(\boldsymbol{\beta}, \sigma)$ of the model parameters given the observed data $\boldsymbol{y}$. In the white noise case, the parameters are $\boldsymbol{\beta}$ and σ. $\boldsymbol{X}$ is assumed deterministic and known and so is not included in the distribution. Bayes rule states that

$$p(\boldsymbol{\beta}, \sigma | \boldsymbol{y}) \propto p(\boldsymbol{y} | \boldsymbol{\beta}, \sigma) p(\boldsymbol{\beta}, \sigma), \tag{9.11}$$

where $p(\boldsymbol{y}|\boldsymbol{\beta}, \sigma)$ is the *likelihood function* corresponding to (7.52) and $p(\boldsymbol{\beta}, \sigma)$ is the *prior distribution*, which weighs the likelihood function by imposing knowledge of the paramter values before any data is collected. For this problem, we can assume that $p(\boldsymbol{\beta}, \sigma) = p(\boldsymbol{\beta})p(\sigma)$, and $p(\sigma)$ may be assumed constant over the region of interest, so it can be ignored. What is the prior distribution $p(\boldsymbol{\beta})$ corresponding to (9.1)? *Hint*: Take negative log of (7.52) and compare to first term of (9.7). This gives the log likelihood function. Then what is a suitable prior pdf for $\boldsymbol{\beta}$ so that its negative log compares to the second term in (9.11), where the 1-norm is assumed?.

2. We have a discrete-time waveform which is obtained as the solution to an LS problem and is known to be a narrowband random process centred around a normalised centre frequency of 0.5 Hz. Describe a regularisation procedure for this scenario.
3. As in Problem 2 above, we have prior knowledge that the recovered waveform is to be confined to a specified subspace. Explain how to incorporate this prior knowledge into a penalty function.
4. The Laplacian distribution is given as

$$p(\boldsymbol{x}|\boldsymbol{\mu}, b) = \frac{1}{2b} \exp\left(- \frac{|\boldsymbol{x} - \boldsymbol{\mu}|}{b} \right).$$

What is the penalty function that corresponds to this distribution, when it is used as a prior distribution?

References

1. B. Blankertz et al., Single-trial analysis and classification of ERP components—a tutorial. NeuroImage **56**(2), 814–825 (2011)
2. C.C. Duncan et al., Event-related potentials in clinical research: guidelines for eliciting, recording, and quantifying mismatch negativity, P300, and N400. Clin. Neurophysiol. **120**(11), 1883–1908 (2009)
3. P.C. Hansen, The L-curve and its use in the numerical treatment of inverse problems (1999)
4. T. Hastie, R. Tibshirani, J. Friedman, *The Elements of Statistical Learning: Data Mining, Inference, and Prediction* (Springer Science & Business Media, Cham, 2009)
5. L.L. Scharf, C. Demeure, *Statistical Signal Processing: Detection, Estimation, and Time Series Analysis* (Prentice Hall, Hoboken, 1991)

Chapter 10
Toeplitz Systems

10.1 Toeplitz Systems [3]

In this section, we study the solution of *Toeplitz*[1] systems of equations, which naturally arise when studying autoregressive systems. Specifically, we focus attention on the normal equations, which yield estimates of the coefficients of a stationary autoregressive (AR) process. Our ostensible objective of this section is to exploit the structure of a Toeplitz system of equations to develop a fast technique for computing its solution that is considerably more computationally efficient than the $\mathcal{O}(n^3)$ flops required when solving using Gaussian elimination or the Cholesky decomposition. We first study AR processes in some detail. Then, our analysis leads to several further insights and developments such as prediction error filters, lattice filters, and efficient methods for computing the inverse and determinants of Toeplitz matrices.

10.2 Autoregressive Processes

The analysis of this section is an extension of Example 2 of Chap. 7, which we briefly review here. Here, we look at *autoregressive processes* and their relationship to Toeplitz systems.

Many random processes, either man-made or naturally occurring, are actually AR processes or can at least be closely approximated by them. Examples are voice and video signals, sinewaves plus noise, a variety of signals in control systems, signals induced by earthquakes, etc. Modelling these signals as AR processes leads to useful processing techniques such as signal compression, parameter estimation,

---

[1] A Toeplitz matrix is one where all elements on a given diagonal are equal. This is a significant form of matrix because covariance matrices of stationary signals are Toeplitz.

J. Reilly, *Fundamentals of Linear Algebra for Signal Processing*,
https://doi.org/10.1007/978-3-031-68915-4_10

identification, spectral estimation, etc. A very successful form of speech and video coding, referred to as *linear predictive coding*, is based on the idea that a human voice signal can be successfully modelled as a time-varying autoregressive process. Thus, being able to analyze and model AR processes is a very important signal processing tool that finds diverse application in many fields of engineering.

There are also *moving average* (MA) processes. These are the output of an all-zero filter in response to a white noise input. There are also autoregressive-moving average (ARMA) processes which are the output of a filter with both poles and zeros, in response to a white noise input. MA and ARMA processes are not directly considered in this chapter. The interested reader is referred to [5] for a deeper examination of these subjects.

Consider an all-pole filter driven by a white noise process $f(n)$ with output $x(n)$, as shown in Fig. 10.1. We define the denominator polynomial $H(z)$ as

$$H(z) = 1 - \sum_{p=1}^{P} h_p z^{-p},$$

where P is referred to as the model order. Thus, the overall filter transfer function is $\frac{1}{H(z)}$. Taking z transforms of the input–output relationships, we have

$$X(z) = \frac{1}{H(z)} F(z) \quad \text{or} \quad X(z)H(z) = F(z). \tag{10.1}$$

Converting the above relationship into the time domain, and realising that multiplication in the z-domain is convolution in time, we have using (10.1)

$$f(n) = x(n) - \sum_{p=1}^{P} h(p)x(n-p), \quad n = P+1, \ldots, N, \tag{10.2}$$

or

$$x(n) = \sum_{p=1}^{P} h(p)x(n-p) + f(n). \tag{10.3}$$

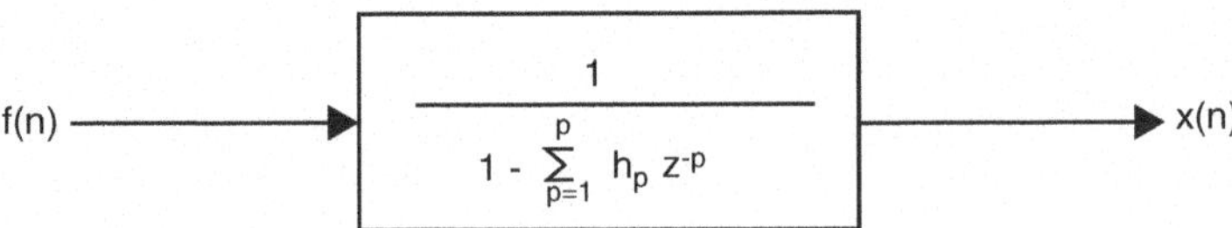

Fig. 10.1 An AR generating filter. A random white noise process $f(n)$ is fed into an all-pole filter to produce the AR process $x(n)$ at its output

Here, we assume the number of available samples N satisfies $N \gg P$. As discussed in Chap. 7, (10.3) offers the interpretation that the present output $x(n)$ is predictable from a linear combination of past outputs within an error $f(n)$, which is referred to as the *forward prediciton error*. This property is derived directly as a consequence of the all-pole characteristic of the filter.

The sequence $x(n)$ defined according to (10.3) is referred to as an *autoregressive process*. The quantity P is referred to as the *order* of the process, whereas the quantities $h(p)$ are referred to as the *coefficients* of the AR process. The input white noise process $f(n)$ is sometimes referred to as the *innovations* process. An AR process is completely characterised by the coefficients $h(\cdot)$ and the variance σ_f^2 of the sequence $f(n)$. In the following analysis, we assume we have available an observation of length N samples of a real, stationary, ergodic AR sequence $x(n)$. For generalisation to the nonstationary or complex case, see [3, 5].

Equation (10.3) gives a clue to determining the AR coefficients $h(p)$ given a sequence of data $x_1, x_2, \ldots, x(N)$. We can write this equation in matrix form as follows, where each row corresponds a specific value of the time index n:

$$\begin{bmatrix} x_{P+1} \\ x_{P+2} \\ \vdots \\ x_N \end{bmatrix} = \begin{bmatrix} x_P & x_{P-1} & \ldots, & x_1 \\ x_{P+1} & x_P & \ldots & x_2 \\ \vdots & \vdots & \ddots & \vdots \\ x_{N-1} & x_{N-2} & \ldots & x_{N-P-1} \end{bmatrix} \begin{bmatrix} h_1 \\ \vdots \\ h_P. \end{bmatrix} + \begin{bmatrix} f_{P+1} \\ f_{P+2} \\ \vdots \\ f_N \end{bmatrix}. \tag{10.4}$$

or

$$\boldsymbol{x} = \boldsymbol{X}\boldsymbol{h} + \boldsymbol{f}, \tag{10.5}$$

where the definitions of the above matrix/vector quantities follow from (10.4). We see that (10.5) is in the form of an LS regression equation, where the variables $\boldsymbol{x}$ are regressed onto themselves. This is the origin of the term "autoregressive".

We can justify the use of an LS procedure for estimating the coefficients using the following argument. From (10.2), the present value of the innovation or prediction error $f(n)$ consists of two terms: the first is the true observed value $x(n)$, while the second is the predicted value $\hat{x}(n)$, which is given by

$$\hat{x}(n) = \sum_{p=1}^{P} h(p)x(n-p). \tag{10.6}$$

Thus, $f(n) = x(n) - \hat{x}(n)$. We refer to (10.6) as a *prediction error filter* (PEF) as shown in Fig. 10.2, where the input is the AR process and the output is the prediction error. It is easily verified that the transfer function of the AR generating filter of Fig. 10.1 is the inverse of that of the PEF of Fig. 10.2. In effect, the PEF operates

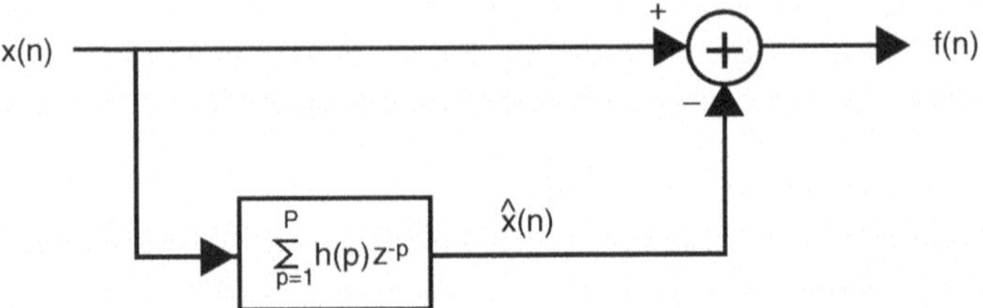

Fig. 10.2 A prediction error filter (PEF). The input is the AR process $x(n)$ and the output is the white noise process $f(n)$. It is readily verified that the transfer function $H(z)$ of this filter is $H(z) = 1 - \sum_{p=1}^{P} h(p)z^{-p}$, which is the inverse of the AR generating filter of Fig. 10.1

by whitening the input AR process. If the prediction error (or innovation) $f(n)$ is small in comparison to $\hat{x}(n)$ most of the time, then we can say the system is well modelled by (10.3). On this basis, we can determine the coefficients $h(\cdot)$ so that the quantity σ_f^2 or equivalently $\frac{1}{N-P}E||\boldsymbol{f}||_2^2$ (i.e. the variance of the sequence $f(n)$), is minimised with respect to $h(\cdot)$, where $\boldsymbol{f}$ is a vector comprising the sequence $f(n)$. In this way, we justify the use of the LS procedure for estimating the AR coefficients $h(p)$, which are found as the solution to the LS normal equations:

$$\boldsymbol{X}^T \boldsymbol{X} \boldsymbol{h}_{LS} = \boldsymbol{X}^T \boldsymbol{x}. \tag{10.7}$$

If expectations of the above are taken, we end up with a Toeplitz system of equations and an approximate Toeplitz system when N is finite. This system of equations can be solved using the LU decomposition or preferably the Cholesky decomposition as discussed in Chap. 5. These each require $\mathcal{O}(P^3)$ arithmetic operations. However, if the observed AR process $x(n)$ is stationary, in which case the corresponding covariance is Toeplitz, we can solve the system in $\mathcal{O}(P^2)$ operations using an algorithm that exploits the Toeplitz structure of the covariance matrix.

10.2.1 The Forward and Backward Prediction Error Equations

Here, we present the *Burg algorithm* [2], which is an extension of the *Levinson–Durbin recursion* (LDR), e.g. [4] for determining the AR prediction coefficients in a recursive manner. These algorithms are based on both the forward and backward prediction errors. We have already discussed the forward prediction case, where we predict the present value $x(n)$ based on a linear combination of past values—i.e. the prediction is in the forward direction. The backward prediction case is considered later.

We can write (10.2) in matrix form in a manner corresponding to (10.4), where each row corresponds to the respective value of the index n:

$$\begin{bmatrix} f(P+1) \\ f(P+2) \\ \vdots \\ f(N-1) \\ f(N) \end{bmatrix} = \begin{bmatrix} x_{P+1} & x_P & \ldots, & x_1 \\ x_{P+2} & x_{P+1} & \ldots & x_2 \\ \vdots & \vdots & \ddots & \vdots \\ x_{N-1} & x_{N-2} & \ldots & x_{N-P-1} \\ x_N & x_{N-1} & \ldots & x_{N-P} \end{bmatrix} \begin{bmatrix} 1 \\ -h_1 \\ \vdots \\ -h_P. \end{bmatrix} \tag{10.8}$$

Here, for every n, the present value $x(n)$ is predicted using P past values of the sequence and then subtracted from the true $x(n)$ to give the respective forward prediction error $f(n)$. Equation (10.8) can also be written in the form

$$\boldsymbol{f} = \boldsymbol{X}\boldsymbol{a}_f \tag{10.9}$$

where $\boldsymbol{a}_f \triangleq [1, -h_1, \ldots, -h_P]^T$. In this context, the sequence $f(n)$ is referred to as the forward *prediction error sequence*.

Backward Prediction Here, a past value is predicted based on a linear combination of future values. Prediction in the backward direction may seem to be a rather odd concept at first, since prediction into the past seems pointless because the predicted past value is already known. But backward prediction is a critical idea in the development of the Burg algorithm to follow. Mathematically, the idea of "predicting" a previous sample from a linear combination of future values (backward prediction) has just as much relevance as predicting a future value from a linear combination of past values (forward prediction).

The mathematical description of the backward prediction operation is given as

$$b(n-P) = x(n-P) - \sum_{p=1}^{P} h(p)x(n-P+p), \qquad n = P+1, \ldots, N. \tag{10.10}$$

where, by analogy to the forward prediction case, $b(n-P)$ is the backward prediction error at time $n-P$. In matrix form, (10.10) can be written as

$$\begin{bmatrix} b(P+1) \\ b(P+2 \\ \vdots \\ b(N-1) \\ b(N) \end{bmatrix} = \begin{bmatrix} x_{P+1} & x_P & \ldots, & x_1 \\ x_{P+2} & x_{P+1} & \ldots & x_2 \\ \vdots & \vdots & \ddots & \vdots \\ x_{N-1} & x_{N-2} & \ldots & x_{N-P-1} \\ x_N & x_{N-1} & \ldots & x_{N-P} \end{bmatrix} \begin{bmatrix} -h_P \\ \vdots \\ -h_1 \\ 1 \end{bmatrix} \tag{10.11}$$

or

$$\boldsymbol{b} = \boldsymbol{X}\boldsymbol{a}_b$$

where $\boldsymbol{a}_b \triangleq [-h_P, -h_{P-1}, \ldots, -h_1, 1]$. In (10.11), the rightmost column of $\boldsymbol{X}$ is predicted from a linear combination of the entries in the first P columns, with an error corresponding to the respective value of the sequence $b(n)$. It is apparent from the development of the LDR algorithm that the expected value of the backward prediction coefficients $b(\cdot)$ is identical to the forward prediction coefficients but flipped top to bottom. In the complex data case, the backward prediction coefficents are also complex conjugated. We note that the $\boldsymbol{X}$-matrix for the forward case in (10.8) is identical to that for the backward case, as in (10.11).

The Order Update Process The idea of the Levinson–Durbin [3, 4] or the Burg [2] recursions is to start with a simple 1×1 system of equations. Then, by induction, we use that result to solve a 2×2 system and recursively iterate until the solution to a $P \times P$ system is obtained. It is assumed the value P, which is the *order* of the underlying AR process, or the number of poles in the all-pole generating filter of Fig. 10.1, is known beforehand. Therefore, we must determine how to increment the forward and backward prediction error equations from order $p - 1$ to p. We first rewrite (10.8) and (10.11) for a prediction order of $p - 1$ instead of order P, where $p - 1$ past or future values, respectively, are used to predict the present value of x. The forward prediction error equations of order $p - 1$ are given as

$$\begin{bmatrix} f_{p-1}(p) \\ f_{p-1}(p+1) \\ \vdots \\ f_{p-1}(N-1) \\ f_{p-1}(N) \end{bmatrix} = \begin{bmatrix} x_p & x_{p-1} & \ldots, & x_1 \\ x_{p+1} & x_p & \ldots & x_2 \\ \vdots & \vdots & \ddots & \vdots \\ x_{N-1} & x_{N-2} & \ldots & x_{N-p} \\ x_N & x_{N-1} & \ldots & x_{N-p+1} \end{bmatrix} \begin{bmatrix} 1 \\ -h_{p-1}(1) \\ \vdots \\ -h_{p-1}(p-1). \end{bmatrix} \tag{10.12}$$

whereas the backward equations are given as

$$\begin{bmatrix} b_{p-1}(p) \\ b_{p-1}(p+1) \\ \vdots \\ b_{p-1}(N-1) \\ b_{p-1}(N) \end{bmatrix} = \begin{bmatrix} x_p & x_{p-1} & \ldots, & x_1 \\ x_{p+1} & x_p & \ldots & x_2 \\ \vdots & \vdots & \ddots & \vdots \\ x_{N-1} & x_{N-2} & \ldots & x_{N-p} \\ x_N & x_{N-1} & \ldots & x_{N-p+1} \end{bmatrix} \begin{bmatrix} -h_{p-1}(p-1) \\ \vdots \\ -h_{p-1}(1) \\ 1 \end{bmatrix}. \tag{10.13}$$

To augment the above two equations to order p, we begin by left appending an appropriate new column to the $\boldsymbol{X}$ matrix to reflect the increased order. The original model is maintained by appending a zero to the coefficient vector in the appropriate position. Then, (10.12) and (10.13) for the forward and backward cases, respectively, become

$$\begin{bmatrix} f_{p-1}(p+1) \\ f_{p-1}(p+2) \\ \vdots \\ f_{p-1}(N-1) \\ f_{p-1}(N) \end{bmatrix} = \begin{bmatrix} x_{p+1} & x_p & \ldots, & x_1 \\ x_{p+2} & x_{p+1} & \ldots & x_2 \\ \vdots & \vdots & \ddots & \vdots \\ x_{N-1} & x_{N-2} & \ldots & x_{N-P-1} \\ x_N & x_{N-1} & \ldots & x_{N-p} \end{bmatrix} \begin{bmatrix} 1 \\ -h_{p-1}(1) \\ \vdots \\ -h_{p-1}(p-1) \\ 0 \end{bmatrix} \tag{10.14}$$

and

$$\begin{bmatrix} b_{p-1}(p) \\ b_{p-1}(p+1) \\ \vdots \\ b_{p-1}(N-2) \\ b_{p-1}(N-1) \end{bmatrix} = \begin{bmatrix} x_{p+1} & x_p & \ldots, & x_1 \\ x_{p+2} & x_{p+1} & \ldots & x_2 \\ \vdots & \vdots & \ddots & \vdots \\ x_{N-1} & x_{N-2} & \ldots & x_{N-p-1} \\ x_N & x_{N-1} & \ldots & x_{N-p} \end{bmatrix} \begin{bmatrix} 0 \\ -h_{p-1}(p-1) \\ \vdots \\ -h_{p-1}(1) \\ 1 \end{bmatrix}. \tag{10.15}$$

Here, we have explicitly included the order $p-1$ as a subscript in order to distinguish the equations for order $p-1$ from those of order p to follow. Notice that in the forward case (10.14), n now varies from $p+1, \ldots, N$ for order p instead of $p, \ldots, N$ as it did in the order $p-1$ case, and in the backward case, n varies from $p \ldots, N-1$ instead of from $p \ldots N$. Thus, the backward prediction errors are delayed by one sample relative to the corresponding forward errors. In each case, the bottom row of $\boldsymbol{X}$ has been deleted, since the value x_{N+1} does not exist.

The method proceeds by multiplying both sides of the backward prediction equations (10.15) at order $p-1$ by a constant ρ_p, where ρ_p is a constant to be determined, and then adding the result to the forward equations. We take advantage of the fact that $\boldsymbol{X}$ is a common factor for both sets of equations. The result is given by

$$\left\{ \begin{bmatrix} f_{p-1}(p+1) \\ f_{p-1}(p+2) \\ \vdots \\ f_{p-1}(N-1) \\ f(N) \end{bmatrix} + \rho_p \begin{bmatrix} b_{p-1}(p) \\ b_{p-1}(p+1) \\ \vdots \\ b_{p-1}(N-2) \\ b_{p-1}(N-1) \end{bmatrix} \right\} =$$

$$\begin{bmatrix} x_{p+1} & x_p & \ldots, & x_1 \\ x_{p+2} & x_{p+1} & \ldots & x_2 \\ \vdots & \vdots & \ddots & \vdots \\ x_{N-1} & x_{N-2} & \ldots & x_{N-p-1} \\ x_N & x_{N-1} & \ldots & x_{N-p} \end{bmatrix} \cdot \left\{ \begin{bmatrix} 1 \\ -h_{p-1}(1) \\ \vdots \\ -h_{p-1}(p-1) \\ 0 \end{bmatrix} + \rho_p \begin{bmatrix} 0 \\ -h_{p-1}(p-1) \\ \vdots \\ -h_{p-1}(1) \\ 1 \end{bmatrix} \right\}. \tag{10.16}$$

The true forward prediction error equations at order p are given as

$$\begin{bmatrix} f_p(p+1) \\ f_p(p+2) \\ \vdots \\ f_p(N-1) \\ f_p(N) \end{bmatrix} = \begin{bmatrix} x_{p+1} & x_p & \dots, & x_1 \\ x_{p+2} & x_{p+1} & \dots & x_2 \\ \vdots & \vdots & \ddots & \vdots \\ x_{N-1} & x_{N-2} & \dots & x_{N-P-1} \\ x_N & x_{N-1} & \dots & x_{N-p} \end{bmatrix} \begin{bmatrix} 1 \\ -h_p(1) \\ \vdots \\ -h_p(p-1) \\ -h_p(p) \end{bmatrix}. \tag{10.17}$$

Given that we have a value for the constant ρ_p (a process which is discussed later), the comparison of (10.16) and (10.17) gives us the desired result that augments the order of the recursion from order $p-1$ to p. It may be seen that the forward prediction error equations at order p are given in terms of the forward and backward equations at order $p-1$ as

$$\begin{bmatrix} f_p(p+1) \\ f_p(p+2) \\ \vdots \\ f_p(N-1) \\ f_p(N) \end{bmatrix} = \begin{bmatrix} f_{p-1}(p+1) \\ f_{p-1}(p+2) \\ \vdots \\ f_{p-1}(N-1) \\ f(N) \end{bmatrix} + \rho_p \begin{bmatrix} b_{p-1}(p) \\ b_{p-1}(p+1) \\ \vdots \\ b_{p-1}(N-2) \\ b_{p-1}(N-1) \end{bmatrix}. \tag{10.18}$$

Again by comparing (10.16) and (10.17), the forward prediction coefficients at order p can be determined as

$$\begin{bmatrix} 1 \\ -h_p(1) \\ \vdots \\ -h_p(p-1) \\ -h_p(p) \end{bmatrix} = \begin{bmatrix} 1 \\ -h_{p-1}(1) \\ \vdots \\ -h_{p-1}(p-1) \\ 0 \end{bmatrix} + \rho_p \begin{bmatrix} 0 \\ -h_{p-1}(p-1) \\ \vdots \\ -h_{p-1}(1) \\ 1 \end{bmatrix}. \tag{10.19}$$

The update on the backward prediction coefficients can be determined in a similar manner. We accomplish this task by multiplying the forward equations at order $p-1$ by ρ_p adding to the backward equations at order $p-1$ and comparing the result to the true backward prediction equations at order p. The result is given as

$$\begin{bmatrix} b_p(p+1) \\ b_p(p+2) \\ \vdots \\ b_p(N-1) \\ b_p(N) \end{bmatrix} = \begin{bmatrix} b_{p-1}(p) \\ b_{p-1}(p+1) \\ \vdots \\ b_{p-1}(N-2) \\ b_{p-1}(N-1) \end{bmatrix} + \rho_p \begin{bmatrix} f_{p-1}(p+1) \\ f_{p-1}(p+2) \\ \vdots \\ f_{p-1}(N-1) \\ f_{p-1}(N) \end{bmatrix}. \tag{10.20}$$

Again, notice that for both the forward and backward prediction error order updates, the backward prediction errors are delayed by one time sample relative to the forward case. As mentioned previously, the backward prediction coefficients at order p may be determined simply by flipping the ordering of the forward coefficients from top to bottom (and conjugating if the data is complex). This gives

$$\begin{bmatrix} -h_p(p) \\ -h_p(p-1) \\ \vdots \\ -h_p(1) \\ 1 \end{bmatrix} = \begin{bmatrix} 0 \\ -h_{p-1}(p-1) \\ \vdots \\ -h_{p-1}(1) \\ 1 \end{bmatrix} + \rho_p \begin{bmatrix} 1 \\ -h_{p-1}(1) \\ \vdots \\ -h_{p-1}(p-1) \\ 0 \end{bmatrix}. \tag{10.21}$$

Thus, given ρ_p, the order update process from $p-1$ to p is now completely specified.

Determination of the Reflection Coefficients ρ_p The coefficients ρ_p are referred to as *reflection coefficients*, for reasons soon to be explained. In the development of the Levinson–Durbin recursion (LDR), it is shown that there exists a value ρ_p such that the order update from $p-1$ to p, as specified by (10.16) and (10.17), is exact—i.e. the prediction coefficients yielded by the LDR are identical (in the asymptotic sense) to those yielded by the respective normal equations at order p. However, in the presence of a finite N, the reflection coefficient values yielded by the LDR may result in AR coefficient values for which the respective all-pole filter of Fig. 10.1 becomes unstable; i.e. the poles of the AR transfer function $\frac{1}{H(z)}$ of Fig. 10.1 may migrate slightly outside the unit circle. The Burg algorithm [2] on the other hand produces a value for ρ_p for which the filter is guaranteed stable, regardless of the quantity of data available. Thus, the Burg method is the preferred approach. The Burg method chooses ρ_p to minimise the following quantity $J(\rho_p)$, given as

$$J(\rho_p) = \frac{1}{2}||\boldsymbol{f}_p||_2^2 + \frac{1}{2}||\boldsymbol{b}_p||_2^2, \tag{10.22}$$

where $\boldsymbol{f}_p$ and $\boldsymbol{b}_p$ are the vectors of the forward and backward prediction errors at order p, corresponding to (10.14) and (10.15), respectively. To minimise (10.22), we set the derivative of $J(\rho_p)$ with respect to ρ_p to zero. The derivative is given as[2]

$$\frac{dJ}{d\rho_p} = \frac{1}{2}\left[\boldsymbol{f}_p^T \frac{d\boldsymbol{f}_p}{d\rho_p} + \frac{d\boldsymbol{f}_p^T}{d\rho_p}\boldsymbol{f}_p\right] + \frac{1}{2}\left[\boldsymbol{b}_p^T \frac{d\boldsymbol{b}_p}{d\rho_p} + \frac{d\boldsymbol{b}_p^T}{d\rho_p}\boldsymbol{b}_p\right]. \tag{10.23}$$

In order to reveal the dependence of J on ρ_p, we must express the quantities $\boldsymbol{f}_p$ and $\boldsymbol{b}_p$ in terms of their representations at order $p-1$, given by (10.19) and (10.21).

[2] This formula follows by recognising that $||\boldsymbol{f}||_2^2 = \sum_i f_i^2$. Each term is a function of ρ. We differentiate each term of the summation using the ordinary chain rule and reassemble the results back into an inner product form.

From these equations, we have

$$\frac{d\boldsymbol{f}_p}{d\rho_p} = \boldsymbol{b}_{p-1} \tag{10.24}$$

and

$$\frac{d\boldsymbol{b}_p}{d\rho_p} = \boldsymbol{f}_{p-1}. \tag{10.25}$$

Substituting these values into (10.23) and rearranging, we obtain

$$\begin{aligned}\frac{dJ}{d\rho_p} &= (\boldsymbol{f}_{p-1}^T + \rho_p \boldsymbol{b}_{p-1}^T)\boldsymbol{b}_{p-1} + (\boldsymbol{b}_{p-1}^T + \rho_p \boldsymbol{f}_{p-1}^T)\boldsymbol{f}_{p-1} \\ &= 2\boldsymbol{f}_{p-1}^T \boldsymbol{b}_{p-1} + \rho_p\left(\boldsymbol{b}_{p-1}^T \boldsymbol{b}_{p-1} + \boldsymbol{f}_{p-1}^T \boldsymbol{f}_{p-1}\right).\end{aligned} \tag{10.26}$$

where we have substituted (10.18) for $\boldsymbol{f}_p^T$ and (10.20) for $\boldsymbol{b}_p^T$ in (10.23). In the last line, we have grouped common terms by noting that $\boldsymbol{f}^T\boldsymbol{b} = \boldsymbol{b}^T\boldsymbol{f}$. Equating this to zero, the optimum value for ρ_p is given as

$$\rho_p = -\frac{2\boldsymbol{f}_{p-1}^T \boldsymbol{b}_{p-1}}{\boldsymbol{f}_{p-1}^T \boldsymbol{f}_{p-1} + \boldsymbol{b}_{p-1}^T \boldsymbol{b}_{p-1}}. \tag{10.27}$$

We see that the value ρ_p can be determined using only the forward and backward prediction errors at order $p-1$. Thus, the specification of the Burg recursion is now complete. The process is initialised for $p = 0$ for which the prediction errors are the original data sequence $x(n)$, and then, the order is successively incremented up until $p = P$. We see that (10.27) is similar in form to the correlation formula between the sequences $\boldsymbol{f}_{p-1}$ and $\boldsymbol{b}_{p-1}$, where the backward sequence has been delayed by one sample, with a modified denominator. Determination of a suitable value for the final prediction order P is discussed later.

The pth-order forward and backward prediction errors can be represented in terms of the $(p-1)$th order prediction errors as shown in Fig. 10.3, which are derived directly from Eqs. (10.18) and (10.20). By cascading these sections, for $p = 0, \ldots, P$, and realising that both the forward and backward zeroth-order prediction error sequences are the original data sequence $x(n)$, we arrive at the structure shown in Fig. 10.4 as an alternative representation of the prediction error filter (PEF) structure of Fig. 10.2.

The Burg iteration procedure may be summarised as follows. Given a stationary AR sequence $x(n)$ of length N samples, we may determine its prediction/AR coefficients by performing the following schema. This schema uses the notation $\boldsymbol{a}_p = [1, -h_p(1), \ldots, -h_p(p)]^T$ as in (10.9); the value $r_0 = \boldsymbol{R}(1, 1)$ is the zero-lag autocorrelation value.

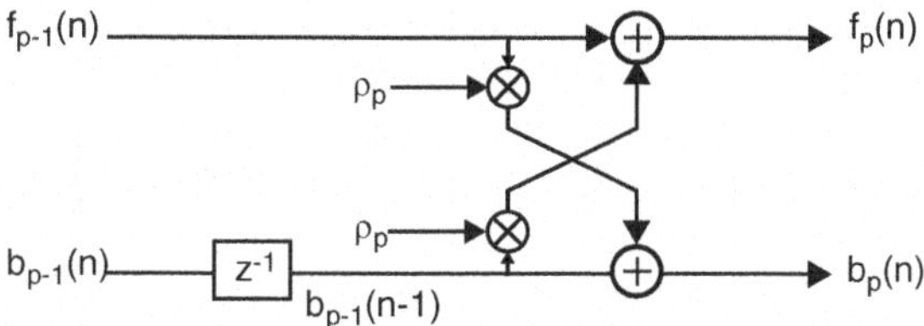

Fig. 10.3 Representation of a single stage of a lattice filter, corresponding to Eqs. (10.18) and (10.20), where the prediction error order is augmented from $p-1$ to p. The delay operator z^{-1} is included because the backward prediction error sequence is delayed by one sample relative to the forward sequence

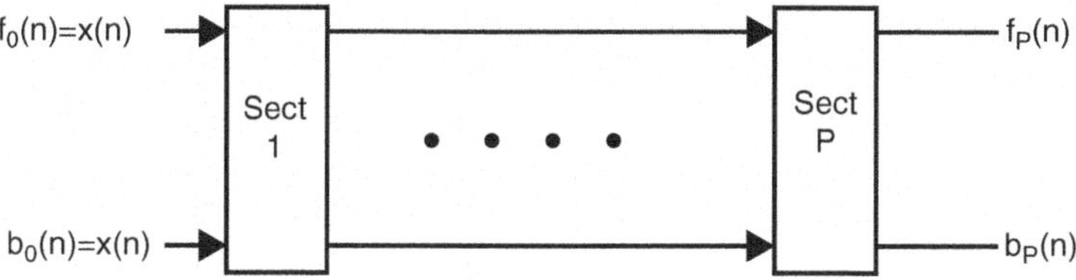

Fig. 10.4 The prediction error filter implemented as a cascaded lattice filter. Each section is implemented as shown in Fig. 10.3

Initialise

$$
\begin{aligned}
&p = 0\\
&a_0(0) = 1\\
&f_0(n) = x(n)\\
&b_0(n) = x(n)\\
&\sigma_0^2 = r_0\\
&p = 1
\end{aligned}
$$

Iterate for $p = 1, \ldots, P$

determine ρ_p from (10.27)
calculate $f_p(n)$ and $b_p(n)$ from (10.18) and (10.20) respectively
calculate the pth-order coefficients $\boldsymbol{a}_p$ from (10.19) and (10.21)
calculate σ_p^2 from (10.31), to follow.

end.

The Levinson–Durbin recursion (LDR) [3], which is not discussed here in detail but is closely related to the Burg algorithm, requires P^2 flops to determine the prediction coefficents $h(p)$, given all the required covariance values. The Burg algorithm on the other hand requires $P^2/2$ flops given the reflection coefficients ρ_p, not counting the number of flops required to evaluate the ρ_p values and the prediction errors $f_p(n)$ and $b_p(n)$. However, the computation of these latter quantities all involve only inner product operations and can therefore be executed very efficiently using floating point accelerator hardware, such as what is found on DSP processors. In contrast, solving the normal equations directly using Gaussian elimination or the Cholesky decomposition requires $\mathcal{O}(P^3)$ flops given the required covariance values. Therefore, given the proper hardware, the Burg algorithm compares favourably in computational terms to the other alternatives.

The issue of the choice of the final prediction order P requires some attention, since usually in practice this quantity is unknown. The typical behaviour of the prediction error variance σ_p^2 (i.e. the variance of either $f_p(n)$ or $b_p(n)$) is that its initial value $\sigma_0^2 = \sigma_x^2$ is relatively high in value but then decreases relatively quickly with increasing p, until p reaches P, whereupon further increases in p result in little or no decrease in σ_p^2. Thus, ideally, the value P can be estimated as that value of p where further decreases in σ_p^2 level off. However, in situations, e.g. where N is relatively small, then this behaviour of the prediction error variance is not as evident. Determination of the value of P is similar in concept to the determination of the rank r in the principal component analysis of Chap. 2. Information theoretic or statistical criteria [1, 6, 7] have been successfully applied to problems such as these. In each case, the method involves a function of order which consists of two terms: the first term is a measure of model fit, which in this case is the prediction error variance, and the second is the model complexity, which in this case is p. The value of p corresponding to the minimum of a weighted sum of these two terms is given as the estimated value of P.

We now investigate the relationship between the prediction error variances at order $p-1$ and order p. We denote the quantities $||\boldsymbol{f}_p||_2^2$ and $||\boldsymbol{b}_p||_2^2$ by the parameter γ_p^2. Since they are equivalent in the asymptotic sense, we do not distinguish between the values of γ_p^2 for the two cases. These quantities are related to the sample prediction error variances σ_p^2 by $\gamma_p^2 = (N-p)\sigma_p^2$. By taking squared 2-norms on each side of (10.18), we can write

$$\gamma_p^2 = \gamma_{p-1}^2 + 2\rho_p \boldsymbol{f}_{p-1}^T \boldsymbol{b}_{p-1} + \rho_p^2 \gamma_{p-1}^2. \tag{10.28}$$

From (10.27), we can write

$$\begin{aligned} \boldsymbol{f}_{p-1}^T \boldsymbol{b}_{p-1} &= -\frac{\rho_p}{2}(\boldsymbol{f}_{p-1}^T \boldsymbol{f}_{p-1} + \boldsymbol{b}_{p-1}^T \boldsymbol{b}_{p-1}) \\ &= -\rho_p \gamma_{p-1}^2, \end{aligned} \tag{10.29}$$

where we have used the fact that $\boldsymbol{f}^T\boldsymbol{f} = \boldsymbol{b}^T\boldsymbol{b} = \gamma^2$. Substituting the above into (10.28), we have

$$\begin{aligned} \gamma_p^2 &= \gamma_{p-1}^2 - 2\rho_p^2 \gamma_{p-1}^2 + \rho_p^2 \gamma_{p-1}^2 \\ &= \gamma_{p-1}^2 (1-\rho_p^2). \end{aligned} \tag{10.30}$$

We can divide each side by $N - p$ (the length of $f(n)$ or $b(n)$) to obtain the prediction error variances σ^2 as follows:

$$\sigma_p^2 = \sigma_{p-1}^2 \left(1 - \rho_p^2\right). \tag{10.31}$$

This equation has the same structure as that from electromagnetics which describes the reflected power from a load on a transmission line. This is the origin of the terminology *reflection coefficient*. We see that the prediction error variance (power) is nonincreasing with increase in order. From (10.31), we see that if $|\rho_p| > 1$ for any p, then σ_p^2 is negative thereby resulting in a failure of the recursion. Furthermore, it may be shown that if $|\rho_m| > 1$, then some of the poles of the transfer function $\frac{1}{H(z)}$ of Fig. 10.1 are outside the unit circle. This of course will lead to instability and a nonstationary process whose covariance matrix does not exist. Therefore, we must have $|\rho_p| \leq 1$ for $p = 1, \ldots, P$. However, with a finite sample of data, it is not guaranteed that the LDR will always satisfy this condition. It is also interesting to note that if at any stage p, for $p = 1, \ldots, P$, if $\rho_p = \pm 1$, then according to (10.31), $\sigma_p^2 = 0$. This means that the prediction error variance at order p is zero, and consequently, the process is perfectly predictable.

As mentioned previously, the Burg algorithm has the advantage that ρ_p satisfies the criterion $|\rho_p| \leq 1$ for any sample size of data. To prove this point, consider the matrix

$$\boldsymbol{W} = \begin{bmatrix} \boldsymbol{f}_p & \boldsymbol{b}_p \\ \boldsymbol{b}_p & \boldsymbol{f}_p \end{bmatrix}$$

Then, the matrix $\boldsymbol{W}^T \boldsymbol{W}$ which is

$$\boldsymbol{W}^T \boldsymbol{W} = \begin{bmatrix} \boldsymbol{f}^T \boldsymbol{f} + \boldsymbol{b}^T \boldsymbol{b} & 2\boldsymbol{f}^T \boldsymbol{b} \\ 2\boldsymbol{f}^T \boldsymbol{b} & \boldsymbol{f}^T \boldsymbol{f} + \boldsymbol{b}^T \boldsymbol{b} \end{bmatrix}$$

is positive semi-definite. Therefore, its determinant, which is the product of the eigenvalues, is nonnegative; hence

$$\left|2\boldsymbol{f}^T \boldsymbol{b}\right| \leq \left|\boldsymbol{f}^T \boldsymbol{f} + \boldsymbol{b}^T \boldsymbol{b}\right|. \tag{10.32}$$

Therefore, by comparing (10.32) with (10.27), we find that $|\rho_p| \leq 1$ as indicated.

It is interesting to note that there is a one-to-one correspondence between the sequence of reflection coefficients and the Pth-order prediction coefficients.[3]. In infinite precision arithmetic, the lattice filter of Fig. 10.4 is identical to the prediction error filter of Fig. 10.2. However, we have seen $|\rho_p| \leq 1, \quad p = 1, \ldots, P$, whereas the magnitude of the prediction coefficients $|h_p(k)|$ can be much larger,

---

[3] The proof is left as an exercise.

especially if some of the poles of the AR generating filter of Fig. 10.1 are near the unit circle. Therefore, for a given number of bits in the representation of the respective quantities, the ρ_p's can be represented with higher precision than the $\boldsymbol{a}_p$ coefficients, or for a given precision, the representation of the ρ_p requires fewer bits than representation of the $\boldsymbol{a}_p$. Thus, the lattice filter of Fig. 10.4 is generally the preferred structure.

10.3 Toeplitz Factorisations

The above analysis leads to several numerically efficient techniques for evaluating several functions of the covariance matrix $\boldsymbol{R}$ corresponding to the observed AR sequence $x(n)$. These include the inverse Cholesky factorisation, the inverse and the determinant. But before we investigate these ideas, we must first develop the so-called *forward prediction equations*. We start in this respect by taking expectations on each side of the normal equations given by (10.7)

$$E(\boldsymbol{X}^T\boldsymbol{X})\boldsymbol{h}_p = E(\boldsymbol{X}^T\boldsymbol{x}_o). \tag{10.33}$$

These are also known as the Yule–Walker equations. We observe from (10.4) that in the present recursive framework, these quantities are dependent on the order p of the prediction. (Here, we use the lowercase p instead of P in the recursive context to denote prediction order.) In the above, it is clear from (10.4) that the dimensions of the quantities $\boldsymbol{X}$, $\boldsymbol{h}$, and $\boldsymbol{x}_o$ all depend on the order p, although in the sequel, except for the quantity $\boldsymbol{h}_p$, we do not explicitly denote the dependence of these quantities on p, for clarity of notation.

We denote the $p \times p$ covariance matrix $E(\boldsymbol{X}^T\boldsymbol{X})$ in (10.33) as $\boldsymbol{R}$. Since the underlying environment is assumed stationary, $\boldsymbol{R}$ is Toeplitz, with diagonals equal to $r_0, r_{-1} \ldots r_{-p}$, where $r_i = E\big(x(n)x(n+i)\big)$. Also, we denote the quantity $E\big(\boldsymbol{X}^T\boldsymbol{x}_o\big)$ as $\boldsymbol{r}_o$, where $\boldsymbol{r}_o = [r_1, r_2, \ldots, r_p]^T$. The normal equations of (10.33) then become

$$\boldsymbol{R}\boldsymbol{h}_p = \boldsymbol{r}_o. \tag{10.34}$$

Since (10.34) involves expectations, it only holds for the ideal case when an infinite amount of data is available to form the covariance matrix $\boldsymbol{R}$ of $x(n)$. In the practical case where we have finite N, the normal equations corresponding to (10.34) are not exactly Toeplitz. However, in the following treatment, we still treat the finite case as if it were exactly Toeplitz. While this form of treatment does not necessarily minimise the prediction error for the finite case, it imposes an asymptotic structure to the finite N solution which tends to produce more stable results.

To develop the forward prediction equations, we need an expression for the quantity $\boldsymbol{f}^T\boldsymbol{f} \triangleq \gamma_p^2$ at order p. We see that γ_p^2 is the minimum of the quantity $||\boldsymbol{f}||_2^2$ from (10.17), obtained for the optimal $\boldsymbol{h}_p$ yielded from the Burg recursion procedure and is equal to $(N-p)\sigma_p^2$ of the sequence $\boldsymbol{f}_p$. The quantity γ_p^2 is given from the regression equations (10.5) by

$$\begin{aligned}\gamma_p^2 &= E(\boldsymbol{f}^T\boldsymbol{f}) = E(\boldsymbol{x}_o - \boldsymbol{X}\boldsymbol{h}_p)^T(\boldsymbol{x}_o - \boldsymbol{X}\boldsymbol{h}_p)\\ &= E(\boldsymbol{x}_o^T\boldsymbol{x}_o - \boldsymbol{x}_o^T\boldsymbol{X}\boldsymbol{h}_p - \boldsymbol{h}_p^T\boldsymbol{X}^T\boldsymbol{x}_o + \boldsymbol{h}_p^T\boldsymbol{X}^T\boldsymbol{X}\boldsymbol{h}_p).\end{aligned}$$

Substituting (10.34) into the above where

$$E(\boldsymbol{X}^T\boldsymbol{x}_o) = \boldsymbol{r}_o = \boldsymbol{R}\boldsymbol{h}_p,$$

and using the fact that $E(\boldsymbol{x}_o^T\boldsymbol{x}_o) = r_0$, we have our desired result:

$$\begin{aligned}\gamma_p^2 &= r_0 - \boldsymbol{r}_o^T\boldsymbol{h}_p - \boldsymbol{h}_p^T\boldsymbol{r}_o + \boldsymbol{h}_p^T\boldsymbol{r}_o\\ &= r_0 - \boldsymbol{r}_o^T\boldsymbol{h}_p. \qquad (10.35)\end{aligned}$$

We can combine (10.34) and (10.35) together into one matrix equation as follows:

$$\begin{matrix}\text{1st row given by (10.35)}\rightarrow \\ \\ \text{remaining rows are given by} \\ \text{(10.34): } \boldsymbol{r}_o - \boldsymbol{R}\boldsymbol{h}_p = \boldsymbol{0}\end{matrix} \left\{ \begin{bmatrix} r_0 & r_{-1} & \dots & r_{-p} \\ r_1 & r_0 & \ddots & r_{-p+1} \\ \vdots & \ddots & \ddots & \\ \vdots & & \ddots & r_{-1} \\ r_p & & r_1 & r_0 \end{bmatrix} \right. \begin{bmatrix} 1 \\ -h_p(1) \\ \vdots \\ \vdots \\ -h_p(p) \end{bmatrix} = \begin{bmatrix} \gamma_p^2 \\ 0 \\ \vdots \\ \vdots \\ 0 \end{bmatrix} \qquad (10.36)$$

These are called the *forward prediction error equations*. They are developed directly from (10.3) where we have predicted $x(n)$ in a forward direction from a linear combination of past values.

Now that we have developed the forward prediction error equations, we can proceed with various factorisations of the Toeplitz matrix $\boldsymbol{R}$ in (10.36). Note that in the case where $\boldsymbol{R}$ is used for AR prediction as in (10.34), $\boldsymbol{R}$ is $p \times p$, whereas the $\boldsymbol{R}$ used to generate the prediction errors as in (10.36) is $(p+1) \times (p+1)$.

We append new columns to each side of (10.36), corresponding to prediction orders $p = P, \dots, 0$. We obtain

$$\begin{bmatrix} r_0 & r_{-1} & \dots & r_{-P} \\ r_1 & r_0 & & \\ \vdots & & \ddots & \\ r_P & & & r_0 \end{bmatrix} \begin{bmatrix} 1 & & & \\ -h_P(1) & 1 & & \\ -h_P(2) & -h_{P-1}(1) & \ddots & \\ \vdots & \vdots & & 1 \\ -h_P(P) & -h_{P-1}(P-1) & \dots & -h_1(1) \; 1 \end{bmatrix}$$

$$
= \begin{bmatrix} \gamma_P^2 & & & & \\ 0 & \gamma_{P-1}^2 & & \boldsymbol{U} & \\ \vdots & \vdots & \ddots & & \\ 0 & 0 & & \gamma_1^2 & \\ 0 & 0 & & 0 & \gamma_0^2 \end{bmatrix} \tag{10.37}
$$

where $\boldsymbol{U}$ is an undetermined upper triangular matrix whose value will soon become apparent. We denote the matrices on the left of (10.37) as $\boldsymbol{R}$ and $\boldsymbol{A}$, respectively, and premultiply each side of (10.37) by $\boldsymbol{A}^T$ to obtain

$$
\boldsymbol{A}^T\boldsymbol{R}\boldsymbol{A} = \begin{bmatrix} \gamma_P^2 & & & & \\ & \gamma_{P-1}^2 & & \boldsymbol{U}' & \\ & & \ddots & & \\ & \boldsymbol{0} & & \gamma_1^2 & \\ & & & & \gamma_0^2 \end{bmatrix} \triangleq \boldsymbol{P}. \tag{10.38}
$$

But the matrix $\boldsymbol{A}^T\boldsymbol{R}\boldsymbol{A}$ is symmetric; therefore, $\boldsymbol{U}'$ must be $\boldsymbol{0}$, since the lower triangular portion of $\boldsymbol{P}$ is known to be zero from (10.36). Thus, the matrix $\boldsymbol{A}^T\boldsymbol{R}\boldsymbol{A} = \boldsymbol{P}$ is *diagonal*; i.e. $\boldsymbol{P} = \text{diag}[\gamma_P^2, \gamma_{P-1}^2, \ldots, \gamma_0^2]$. This is a significant result since it forms the basis of the following factorisations.

From (10.38), considering that $\boldsymbol{U}'$ is zero, we have

$$
\boldsymbol{R} = \boldsymbol{A}^{-T}\boldsymbol{P}\boldsymbol{A}^{-1}. \tag{10.39}
$$

Since the diagonal entries of $\boldsymbol{P}$ are nonnegative, we can write

$$
\boldsymbol{R} = \boldsymbol{B}^T\boldsymbol{B} \tag{10.40}
$$

where $\boldsymbol{B} = \boldsymbol{S}\boldsymbol{A}^{-1}$, where $\boldsymbol{S} = +\boldsymbol{P}^{1/2}$. Thus, $\boldsymbol{B}^{-1} = \boldsymbol{A}\boldsymbol{S}^{-1}$ is the inverse Cholesky factor of $\boldsymbol{R}$. The computation of $\boldsymbol{S}^{-1}$ requires relatively few FLOPS since $\boldsymbol{P}$ is diagonal. The inverse Cholesky factor plays an important role in signal processing and related fields, due to its role in whitening random processes. To illustrate further, from (10.40), we can write

$$
\boldsymbol{B}^{-T}\boldsymbol{R}\boldsymbol{B}^{-1} = \boldsymbol{I},
$$

In the finite N case, $\boldsymbol{R} = \frac{1}{N-p}\boldsymbol{X}^T\boldsymbol{X}$, where $\boldsymbol{X}$ is given in (10.8) and therefore the matrix $\sqrt{\frac{1}{N-p}}\boldsymbol{X}\boldsymbol{B}^{-1}$ has orthonormal columns, i.e. $\sqrt{\frac{1}{N-p}}\boldsymbol{X}\boldsymbol{B}^{-1} = \boldsymbol{Q}$, where $\boldsymbol{Q}$ is orthonormal. Therefore, $\boldsymbol{X} = \sqrt{N-p}\ \boldsymbol{Q}\boldsymbol{B}$ is the QR decomposition of $\boldsymbol{X}$. This is a very efficient and stable method of calculating the QR decomposition.

A further point is that by comparing the columns of the matrix $\boldsymbol{X}\boldsymbol{B}^{-1}$ from (10.37) to the forward prediction error sequences $f_p(n)$ given by (10.8), we see that these columns are in fact the unnormalised forward prediction error sequences at the order corresponding to the respective column. It therefore follows that the forward prediction error sequences at different orders are orthogonal.

Furthermore, using (10.39), it is straightforward to calculate $\boldsymbol{R}^{-1}$. This is given simply as

$$\boldsymbol{R}^{-1} = \boldsymbol{A}\boldsymbol{P}^{-1}\boldsymbol{A}^{T}. \tag{10.41}$$

This is a very efficient method of computing $\boldsymbol{R}^{-1}$, since as we have seen previously, $\boldsymbol{P}$ is diagonal.

Finally, we can also use (10.39) to calculate det$\boldsymbol{R}$. It is clear that $\boldsymbol{A}$ is unit lower triangular (ULT) (i.e. with one's along the diagonal). The inverse of a ULT is also ULT. Since the determinant of a ULT matrix is one, it follows that

$$\det\boldsymbol{R} = \prod_{p=0}^{P} \gamma_p^2 = r_o \prod_{p=1}^{P} \gamma_p^2. \tag{10.42}$$

It is interesting to note from (10.31) that if $\rho_p = 1$ for any p, then the respective σ_p^2 (i.e. γ_p^2) is zero and the corresponding prediction error sequence is then zero, with the result that the AR process is thus perfectly predictable. Therefore, according to (10.42), the determinant of the covariance matrix of a perfectly predictable AR process is zero and is thus rank deficient.

Problems

1. What happens to the Burg recursion when $\boldsymbol{R}$ is poorly conditioned or even rank deficient? Does the condition number bound of Chap. 5 still apply in this case? Discuss fully.
2. Describe characteristics of the sequence $x(n)$ such that the $\rho_p = 0$.
3. What happens to the solution when the final prediction order P is a) too small and b) too large?
4. Show that the forward prediction error sequences at different orders resulting from (10.37) are orthogonal. Explain how to alter (10.37) so that the backward sequences at different orders are orthogonal.
5. Compare the number of flops required for computing the inverse Cholesky factor using Gaussian elimination methods to that required using the Burg method.
6. Show there is a one-to-one correspondence between the sequence $[\rho_0, \ldots \rho_P]$ and the AR prediction coefficients $[h_1, \ldots, h_P]$. Given one, explain how to find the other.

References

1. H. Akaike, A new look at the statistical model identification. IEEE Trans. Autom. Control **19**(6), 716–723 (1974). https://doi.org/10.1109/TAC.1974.1100705
2. J.P. Burg, *Maximum Entropy Spectral Analysis* (Stanford University, Stanford, 1975)
3. S. Haykin, *Nonlinear Methods of Spectral Analysis*, vol. 34 (Springer Science & Business Media, New York, 2006)
4. H. Krishna, S. Morgera, The Levinson recurrence and fast algorithms for solving Toeplitz systems of linear equations. IEEE Trans. Acoust. Speech Signal Process. **35**(6), 839–848 (1987)
5. S.L. Marple Jr., W.M. Carey, *Digital Spectral Analysis with Applications* (Prentice-Hall, Englewood Cliffs, NJ, 1989)
6. A.A. Neath, J.E. Cavanaugh, The Bayesian information criterion: background, derivation, and applications. Wiley Interdiscip. Rev. Comput. Stat. **4**(2), 199–203 (2012)
7. J. Rissanen, Modeling by shortest data description. Automatica **14**(5), 465–471 (1978)

Appendix A
Review of Linear System Theory

In this chapter, we present a concise summary of the fundamentals of linear system theory. The material in this appendix is not part of the mainstream of this text per se. However, the topic of signal processing at large is intimately connected with linear system theory, and as such, there are numerous examples throughout the text which assume knowledge of this topic. This appendix is included so that readers who have no background in this area can at least gain sufficient knowledge to appreciate the message of the examples. It also gives an opportunity for those readers who have already taken a course in this area but could do with a brushup. The presentation of this appendix is intended to be very brief; the interested reader may consult the references for a more detailed treatment. A basic background in complex variables is assumed.

We first look at the Fourier transform and its variations arising from sampling the frequency and time domain signals. These include the discrete-time Fourier transform (DTFT) and the discrete Fourier transform (DFT). Various properties of the Fourier transform are presented, as well as many examples, to aid in the understanding of this subject. We then describe how difference equations are used to model linear, time-invariant (LTI) systems and look at how the z-transform is used to characterise the difference equation in the LTI case. With this background in place, we review the implementation and operation of various types of digital filters.

A.1 The Fourier Transform

First, we review a few preliminaries. Recall that a complex number z may be represented in rectangular form; i.e. $z = a + jb$, where a, b are the real and imaginary parts respectively, and $j = \sqrt{-1}$. The quantity z can also be represented in polar form $z = r\exp(j\theta)$, where r is the magnitude and θ is the angle or phase. For our purposes, the polar form is more convenient since it offers a simpler

J. Reilly, *Fundamentals of Linear Algebra for Signal Processing*,
https://doi.org/10.1007/978-3-031-68915-4

interpretation of the underlying math. A complex exponential $e(t) = \exp(j2\pi f_o t)$ is a complex, time-varying vector with unit magnitude and an angle that increases linearly at the rate of f_o cycles per second. Thus, z rotates in the complex plane at a rate of f_o Hz. Note that f_o can be either positive or a negative—in the positive case, the exponential rotates in the counterclockwise direction and in the clockwise direction otherwise. In the continuous version of Fourier analysis, the time domain signal $x(t)$ is represented in a basis consisting of an orthogonal set of complex exponential basis functions over a continuum of frequencies, both positive and negative. The weights or coefficients of $x(t)$ in this complex exponential basis are the quantities $X(f)$ given by (A.1) below. The units of f are cycles per second, i.e. Hz.

We now consider a nonperiodic, deterministic, continuous-time signal $x(t)$ with finite energy. The Fourier transform $X(f)$ of $x(t)$ is defined as

$$X(f) = \int_{-\infty}^{+\infty} x(t) \exp(-j2\pi f t) dt. \tag{A.1}$$

The quantity $X(f)$ is referred to as the *frequency domain* representation of $x(t)$, whereas $x(t)$ itself is referred to as the *time domain* representation. The quantity $X(f)$ is also referred to as the *spectrum* of $x(t)$. There are many good references on the application of the Fourier transform, e.g. [2, 5] that the reader may refer to for further details and examples.

Given $X(f)$, we can recover the time domain representation using the following integral, which is referred to as the inverse Fourier transform:

$$x(t) = \int_{-\infty}^{+\infty} X(f) \exp(j2\pi f t) df. \tag{A.2}$$

Thus, there is a one-to-one correspondence between $x(t)$ and $X(f)$, i.e. knowing one completely specifies the other. The Fourier transform of any signal $x(t)$ is in general a complex quantity; i.e. in polar coordinates, $X(f)$ has both a magnitude response $R(f)$ and phase response $\theta(f)$ which are both functions of frequency. Therefore, we may write the complex quantity $X(f)$ using the polar repesentation as $X(f) = R(f) \exp\big(j\theta(f)\big)$. It is worthy of note that any real-life signal $x(t)$ that can be measured using some form of physical instrumentation must be pure real in the time domain. We show later in Property 13 that if $x(t)$ is indeed pure real, then $X(f)$ must be conjugate symmetric; i.e., $X(-f) = X^*(f)$, where $(\cdot)^*$ indicates complex conjugation. This relation implies that $R(f)$ is symmetric with frequency; i.e., $R(f) = R(-f)$, and that $\theta(f)$ is anti-symmetric with frequency; i.e., $\theta(-f) = -\theta(f)$.

The inner product p of two continuous-time functions $a(t)$ and $b(t)$ is given as $p = \int a(t)b^*(t)dt$, where in the complex case the second argument is complex conjugated (as indicated by the * symbol). Thus, the quantity $X(f)$ in (A.1) may be seen to be the inner product (correlation) between the complex exponential term $\exp(j2\pi f t)$ and $x(t)$, as a function of frequency. In this manner, we see that $X(f)$ is indeed the coefficient (weighting) of the basis function $\exp(j2\pi f t)$ in the

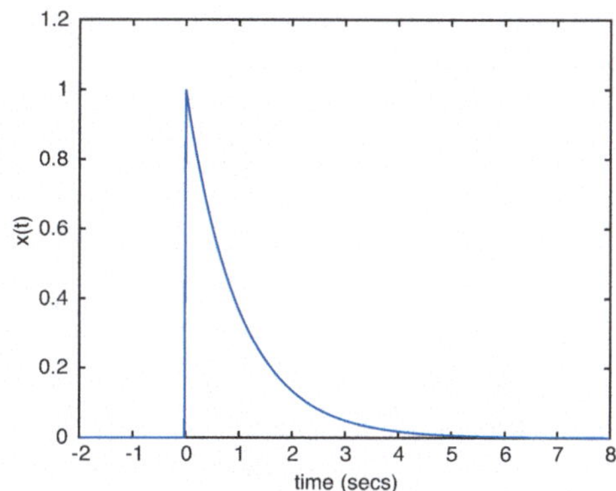

Fig. A.1 The decaying exponential pulse $x(t) = u(t)\exp(-at)$ whose spectrum we wish to evaluate. In this example, $a = 1$

representation of $x(t)$. Now that we know the coefficients $X(f)$, we can reconstruct $x(t)$ as a linear combination of basis functions $\exp(j2\pi ft)$, each weighted by their respective coefficient $X(f)$. This is the operation implicit in Eq. (A.2).

A.1.1 Examples of Some Selected Fourier Transform Pairs

Here, we present a few examples of Fourier transform pairs that are commonly used in signal processing. The notation $x(t) \rightleftharpoons X(f)$ indicates that $x(t)$ and $X(f)$ are a Fourier transform pair.

Example A.1 (The Exponential Pulse) We wish to find the Fourier transform of the decaying exponential pulse given by $x(t) = \exp(-at)u(t)$ shown in Fig. A.1, where a is a positive real number and $u(t)$ is the unit step function.[1] Substituting this definition of $x(t)$ into (A.1) and recognising that $x(t) = 0, t < 0$ because of the step function, we have

$$\begin{aligned} X(f) &= \int_0^{+\infty} \exp(-at)\exp(-j2\pi ft)dt \\ &= \int_0^{+\infty} \exp\big[-t(a+j2\pi f)\big]dt \\ &= \frac{1}{a+j2\pi f} \\ &= \frac{1/a}{1+j2\pi(f/a)}. \end{aligned} \tag{A.3}$$

[1] The unit step function $u(t)$ is defined such that

$$u(t) = \begin{cases} 1, & t \geq 0, \\ 0 & \text{otherwise.} \end{cases}$$

The magnitude response $R(f)$ and phase response $\theta(f)$ (in degrees) are given from (A.3), respectively, as

$$R(f) = \frac{1/a}{\left[1 + \left(2\pi(f/a)\right)^2\right]^{\frac{1}{2}}} \tag{A.4a}$$

$$\theta(f) = -\frac{180}{\pi} \arctan \frac{2\pi f}{a}. \tag{A.4b}$$

These functions are plotted in Figs. A.2 and A.3, respectively. It is indeed a remarkable phenomenon that the weighted sum of the complex exponentials, where the coefficients are given by $X(f)$, has the following properties: it adds up to zero

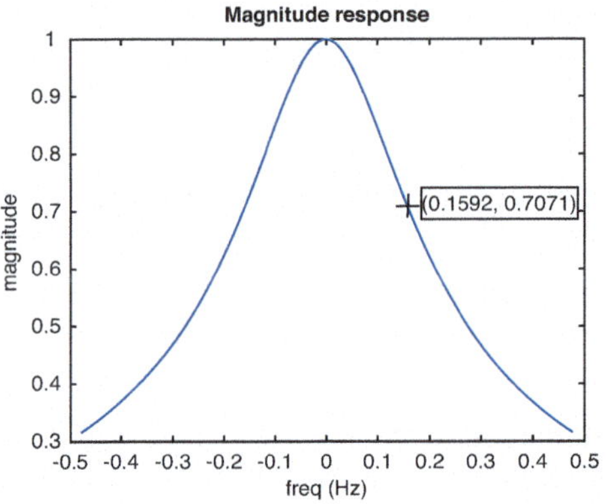

Fig. A.2 The magnitude response $R(f)$ given by (A.4a) corresponding to the exponential pulse of Fig. A.1. Note that since $x(t)$ is pure real, $X(f)$ is conjugate symmetric about $f = 0$ as stated by Property 13 below, and therefore $R(f)$ is symmetric about $f = 0$. At $f = \pm a/2\pi = \pm 0.1592$, $R(f) = 1/\sqrt{2} = 0.7071$ as shown, where in this example $a = 1$. Note that $R(f) \to 0$ as $f \to \pm\infty$

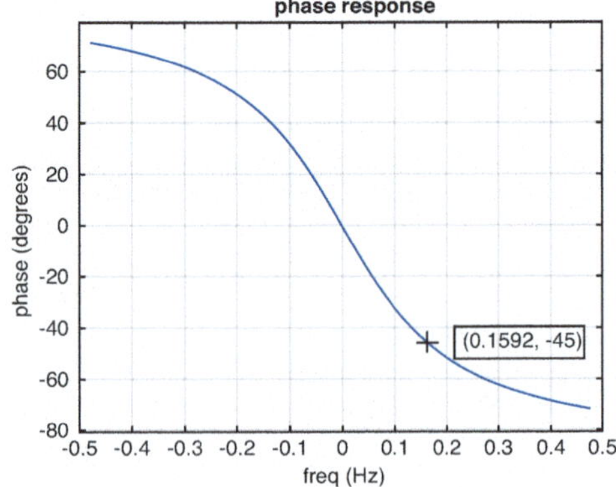

Fig. A.3 The phase response $\theta(f)$ given by (A.4b) corresponding to the exponential pulse of Fig. A.1. Note the antisymmetry of $\theta(f)$ about $f = 0$ due to the conjugate symmetry of $X(f)$, a fact that arises since $x(t)$ is pure real. At $f = \pm a/2\pi, \theta(f) = \mp 45°$ as shown. The phase response asymptotically approaches $\mp 90°$ as $f \to \pm\infty$

for $t \leq 0$; it exhibits a discontinuous edge from the value 0 to 1 at $t = 0$; it is pure real and exhibits the appropriate exponential decay for $t > 0$. We note that the spectrum $X(f)$ contains both positive and negative frequencies. This may appear to be in contravention of the usual definition of frequency, which is usually taken to be a positive value only. However, in the present context, we are dealing with complex exponentials, not the regular sinusoids as basis functions. Complex exponentials may have either a positive frequency, in which case they rotate in the counterclockwise direction in the complex plane, or a negative frequency in which case they rotate in the clockwise direction.

In fact, we now show that the existence of negative frequency components is one of the necessary conditions for $x(t)$ to be pure real. We take the integral in (A.2) and divide it into its positive and negative regimes:

$$\begin{aligned} x(t) &= \int_{-\infty}^{+\infty} X(f) \exp(j2\pi f t) df \\ &= \int_{-\infty}^{0} X(f) \exp(j2\pi f t) df + \int_{0}^{+\infty} X(f) \exp(j2\pi f t) df \end{aligned}$$

Replacing f with $-f$ in the first integral on the right, we have

$$x(t) = \int_{0}^{+\infty} X(-f) \exp(-j2\pi f t) df + \int_{0}^{+\infty} X(f) \exp(j2\pi f t) df.$$

If $x(t)$ is to be pure real for all t, then the arguments in these two integrals must be complex conjugates of each other for all values of f and t. In this way, the imaginary components cancel each other for all time and at each frequency. We see that the exponential terms $\exp(j2\pi f t)$ and $\exp(-j2\pi f t)$ already form a complex conjugate pair for all t and f. For the complete arguments to be complex conjugates, we must also have $X(-f) = X^*(f)$. If we write $X(f)$ in polar form as discussed already, we have $X(f) = R(f) \exp\big(j\theta(f)\big)$, where $R(f)$ is the magnitude function and $\theta(f)$ is the angle function. It is then clear that $X^*(f) = R(f) \exp\big(-j\theta(f)\big)$. Substituting these quantities into the above, we have

$$\begin{aligned} x(t) &= \int_{0}^{+\infty} R(f) \exp(-j\theta(f)) \exp(-j2\pi f t) df \\ &\quad + \int_{0}^{+\infty} R(f) \exp(+j\theta(f)) \exp(j2\pi f t) df \\ &= \int_{0}^{+\infty} R(f) \Big[\exp\big[-j\big(\theta(f) + 2\pi f t\big)\big] + \exp\big[+j\big(\theta(f) + 2\pi f t\big)\big] \Big] df \\ &= \int_{0}^{+\infty} 2R(f) \cos\big[\theta(f) + 2\pi f t\big] df. \end{aligned}$$

where in the last line the familiar identity for $\cos(\theta)$ has been used. Since the above integrand is always pure real under the stated conditions, $x(t)$ is pure real. Thus, the

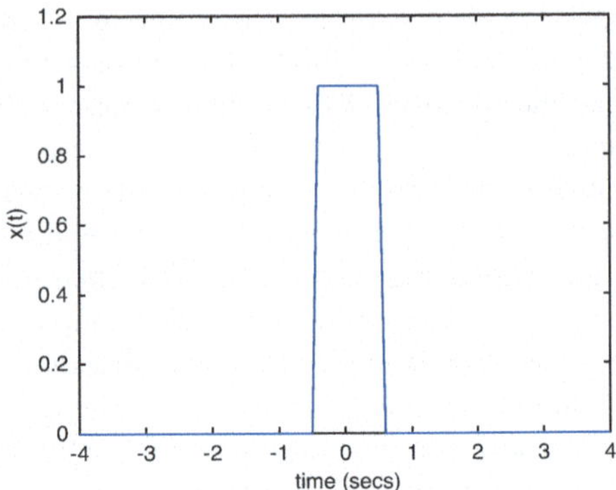

Fig. A.4 A rectangular pulse $x(t)$ in the time domain of duration $T = 1$ second and amplitude $A = 1$. The corresponding Fourier transform $X(f)$ is shown in Fig. A.5. Note that this function is real and symmetric about $t = 0$, which, according to Properties 13 and 20 discussed below, results in $X(f)$ being both real and symmetric as well

negative frequency components are necessary for $x(t)$ to be pure real, which is what any signal existing in real life must be. An additional necessary condition is that $X(-f) = X^*(f)$; i.e. the positive and negative coefficients at any given frequency must be complex conjugates of each other. Note that the above treatment serves as a proof for Property 13 to follow.

Example A.2 (The Rectangular Pulse) Here, we consider a rectangular pulse of amplitude A and duration T in the time domain, symmetrically spaced about $t = 0$, as shown in Fig. A.4. The Fourier transform $X(f)$ is readily evaluated from (A.1) as

$$\begin{aligned} X(f) &= \int_{-\infty}^{+\infty} x(t) \exp(-j2\pi f t) dt \\ &= A \int_{-T/2}^{T/2} \exp(-j2\pi f t) dt \\ &= \frac{-A}{j2\pi f} \exp(-j2\pi f t) \Big|_{-T/2}^{T/2} \\ &= \frac{-A}{j2\pi f} \Big[\exp(-j\pi f T) - \exp(\pi f T) \Big] \\ &= \frac{A}{\pi f} \left[\frac{\exp(\pi f T) - \exp(-\pi f T)]}{2j} \right] = AT \frac{\sin(\pi f T)}{\pi f T} \\ &= AT \operatorname{sinc}(fT), \end{aligned} \tag{A.5}$$

where $\operatorname{sinc}(x) \triangleq \sin(\pi x)/(\pi x)$ and where we have multiplied top and bottom in the second last line by the factor T to reveal the sinc function inherent in $X(f)$. Note that by applying l'Hopital's rule [6] we have $\operatorname{sinc}(0) = 1$. This function has zeros at values $f = n/T$, where n is an integer. The time and frequency domain functions are plotted in Figs. A.4 and A.5, respectively. Note that the sinc function

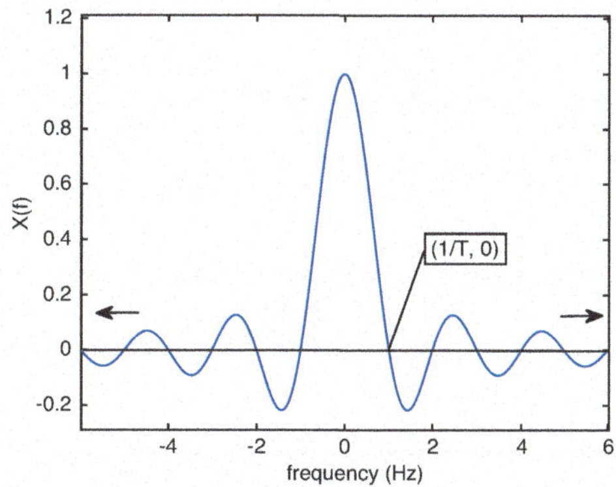

Fig. A.5 The Fourier transform $X(f) = AT\,\mathrm{sinc}(fT)$ corresponding to the rectangular pulse $x(t)$ of Fig. A.4 in the time domain. The sinc function is zero at frequencies equal to any multiple of $1/T$ Hz, where in this case, $T = 1$

decays relatively slowly with frequency at the rate $1/f$. Thus, $X(f)$ decays slowly enough to contain the high-frequency components that are necessary to model the instantantaneous transitions that are present in $x(t)$.

Example A.3 (The δ-Function and the Periodic Pulse Train) The Dirac δ-function $\delta(t)$ is defined as a function which is zero everywhere except at $t = 0$, such that $\delta(0) = \infty$ and the total area under the curve is one. Thus, the δ-function is a pulse with infinite height at $t = 0$ and infinitesimal width and zero everywhere else. It is difficult to determine the Fourier transform of the δ-function by direct evaluation of the integral (A.1) due to the discontinuous nature of $\delta(t)$. However, it is possible to evaluate the Fourier transform $\Delta(f)$ of $\delta(t)$ by indirect methods—details are given in [2]. The result is

$$\delta(t) \rightleftharpoons 1. \tag{A.6}$$

An additional Fourier transform pair that is of relevance in the discretisation of a continuous-time signal $x(t)$ is the periodic pulse train $p(t)$, where a δ-function occurs periodically in time every T seconds. The function $p(t)$ is given by

$$p(t) = \sum_{n=-\infty}^{+\infty} \delta(t - nT). \tag{A.7}$$

Again, the Fourier transform $p(f)$ of $p(t)$ is difficult to evaluate directly, but an indirect analysis yields the following result:

$$p(t) \rightleftharpoons f_o \sum_{n=-\infty}^{\infty} \delta(f - nf_o) \triangleq P(f), \tag{A.8}$$

where $f_o = 1/T$. These functions are plotted in Fig. A.6. We see that the Fourier transform of a periodic pulse train $p(t)$ with period T in time is another periodic pulse train $P(f)$ with period f_o in frequency, weighted by the factor f_o.

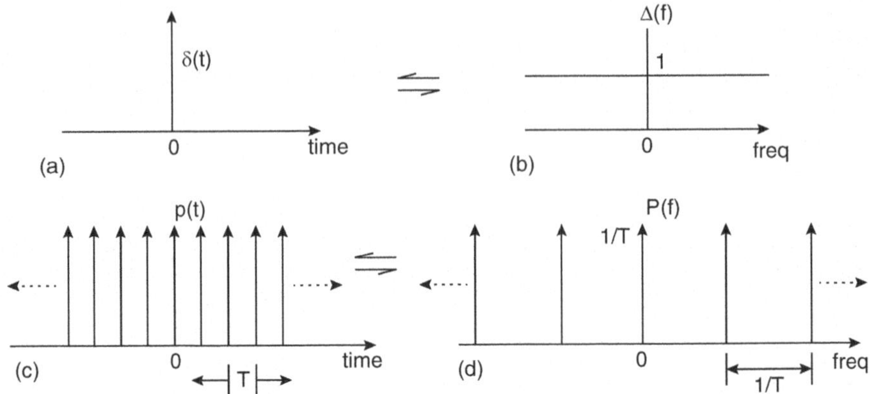

Fig. A.6 Fourier transforms relating to δ-functions. Note that all δ-functions in this figure have infinite height but finite area. (**a**) The δ-function itself. (**b**) Its Fourier transform $\Delta(f)$. (**c**) The periodic pulse train $p(t)$ with period T. (**d**) The Fourier transform $P(f)$ of $p(t)$. When we refer to the "weight" of a δ-function which is $1/T$ in this case, we imply its area and not its amplitude, since that is infinite

There are many good references on the application of the Fourier transform, e.g. [1, 2, 5] that the reader may refer to for further details and examples on this topic.

A.1.2 Properties of the Fourier Transform

Here, we outline a few of the most important properties of the Fourier transform that are important for its interpretation and understanding. The properties are stated without proof; for a more extended treatment on this topic, the reader is referred to [2].

Property 12 (Scaling) *If $x(t) \rightleftharpoons X(f)$, then $x(at) \rightleftharpoons \frac{1}{|a|} X\left(\frac{f}{a}\right)$, where a is a real number.*

This property may be interpreted in the following manner. For example, if $x(t)$ is compressed in time, then $|a| > 1$ and therefore $x(at)$ varies more quickly in time. Faster variation in time requires higher-frequency components to be present in the corresponding spectrum. The spectrum corresponding to $x(at)$ is given above as $\frac{1}{|a|} X\left(\frac{f}{a}\right)$, indicating that the frequency domain is expanded and therefore contains frequency components over a wider range of frequencies. This is turn implies the presence of higher-frequency components in the spectrum corresponding to $x(at)$ that allow for its faster variation. Note that if $|a| < 1$, the time domain is expanded and the frequency domain is compressed; i.e. the behaviour in the two domains is interchanged.

Property 13 *If $x(t)$ is pure real, then $X(-f) = X^*(f)$.*

i.e. if $x(t)$ is pure real (which it must be in real life), then $X(f)$ must exhibit conjugate symmetry about the vertical axis. Therefore, using the familiar rules of complex conjugation, we must have $X(-f) = X^*(f)$, or $R(-f) = R(f)$ and $\theta(-f) = -\theta(f)$, where as before $R(f)$ and $\theta(f)$ denote the magnitude and phase responses of $X(f)$, respectively. A justification of this property is given in the discussion of Example A.1.

Property 14 (Convolution in Time) *Consider two Fourier transform pairs* $x_1(t) \leftrightharpoons X_1(f)$ *and* $x_2(t) \leftrightharpoons X_2(f)$. *Then*

$$X_1(f)X_2(f) \leftrightharpoons \int_{-\infty}^{+\infty} x_1(\tau)x_2(t-\tau)d\tau \tag{A.9}$$

where the integral on the right is the continuous-time convolution integral.

We discuss the convolution operation later in Sect. A.3 for the discrete-time case, since this version of the convolution operation is much easier to understand.

Thus, the multiplication of two spectra in the frequency domain corresponds to convolution of their respective time domain representations. This property has relevance to filtering operations. A filter may be considered to be a black box with an input $x(t)$ and an output $y(t)$, whose purpose is to create an output $y(t)$ where the bands of unwanted frequencies that are present in the input $x(t)$ are rejected at the output. This means that the magnitude response of the filter is close to one in the frequency bands that are to be preserved at the output and small in the bands that are to be rejected. An example of an (ideal) frequency response is shown in Fig. A.7, where it is seen that only low-frequency components below the frequency f_c are to be passed through to the output. It should be noted that it is not possible in practice to implement a filter with the ideal frequency response of Fig. A.3; however, it is possible to build filters that are close approximations to it.

Analog or digital filters operate by creating an output $y(t)$ that is inherently a convolution operation of the intrinsic *impulse response* $h(t)$ of the filter with the input signal $x(t)$. The impulse response is internally generated by the filter's hardware configuration. We briefly discuss digital filters in more detail later in Sect. A.3. The impulse response $h(t)$ is by definition the inverse Fourier transform of the frequency response $H(f)$ of the filter. Thus, in accordance with (A.9), the convolution operation implicitly performed by the filter in the time domain has the effect of multiplying the spectrum $X(f)$ of the input signal with the frequency response $H(f)$ of the filter to give the output spectrum $Y(f)$. Therefore, with respect to the example of Fig. A.7, any frequency components in the input spectrum $X(f)$ that are above the cutoff frequency f_c are rejected in forming the output $Y(f)$. For obvious reasons, the type of frequency response shown in Fig. A.3 is referred to as *low pass*.

Property 15 (Convolutions Involving the δ-Function) *Further to Property 14 above, consider* $x_2(t)$ *to be a δ-function and* $x_1(t)$ *to be an arbitrary signal. Then, from (A.6),* $X_2(f) = 1$. *Therefore, from (A.9), the spectrum resulting from the*

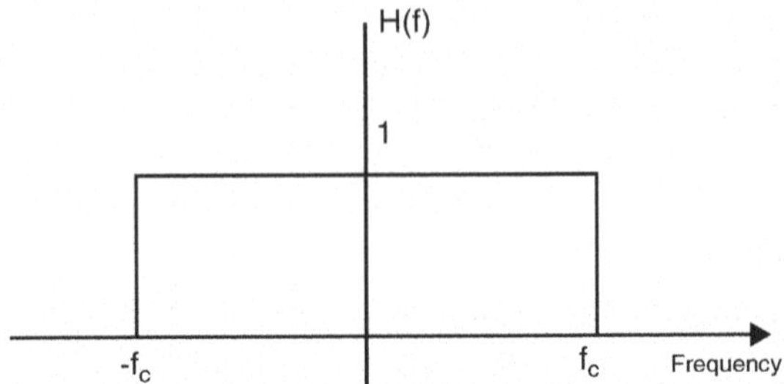

Fig. A.7 An ideal low-pass filter magnitude response. The passband is the region of frequencies which satisfy $|f| < |f_c|$, i.e. the region where $|H(f)| = 1$. The stop band is the region where $|f| > |f_c|$ where $|H(f)| \approx 0$. The transition band is the region of frequencies between the pass and stop bands. In the ideal case, its width is zero, but in the practical case, it has finite width

convolution $x_1(t) \otimes x_2(t) = x_1(t) \otimes \delta(t)$ is $X_1(f) \times 1 = X_1(f)$, where $\otimes$ denotes the convolution operator. Correspondingly, we have $x_1(t) \otimes \delta(t) = x_1(t)$.

Thus, the convolution of any function with a δ-function is the function itself. This property is used in Sect. A.2 to develop the spectrum of a sampled signal.

Property 16 (Translation in Time) *If $x(t) \rightleftharpoons X(f)$, then*

$$x(t-\tau) \rightleftharpoons X(f)\exp(-j2\pi f\tau), \tag{A.10}$$

where τ is the amount of time shift in seconds and f is frequency in Hertz.

Since $X(f) = R(f)\exp[j\theta(f)]$, the spectrum of the shifted version $x(t-\tau)$ of $x(t)$ is given from (A.10) as $X(f) = R(f)\exp[j\theta(f)]\exp[-j2\pi f\tau] = R(f)\exp[j(\theta - 2\pi f\tau)]$. Thus, shifting $x(t)$ by τ seconds in time corresponds to adding a phase term of $-2\pi f\tau$ radians to $X(f)$, that is, linear in frequency with slope is proportional to the time shift τ. It may be observed that the quantity $-2\pi f\tau$ radians is precisely the phase shift required to time shift each frequency component present in $x(t)$ by the amount τ.

Property 17 (Discretisation (Sampling) in Time) *Let $x(t) \rightleftharpoons X(f)$. We form a sampled version $x(nT)$ of $x(t)$ by uniformly sampling $x(t)$ at a rate of f_o samples per second, with a corresponding sampling period of $T = 1/f_o$ seconds. Let the sampled signal $x(nT)$ be defined as*

$$x(nT) = \begin{cases} x(t), \ t = nT, & n = \ldots -2, -1, 0, 1, 2, \ldots \\ 0, & \textit{otherwise}. \end{cases} \tag{A.11}$$

Then, the spectrum of $x(nT)$, denoted by the quantity $X(e^{j2\pi fT})$, is given as

$$X(e^{j2\pi fT}) = \frac{1}{T}\sum_{n=-\infty}^{+\infty} X(f - nf_o). \tag{A.12}$$

For a more detailed treatment of this topic, refer to, e.g. [7]. It is clear from (A.12) that $X(e^{j2\pi fT})$ is periodic in frequency with period f_o Hz. An interpretation of this property therefore is that discretisation in time results in periodicity in frequency.

Property 18 (Duality) *If* $x(t) \rightleftharpoons X(f)$*, then*

$$X(t) \rightleftharpoons x(-f). \tag{A.13}$$

This property follows due to the similarity of the integrals of (A.1) and (A.2). For example, we have seen from Example A.2 above that the inverse Fourier transform of a rectangle function in time is a sinc function in frequency. According to the property of duality above, if $x(t)$ is a sinc function in time, then its Fourier transform $X(f)$ is a rectangular function in frequency.

In words, we may interpret this property as follows. Suppose we apply some form of operation A on the time domain signal $x(t)$. For example, A can be discretisation, translation, scaling, etc. Operation A in the time domain produces some result B on $X(f)$ as we have seen from the above properties. Then, if the operation and result are interchanged, i.e. we apply B to the time domain signal $x(t)$, then the result on $X(f)$ becomes A (after suitable consideration is given to the negative sign in the argument of $x(-f)$ in (A.13). For example, Property 16 above states that if $x(t)$ is translated (operation A) in the time domain time by a delay τ, then a linear phase term is added to $X(f)$ (result B). Now, if B is applied to $x(t)$, i.e. the linear phase term $\exp(j2\pi f_o t)$ is applied to $x(t)$, then $X(f)$ is translated by f_o Hz.

A.1.3 Application of the Duality Property to Properties 12–17 Above

Applying the duality property on the properties given above yields the following additional properties. In the application of these properties, recall that the frequency variable f must be replaced with $-f$ to yield the quantity $x(-f)$ as indicated by (A.13).

Property 19 *With reference to Property 12, compression in one domain causes expansion in the other.*

Property 20 *With reference to Property 13, if one domain is real, the other is (conjugate) symmetric. If follows that if one domain is both real and symmetric, the other is also real and symmetric.*

Property 21 *Generalising the convolution Property 14, we see that convolving two signals in one domain corresponds to the multiplication of their representations in the other domain.*

Property 22 *As implied above, applying duality to Property 16 implies that if an exponential phase term* $\exp(j2\pi f_o t)$*, which is linear in time with slope proportional* *to f_o, is applied to $x(t)$, then the respective spectrum $X(f)$ is translated in frequency to give $X(f - f_o)$.*

Assume $x(t)$ is a real bandlimited signal whose spectrum is conjugate symmetric about $f = 0$. This is a typical characteristic of real-world signals such as audio and video signals. Such signals are referred to as *baseband* signals. It is interesting to consider the expression

$$x(t)\exp(j2\pi f_o t) + x(t)\exp(-j2\pi f_o t) = 2x(t)\cos(2\pi f_o t)$$
$$\rightleftharpoons X(f - f_o) + X(f + f_o).$$

Thus, applying the present property to each of the above terms, we see that the effect of multiplying $x(t)$ by a sinusoidal function in time translates the respective spectrum both upwards and downwards by f_o Hz. The resulting spectrum is seen to be conjugate symmetric, as required for the corresponding time domain version to be pure real. This is the underlying principle behind modulation, which is a required process in the propagation of signals through transmission media such as the internet.

Property 23 *With reference to Property 17, duality implies that a spectrum which is discrete in frequency with sample period f_o Hz is periodic in time with period $1/f_o$ seconds. Thus, discreteness in one domain implies periodicity in the other domain.*

A.2 Discretisation in Time and Frequency

The analog or continuous-time representation of signals as we have discussed so far is not very useful in modern applications of signal processing methods, since signals must be representable in digital form in order to be processed by a computer or sent over a digital communications link. There are two steps required in the conversion of a continuous-time analog signal $x(t)$ into a digital format. The first involves uniformly sampling (discretising) $x(t)$ at times nT, where T is the sampling period, to produce a sequence of samples $x(nT)$, where n is an integer. This is the process described by (A.11) and depicted in Fig. A.8. In executing the sampling process, we must ensure that the original signal $x(t)$ is recoverable from its sequence of samples $x(nT)$—more on this later. Once we have $x(nT)$, in the second stage, we quantise each sample value of the sequence $x(nT)$ into a binary number containing a finite number of bits. We can then concatenate the bits representing each sample into a digital bit stream, consisting of a prescribed number of bits per second. The signal is now in a digital format, ready to be transmitted through a digital channel or processed by computer. The second stage is referred to as *analog-to-digital conversion* (ADC), a process which is well described in [3]. Because the second stage is not as relevant to the current signal processing discussion, in the following, we discuss only the first stage in detail. As an example, ordinary landline telephone

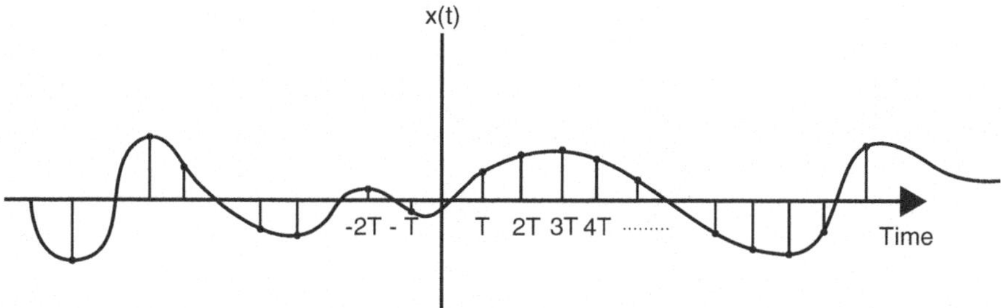

Fig. A.8 The sampling process in time. The continuous curve represents the original signal $x(t)$, whereas the sequence of spikes constitutes the corresponding discrete-time signal $x(nT)$ according to (A.11). The result is a discrete pulse train in time

speech is sampled at the rate of 8000 samples per second. Each sample is converted into an 8-bit number, yielding a 64 Kbits/second bit stream that represents the speech. There are numerous texts available of the topic of *digital signal processing* (DSP) and related subjects, such as [1, 7, 8]. The reader may refer to these texts for a more detailed discussion on DSP methods.

A.2.1 Discretisation in Time and the Nyquist Condition

For reasons which will soon become apparent, the spectrum of the discretised version of the continuous-time signal $x(t)$ is denoted as $X(e^{j2\pi fT})$. Once the samples are stored into memory, the value of the sampling period T loses relevance, in that once the samples are in memory, it is not possible to distinguish whether the sampling period is, e.g. 1 year or 1 nsec. For this reason, from now on, we normalise the value T to $T = 1$. Then, in this case, the complex exponential $e^{j2\pi fT} \equiv e^{j2\pi f}$ is periodic in f with period 1 Hz. Thus, f must be contained within the interval of one period; e.g. $-1/2 < f \leq 1/2$ so that there is a one-to-one correspondence between f and $e^{j2\pi f}$. (In principle, the range of f can be any interval of one period duration; the interval $0 < f \leq 1$ is another convenient choice.) This normalisation implies that real-world frequency values are divided by the sampling frequency f_o to give the equivalent normalised digital frequency. Likewise, real-world time intervals are multiplied by f_o to give their corresponding time values in the normalised and discretised framework. Since $T = 1$, it is customary to represent the discrete-time sequence as $x[n]$ and its Fourier transform as $X(e^{j2\pi f})$.

The process of sampling a signal $x(t)$ in time may be viewed as the point-by-point multiplication of the original analog signal $x(t)$ by the pulse train $p(t)$ given by (A.7) in Example A.3 above and as shown in Fig. A.8. We see from Property 21 that multiplication in time corresponds to convolution in frequency. Thus, we wish to evaluate $X(e^{j2\pi f}) = P(f) \otimes X(f)$. We have also seen that the convolution of some arbitrary function with a δ-function is the function itself. This convolution in frequency is depicted in Fig. A.9, where we see that the nth δ-function component of $p(f)$ contributes one periodic repetition in the form $X(f - nf_o)$, where n is an integer, to the spectrum $X(e^{j2\pi f})$. The result is a function with periodic repetitions of the original spectrum $X(f)$ spaced at intervals of f_o Hz.

Figure A.9c shows the preferred case where it is seen that $f_o > 2f_c$, where f_c is the highest-frequency component of $X(f)$. In this case, the spectral repetitions do not overlap, and it is possible to recover $x(t)$ with no error by a low-pass filtering operation, such that the repetition around 0 Hz is isolated at the output. The condition $f_o > 2f_c$ is referred to as the *Nyquist condition* and determines the minimum sampling frequency f_o for error-free recovery of the original continuous-time signal $x(t)$. For more detail, refer to [7].

If the sampling frequency f_o is reduced so that the Nyquist criterion is not satisfied, then $f_o < 2f_c$ as shown in Fig. A.9d and (e), causing the spectral repetitions overlap and therefore $x(t)$ cannot be recovered without error. The error resulting from the spectral overlap is referred to as *aliasing error*.

The presence of higher-frequency components in $X(f)$ implies that $x(t)$ can vary more quickly in time. The fact that $X(f)$ is bandlimited thus limits the rate at which $x(t)$ can vary between its sampling instants. Thus, when an $x(t)$ is reconstructed where the Nyquist condition has been satisfied, the lack of high-frequency components creates the condition where there can be only the one path (corresponding to the true $x(t)$) between the samples of $x[n]$ in the reconstruction of the original version of $x(t)$.

A.2.2 The Discrete-Time Fourier Transform (DTFT)

We can now combine the above discussions on discretisation to determine a Fourier transform relationship corresponding to (A.1) and (A.2) for the case where $x[n]$[2] is discrete in time with sampling period $T = 1$sec. and where as we have seen from Property 17, the frequency domain is continuous as well as periodic with period 1 Hz. As discussed earlier, we choose the interval for f as $-1/2 < f \leq 1/2$. The resulting equations for this modified version of the Fourier relationship, which accommodates these conditions, are given as

$$x[n] = \int_{-1/2}^{1/2} X(e^{j2\pi f}) \exp(j2\pi fn)df. \tag{A.14}$$

[2] When a continuous-time signal $x(t)$ is discretised in time and the sampling period normalised to one second, the standard notation for the resulting signal becomes $x[n]$.

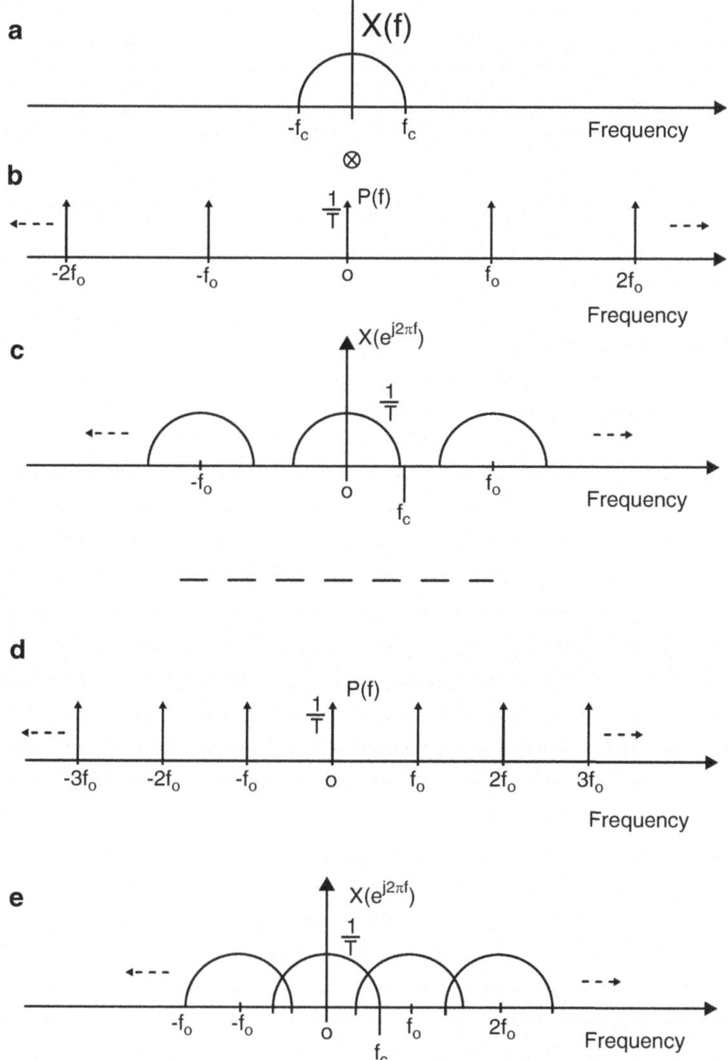

Fig. A.9 The derivation of the spectrum $X(e^{j2\pi f})$ of a sampled signal. (**a**) The spectrum $X(f)$ of the original continuous-time signal. (**b**) The spectrum $P(f)$ of the sampling pulse train $p(t)$, for the case when the Nyquist criterion is satisfied, i.e. when $f_o > 2f_c$, where f_c is the highest-frequency component of $X(f)$. (**c**) The spectrum $X(e^{j2\pi f})$ of the sampled signal, where $X(e^{j2\pi f}) = X(f) \otimes P(f)$. In this case, the sampling rate f_o is fast enough so the spectral repetitions do not overlap and $X(f)$ is recoverable through a low-pass filtering operation without error. (**d**) The spectrum $P(f)$, where in this case $f_c < 2f_o$ and so the signal is undersampled, with the result that the Nyquist condition is violated. (**e**) The corresponding sampled spectrum $X(e^{j2\pi f})$. Here, the spectral repetitions overlap and $X(f)$ cannot be recovered without error. This type of error due to undersamplng is referred to as *aliasing error*

To evaluate the frequency domain representation $X(e^{j2\pi f})$ from the discrete-time sequence $x[n]$, the integral in (A.1) is replaced by a summation to yield

$$X(e^{j2\pi f}) = \sum_{n=-\infty}^{\infty} x[n] \exp(-j2\pi f n). \tag{A.15}$$

The above relations are referred to as the *DTFT*. A detailed derivation of these relationships is presented in [7]. Notice that in these equations, the time domain is discrete, whereas the frequency variable f is continuous and periodic, with one period varying over the range $-1/2 < f \leq 1/2$. It is interesting to note that the DTFT is the dual of the Fourier series representation of a periodic signal in time. In this latter case, the time domain is continuous and periodic, whereas the frequency domain representation is discrete. In the DTFT case, the frequency domain is periodic and continuous, whereas the time domain is discrete.

The time domain signal corresponding to the DTFT can now be processed by computer, but this is not possible for the frequency domain representation, since it is a continuous entity. In Sect. A.2.3, we present a Fourier transform relationship referred to as the *discrete Fourier transform* (DFT) that is discrete in both domains.

Aliasing and the DTFT It is to be noted that if the original spectrum $X(f)$ contains frequency components that violate Nyquist's condition, i.e. f_c from Fig. A.9 satisfies $f_c > 1/2$ in normalised values, then $X(e^{j2\pi f})$ will contain aliasing error, since one period of $X(e^{j2\pi f})$ consists of overlapped segments of $X(f)$ in that case. Then, reconstruction of $x[n]$ through (A.14) at the sampling points will be exact, but interpolated values between the sample points will contain error due to aliasing.

Example A.4 (A Discrete-Time Rectangular Pulse) This example is similar to that of Example A.2 above, except that the waveform in time is now discrete. The discrete-time pulse is shown in Fig. A.10, where it is seen we have a rectangular pulse of duration of $N = 11$ samples with amplitude one, symmetrically spaced about $n = 0$. A straightforward evaluation of the DTFT formula of (A.15) yields the following closed-form result:

$$X(e^{j2\pi f}) = \frac{\sin(\pi f N)}{\sin(\pi f)}, \tag{A.16}$$

where N is the duration of $x[n]$ in samples. The discrete-time rectangular pulse and its corresponding DTFT $X(e^{j2\pi f})$ are shown in Figs. A.10 and A.11, respectively. This function of (A.16) is similar in some respects to the sinc function of Example A.2, especially close to the origin. Its value at $f = 0$ is N through l'Hopital's rule, and there are zero crossings at intervals of $f = 1/N$ Hz, corresponding to the values where the argument in the numerator in (A.16) is a multiple of π radians. However, we have seen that the function $X(e^{j2\pi f})$ yielded by the DTFT consists of periodic repetitions of $X(f)$, which is the Fourier transform of the continuous-

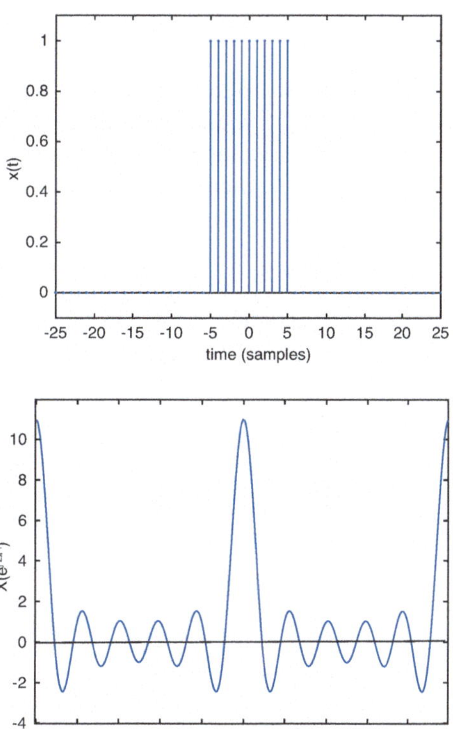

Fig. A.10 A discrete-time rectangular pulse $x[n]$ of length 11 samples symmetrically spaced about the time origin

Fig. A.11 The DTFT $X(e^{j2\pi f})$ of $x[n]$ shown in Fig. A.10 that has the form given by (A.16). Notice it is continuous in frequency. By convention, we would normally plot only one period of the function over the range $-0.5 < f \leq 0.5$ Hz; however, here we have plotted two periods to emphasise its periodic property

time version $x(t)$. We have seen from Example A.2 that the $X(f)$ in this case is a sinc function. However, we have seen from Fig. A.5 that the sinc function decays slowly with frequency, and as such, it can be observed from this figure that its value at $f = 1$ Hz is still quite significant. Therefore, $X(e^{j2\pi f})$ contains substantial overlap between the periods of $X(f)$, giving rise to aliasing error. This explains the differences between the sinc function form of $X(f)$ from (A.5) and functional form of $X(e^{j2\pi f})$ given by (A.16) for this specific case.

A.2.3 The Discrete Fourier Transform (DFT)

The DFT is an extension of the DTFT in that we discretise the frequency domain signal by extracting N samples of $X(e^{j2\pi f})$ over the range $0 < f \leq 1$. As we have seen from Property (A.12) and its dual, Property 23, discreteness in one domain implies periodicity in the other. Therefore, the result of discretising $X(e^{j2\pi f})$ in frequency is that the time domain $x[n]$ now becomes periodic with period N

samples. Consequently, both domains in the DFT representation are discrete and periodic.

To evaluate the frequency domain representation, which we denote as $X[k]$, we sample $X(e^{j2\pi f})$ at the uniformly spaced set of N frequency values $\frac{2\pi k}{N}, k = 0, 1, \ldots, N-1$. For the purposes of notational convenience with the DFT, we define the quantity W_N^m, where m is integer, as

$$W_N^m = e^{j\frac{2\pi}{N}m}.$$

To form the definition of the DFT, the integral in (A.14) becomes a summation to yield the following DFT pair of relations:

$$X[k] = \sum_{n=0}^{N-1} x[n] W_N^{-kn} \tag{A.17}$$

and

$$x[n] = \frac{1}{N} \sum_{k=0}^{N-1} X[k] W_N^{kn}. \tag{A.18}$$

It is to be noted that both these sequences are periodic with period N and both $x[n]$ and $X[k]$ represent only one N-length period of the infinite sequence. Equation (A.17) is equivalent to (A.15) where the range of $x[n]$ has been replaced with the finite duration of a single period.

It is clear that Eq. (A.18) can be written in matrix form in the following way:

$$\boldsymbol{x} = \frac{1}{N} \boldsymbol{W} \boldsymbol{X}, \tag{A.19}$$

where

$$\boldsymbol{W} = \begin{bmatrix} 1 & 1 & 1 & \ldots & 1 \\ 1 & W^1 & W^2 & \ldots & W^{(N-1)} \\ 1 & W^2 & W^4 & \ldots & W^{2(N-1)} \\ \vdots & \vdots & \vdots & \ddots & \vdots \\ 1 & W^{(N-1)} & W^{2(N-1)} & \ldots & W^{(N-1)^2} \end{bmatrix} \tag{A.20}$$

where the subscripts N on the W's have been omitted for notational clarity and $\boldsymbol{X} = \left[X[0], X[1], \ldots, X[N-1]\right]^T$ and $\boldsymbol{x} = \left[x[0], x[1], \ldots, x[N-1]\right]^T$. Likewise, (A.17) can be written in the form

$$\boldsymbol{X} = \boldsymbol{W}^H \boldsymbol{x}, \tag{A.21}$$

where superscript H denotes the Hermitian transpose, as defined in Chap. 1. Since $\boldsymbol{W}$ is symmetric, the Hermitian operation only changes the sign of the exponent of $\boldsymbol{W}$.

The columns of $\boldsymbol{W}$ form a basis of orthogonal complex exponentials with frequencies which are multiples of $1/N$ Hz. It is apparent from (A.19) that $\boldsymbol{x}$ is a linear combination of these complex exponential basis functions, whose coefficients are the elements of $\boldsymbol{X}$. Equation (A.19) can be interpreted as a change of basis. In this case, $\boldsymbol{X}$ is the representation of $\boldsymbol{x}$ in this complex exponential basis. Likewise, (A.21) is the inverse transform. Element $X[k]$ is the coefficient of the kth complex exponential (i.e. the kth "frequency component") of $\boldsymbol{x}$ in this complex exponential basis, and therefore, $X[k]$ indicates how much of the kth exponential is contained in $\boldsymbol{x}$.

The direct evaluation of the DFT using (A.19) or (A.21) requires N^2 add/multiply arithmetic operations to compute. On the other hand, the DFT can be evaluated exactly with only $\frac{N}{2}\log_2(N)$ arithmetic operations using the *fast Fourier transform* (FFT) algorithm. This algorithm requires N to be a power of 2; i.e. N must be expressed in the form $N = 2^m$ where m is integer. For moderate and large values of N, the FFT is indeed a much faster means of computing the DFT than the direct approach. The FFT algorithm and its derivation is discussed in many references, e.g. [7, 8].

Both the DTFT and the DFT share properties that are similar to those of the continuous version of the Fourier transform discussed above, although they have to be modified somewhat to account for the discrete nature of the respective domains. See [7] for more detail.

Example A.5 (The DFT of a Discrete Rectangular Pulse) Here, we discuss the DFT version where $x[n]$ is a discrete, rectangular pulse symmetrically spaced about the origin, for the case where the pulse duration is 11 samples and the period $N = 31$ samples. One period of the infinite periodic time domain sequence is shown in Fig. A.12. In this case, the original analog pulse is actually "on" over the interval $t = -5$ samples to $t = 5$ samples. The DFT time interval corresponding to one period must be chosen to extend from 0 to $N - 1$ samples. As such, by considering the periodic nature of the sequence, the "on" values at samples 26–30 in Fig. A.12 are actually the samples corresponding to the negative time instants of the next period.

The DFT version of the frequency domain is shown in Fig. A.13. As may be observed, it is a sampled version of the DTFT response shown in Fig. A.11, given by Eq. (A.16). The negative frequency values appear in the range $f > 0.5$ Hz.

A.3 Digital Filters

Filters are a critical component of many forms of electronic system. They appear in every form of communication device, including cell phones. They are used for such purposes as noise reduction, equalisation, modulation, and multiplexing. Since

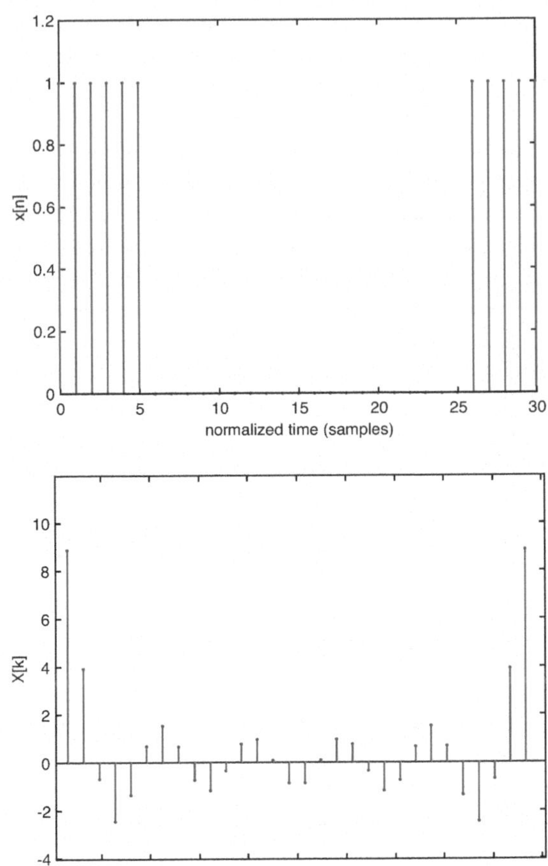

Fig. A.12 The DFT representation of the symmetric rectangular pulse in the time domain, for a pulse duration of 11 samples and a period $N = 31$ samples. The samples from $n = 26$–30 are the values corresponding to negative time of the next period

Fig. A.13 The frequency domain representation $X[k]$ of the rectangular pulse of Fig. A.12. The figure shows a sampled version of the $\sin(\pi f N)/\sin(\pi f)$ response of (A.16) of length N samples. The function $X[k]$ shown is understood to be one period of what is actually an infinite periodic sequence

filters play such an important role in electronics, there is a vast literature on this topic, along with many different methods available for their implementation. The topic of filter design is covered in detail in, e.g. [4]. Digital filters are a preferred form of filter implementation for many applications, in that they provide a cost-effective and flexible approach, provided the frequencies involved are not too high. A filter is a black box with an input $x[n]$, an output $y[n]$ whose purpose is to generate $y[n]$ such that unwanted frequency components that are present in $x[n]$ are suppressed at the output. An important intrinsic component of the filter is its *impulse response* $h[n]$, which is an embedded function of the filter's hardware. The impulse response, as the name suggests, is the response $y[n]$ when $x[n] = \delta[n]$ and is the inverse Fourier transform of the filter's frequency response $H(f)$. The filter's operation, as discussed previously in Property 14, produces an output $y[n]$ which is the convolution of $x[n]$ with $h[n]$. Thus, from Property 14, the corresponding frequency domain representation of the filter output is $Y(f) = H(f)X(f)$, so that the frequencies corresponding to the region where $|H(f)|$ is low are suppressed at the output. There are two common types of digital filter; these are the infinite impulse response (IIR) and the finite impulse response (FIR) filters. Both of these are developed in the sequel.

The classic method of modelling the behaviour of discrete-time dynamic systems is the *difference equation*, which is analogous to the use of the differential equation for modelling continuous-time systems. Conventional digital filters are modelled as *linear, time-invariant* (LTI) systems, which are a subclass of discrete-time dynamic systems. The present value of the output $y[n]$ of any LTI discrete-time system can be modelled in the form

$$\begin{aligned} y[n] &= b_0 x[n] + b_1 x[n-1] + \ldots + b_p x[n-p] - a_1 y[n-1] \\ &\quad - a_2 y[n-2] - \ldots - a_q y[n-q] \end{aligned} \tag{A.22}$$

where $y[n], x[n]$ are the output and input sequences, respectively, $b_0, \ldots, b_p$ are the numerator coefficients where p is the *order* of the numerator, and $a_1, \ldots, a_q$ are the denominator coefficients of order q. The context of the terms "numerator' and "denominator" coefficients will become clear shortly. By choosing the numerator and denominator coefficients carefully, filters with a wide range of frequency responses can be realised.

It is possible to solve the LTI difference equation of (A.22) recursively. That is determining the present value of the output $y[n]$, knowing the present and p past values of the input, and the past q values of the output. However, it is desirable to develop a closed-form expression for the output. In the LTI case, a closed-form solution is indeed possible to develop, using a theoretical construct referred to as the *z-transform*. Given a discrete-time sequence $w[n]$, its z-transform $W(z)$ is defined as

$$W(z) = \sum_{n=-\infty}^{\infty} w[n] z^{-n}, \tag{A.23}$$

where z is a continuous complex variable. It is interesting to note that, except for a few cases, the actual value of the variable z is irrelevant. Exceptions are that the value of z must be chosen so that the infinite summation in (A.23) converges to a sensible value. This matter is discussed in the many texts that are available on digital signal processing, e.g. [7]. The z-transform may be viewed as a transformation of an entire sequence into a single function (given by (A.28) below) in the argument z. Thus, the complete behaviour of any LTI discrete-time system can be represented by the (relatively few) set of coefficients $b_0, \ldots, b_p$ and $a_1 \ldots, a_q$.

The z-transform has at least three properties that are relevant to solving the LTI difference equation. These are:

Property 24 (Convolution) *If $x[n] \leftrightharpoons X(z)$ and $y[n] \leftrightharpoons Y(z)$ are two sets of z-transform pairs, then $x[n] \otimes y[n] \leftrightharpoons X(z)Y(z)$, where the symbol $\otimes$ denotes* *discrete-time convolution; i.e.*

$$\sum_{k=-\infty}^{\infty} x[k]y[n-k] \leftrightharpoons X(z)Y(z), \tag{A.24}$$

where the summation on the left is the discrete-time convolution operation.

This property is analogous to Property 14 for the continuous-time case.

Property 25 (Time Shift) *If $x[n] \leftrightharpoons X(z)$ are a z-transform pair, then $x[n-k] \leftrightharpoons z^{-k}X(z)$.*

Property 26 (Relationship Between the DTFT and the z-Transform) *If $x[n] \leftrightharpoons X(z)$ are a z-transform pair, then the DTFT $X(e^{j2\pi f})$ of $x[n]$ is obtained by evaluating $X(z)$ around the unit circle in the z-plane; i.e. at values $z = e^{j2\pi f}$; i.e.*

$$X(e^{j2\pi f}) = X(z)\big|_{z=e^{j2\pi f}}. \tag{A.25}$$

This follows by comparing (A.23) with (A.15) and noticing they are identical in form for $z = e^{j2\pi f}$.

We now digress to discuss the discrete-time convolution property (Property 24) in more depth, since it is such an important concept in digital signal processing. A more detailed description of convolution is available in [7]. Since the convolution sum in (A.24) is commutative, the convolution operation which produces an output $y[n]$ in response to an input $x[n]$ and an impulse response $h[n]$ can be written in either of the following two forms:

$$\begin{aligned} y[n] &= \sum_{k=-\infty}^{\infty} h[k]x[n-k] \\ &= \sum_{k=-\infty}^{\infty} x[k]h[n-k]. \end{aligned} \tag{A.26}$$

Here, we consider only (A.26). The kth term in (A.26) represents what we refer to as the kth *partial response*, which we view as a function of n. It is evident from the equation that the kth partial response is the impulse response response $h[n]$ due to the input $x[k]$ acting in isolation, as a function of n, delayed by k samples, and weighted by the input value $x[k]$. Since the convolution operation only applies to the LTI case, we must assume that the impulse response does not change with time and that linearity applies. Under these conditions, the total output $y[n]$ is given by the sum of partial responses.

A simple graphic description of the convolution operation is given in Figs. A.14 and A.15. The impulse response in this case is a sequence of 5 one's, starting at time 0. The input sequence is [1, 0.5, 2], again starting at time 0. The sequence of

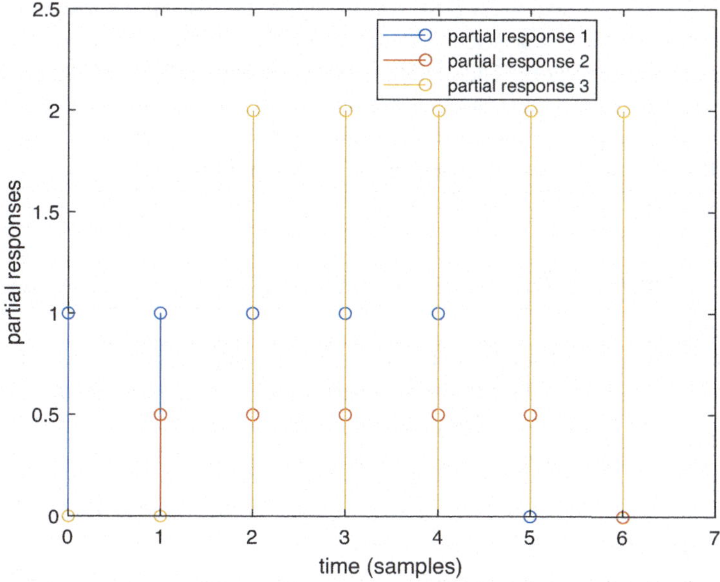

Fig. A.14 The individual partial responses corresponding to the case where $x[k] = [1, 0.5, 2]$ and $h[k] = [1, 1, 1, 1, 1]$. The partial response in blue corresponds to $k = 0$ in (A.26) and is the impulse response due the input $x[0]$ acting alone. A similar situation holds for the partial response in red, but it corresponds to $k = 1$ where the partial response is delayed by 1 sample and is weighted by the value $x[1] = 0.5$. Likewise, the orange partial response corresponds to $k = 2$, is delayed by 2 samples, and is weighted by $x[k] = 2$

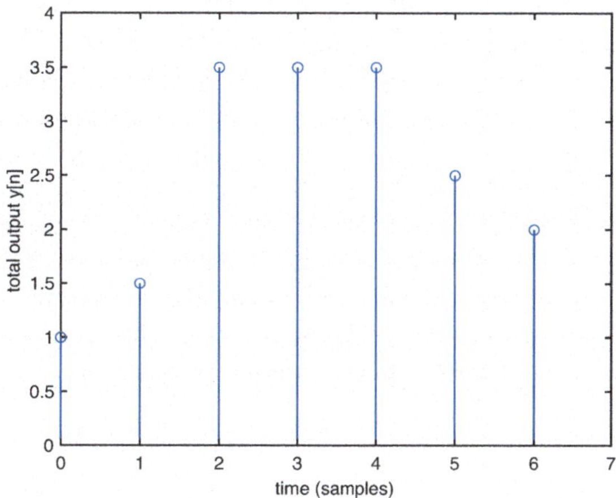

Fig. A.15 The total response $y[n]$, which is the sum of the partial responses

corresponding partial responses is shown in Fig. A.14. The first partial response is in response to $h[0] = 1$, starts at $t = 0$, and is 5 pulses in duration. It is indicated by the blue circles in the figure. The second is in response to $h[1] = 0.5$, starts at $t = 1$, and extends 5 pulses, as indicated by the red circles. A similar situation applies to the third partial response, shown in the orange circles. The combined output $y[n]$, which is the sum of the partial responses, is shown in Fig. A.15.

A.3.1 Solving the Difference Equation Using the z-Transform

Now that the necessary background is in place, we can discuss how the z-transform can be used to solve the LTI difference equation to yield a closed–form solution for the output of the LTI system. By taking z-transforms on both sides of (A.22) and using Property 25, we have

$$X(z)\big[b_0 + b_1 z^{-1} + \ldots + b_p z^{-p}\big] = Y(z)\big[a_1 z^{-1} + \ldots a_q z^{-q}\big]. \tag{A.27}$$

The above can be written in the form

$$H(z) \triangleq \frac{Y(z)}{X(z)} = \frac{b_0 + b_1 z^{-1} + \ldots + b_p z^{-p}}{1 + a_1 z^{-1} + \ldots + a_q z^{-q}}. \tag{A.28}$$

The quantity on the far left is referred to as the *transfer function* of the system and is given the notation $H(z)$ as indicated. Its inverse z-transform $h[n]$ is the system *impulse response*. The *frequency response* $H(e^{j2\pi f})$ of the LTI system is determined by evaluating the DTFT of the transfer function. This is easily calculated by evaluating $H(z)$ on the unit circle, i.e. $H(e^{j2\pi f}) = H(z)\big|_{z=e^{j2\pi f}}, 0 < f \leq 1$. Each point on the unit circle corresponds to a unique frequency value; i.e. $f = 0$ corresponds to the value $z = e^0 = 1$, whereas $f = \pm 1/2$ corresponds to the point $z = e^{j\pi} = -1$. Also, e.g. the $+30°$ point on the unit circle corresponds to a frequency of $1/12$ Hz. The upper half of the unit circle corresponds to positive frequencies and the bottom half to negative frequencies.

The key point in understanding the operation of the LTI system is that it generates its output by inherently convolving the input signal $x[n]$ with the embedded impulse response $h[n]$ to give the output $y[n]$. Therefore, the z-transform $Y(z)$ of the output can be obtained directly using Property 24 as

$$Y(z) = H(z)X(z). \tag{A.29}$$

A closed-form version of the desired time domain signal $y[n]$ can then be obtained by inverting $Y(z)$. The means to do this are discussed shortly. The LTI difference equation is then solved in closed form.

It is customary to represent $H(z)$ in the argument z rather than z^{-1}. Therefore, assuming $q > p$, we multiply the numerator and denominator of (A.28) by z^q to give

$$H(z) = z^{q-p} \frac{b_0 z^p + b_1 z^{p-1} + \ldots + b_p}{z^q + a_1 z^{q-1} + \ldots + a_q}. \tag{A.30}$$

Thus, the entire dynamic behaviour of the underlying LTI system is encapsulated into the rational polynomial form of (A.30). It is indeed interesting to observe that the relatively low number of $p + q$ coefficients of (A.30) is enough to completely specify the entire (possibly infinite length) impulse response of the LTI system. This observation is a consequence of the fact that the impulse response of an LTI system is constrained (under mild conditions) to consist of a linear combination of simple functions, such as decaying exponentials, exponentially decaying sinusoids, δ-functions, and step functions. Each of these functions require few parameters to specify them.

The roots of the numerator and denominator polynomials of (A.30) are referred to, respectively, as the *zeros* and *poles* of the transfer function. The zeros and poles may be plotted as points in the so-called z-plane, which is a two-dimensional complex plane whose axes are the real and imaginary parts of z. As may be seen from (A.22), the LTI difference equation is recursive (i.e. with feedback from the output), and so the output signal has the potential to grow to infinite amplitude if the system is inherently unstable. This condition is avoided if the poles are located within the unit circle in the z-plane. (The unit circle has a radius of 1, centred at the origin.)

As implied earlier, in order to determine the output sequence $y[n]$ in response to an input $x[n]$, we find $Y(z)$ through (A.29) and then invert $Y(z)$ to obtain $y[n]$. There are many ways to perform the inversion process, but the most direct is through the so-called *partial fraction expansion* (PFE) method. The PFE yields a closed-form expression for the time domain response $y[n]$ corresponding to $Y(z)$. We do not discuss this topic in detail here; however, there are many texts giving the theory of the PFE, with examples of its use [7, 8]. The PFE is simplest when there are no repeated roots and $q > p$, although modifications are straightforward when these conditions are not met.

A.3.2 The Infinite Impulse Response (IIR) Digital Filter

The implementation of the IIR digital filter is simply a matter of implementing the difference equation (A.22). In the IIR case, both the b and a coefficients from (A.30) are generally nonzero. As we see later, this is in contrast to the FIR case, where the coefficients $a_1, \ldots, a_q$ are set to zero. The pertinent issue with IIR digital filters is the selection of the a and b coefficients, which determine everything about the characteristics of the filter.

In this vein, there is a very large body of knowledge available on analog filter[3] implementations which can be put to good use in the design of IIR digital filters. This topic is discussed in detail in [4]. Here, we concentrate on the low-pass filter configuration, since other filter configurations (such as high-pass, band pass) can be realised by simple transformations of the low-pass prototype. Because it is not possible to implement the ideal rectangular low-pass filter characteristic exactly, various types of approximations are adopted. Commonly used approximations in this respect are Butterworth, Chebychev, elliptic, and Bessel. In this treatment, we concentrate on the Butterworth design; however, the procedure for implementing other filter forms follows the same procedure as that of the Butterworth.

Analog filters are governed by LTI differential equations rather than LTI difference equations as is the case with digital implementations. The analog case uses the Laplace transform in the complex variable s, instead of the z-transform, to convert the differential equation into a transfer function $H(s)$ in the form of rational polynomial that is analogous in structure to (A.30). Thus, the analog filter also has poles and zeros, but in the s–plane rather than the z-plane. The imaginary part of s is $j\omega$, where ω is radian frequency. The only difference is that in the analog case, the analog frequency response is determined by evaluating $H(s)\big|_{s=j\omega}$; i.e. the frequency response is determined by evaluating $H(s)$ along the entire imaginary axis, which is an infinite straight line in the s-plane, rather than a circle as in the z-plane case.

To design an IIR digital filter, we start off by designing a corresponding analog filter that suits our needs. (This involves determining the filter order (i.e. the parameter q in (A.30), the cutoff frequency f_c, and the filter type, i.e. Butterworth, Chebychev, etc.) Once the analog filter has been specified, there are ample tables available online or in texts which provide the a and b coefficients for our specified filter. We then must transform its analog response, which is a function evaluated along the straight-line $j\omega$ axis in the s-plane, into an equivalent response which results when the function $H(z)$ is evaluated around the unit circle in the z-plane. Fortunately, there is a simple, nonlinear transformation that is exactly suited to our needs. This is the *bilinear transformation* and is given by

$$s = 2\frac{1 - z^{-1}}{1 + z^{-1}}. \tag{A.31}$$

The value $z = -1$ maps into $s = \pm\infty$ where $z = -1$ corresponds to a digital frequency of $\pm 1/2$ and $z = 1$ maps into the value $s = 0$, for which z corresponds to the digital frequency 0. This equation assumes a normalised sampling period of $T = 1$ second. This value above is substituted for s in the analog transfer function $H(s)$ to yield an equivalent transfer function $H(z)$. The result is then simplified into the form of (A.30), whereupon the a and b coefficients are revealed. Because the

[3] Here, we adopt the commonly used terminologies *analog* and *digital* to refer to continuous-time and discrete-time systems, respectively.

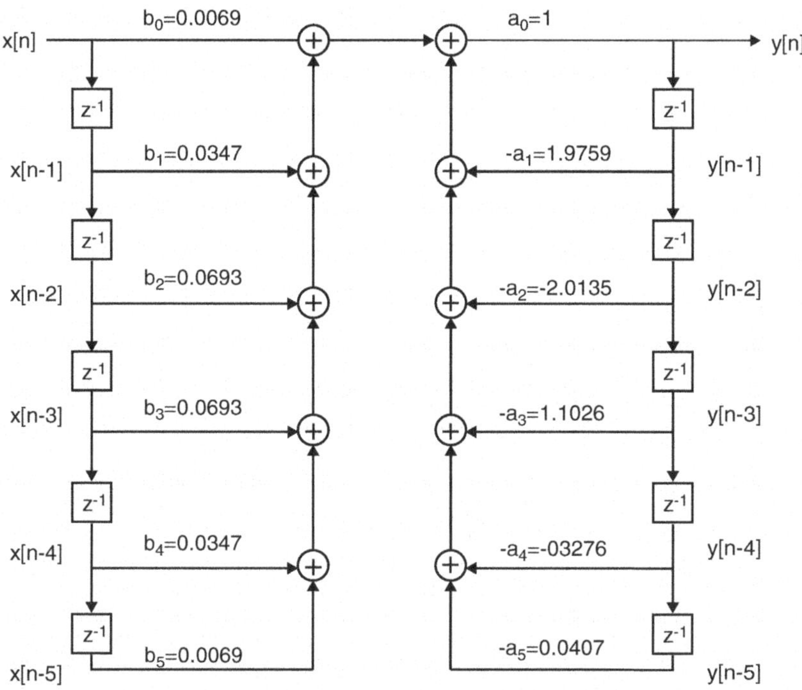

Fig. A.16 Flowchart for implementation of the low-pass, digital Butterworth filter with normalised cutoff frequency of 0.15 Hz. The coefficient values shown are the actual values for this example. The z^{-1} boxes represent delays of one sample value

bilinear transformation is nonlinear, the resulting cutoff frequency of the digital filter is not equivalent to the specified value of the analog filter. Therefore, the cutoff frequency value of the analog filter must be "pre-warped" to provide the specified value of the digital filter. A simple formula exists for doing this. The reader is referred to the various references for more detail. It is of interest to note that software packages such as Matlab® contain built-in functions which provide the filter coefficients directly, for all the filter types mentioned.

Example A.6 (The IIR Butterworth Frequency and Impulse Responses) Here, we demonstrate the design process for a fifth-order, low-pass Butterworth digital filter, with a normalised cutoff frequency of 0.15 Hz. The design process involves only the specification of the coefficients a and b in (A.30), once the filter type, filter order, and cutoff frequency have been determined. We could have looked up the values of the corresponding analog coefficients and converted them to digital form using the bilinear transform of (A.31) to yield the digital coefficients. However, in this example, we used the much simpler method of determining the coefficients by using the built-in function "butter" in Matlab® for this purpose.

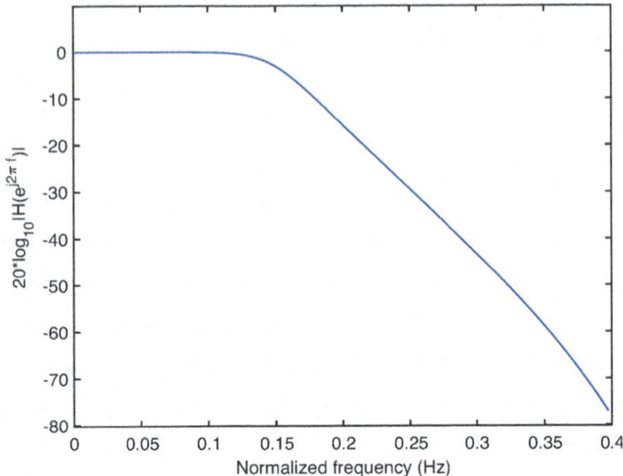

Fig. A.17 The magnitude response of the q = fifth-order digital Butterworth filter, with a normalised cutoff frequency of 0.15Hz. Only the positive frequency values are shown. The vertical axis is represented using a deciBel (dB) scale; i.e., in the form $20\log_{10}\left|H(e^{j2\pi f})\right|$. The logarithmic function expands the smaller values of the response so that the attenuation in the stop band is displayed with higher precision. The cutoff frequency of 0.15 Hz marks the transition between the passband (low frequencies in this case) and the stop band (high frequencies). Once the coefficients have been determined, the respective transfer function $H(z)$ is given through (A.30). The frequency response shown was then formed from the DTFT, by evaluating $H(z)$ at values around the unit circle, i.e. at values $z = e^{j2\pi f}$, $f = [0, 0.5]$

The IIR digital filter is implemented by direct computation of the respective LTI difference equation, once the coefficients are determined. The IIR difference equation is implemented using a flowchart in the form of that shown in Fig. A.16. This figure shows the fifth-order flowchart configuration for the example at hand, showing the actual coefficient values supplied by Matlab® for this example. This flowchart may be adapted to other filter orders in an obvious way. There exist more efficient realisations of this flowchart that require only one delay chain instead of two. Details are provided in the references provided in this section.

The initial values in the delay chain of the flowchart are set to some suitable values, e.g. zero. During the first sample period where $n = 1$, the value $x[1]$ is fed into the structure and the output $y[n]$ for the present conditions is evaluated. Then, during the next sample period where $n = 2$, the values in the two delay chains all fall down one level, and the value $x[2]$ appears at the input. The respective output $y[2]$ is then calculated. The process then repeats for $n = 3, 4, \ldots$.

The magnitude frequency response and the corresponding impulse response for this example are shown in Figs. A.17 and A.18, respectively. It is to be noted that the order q of the filter controls the roll-off rate[4] of the magnitude response of the filter. The higher the value of q, the faster the roll-off rate and the more effective is

[4] Roll-off rate is the rate of increase of attenuation between the passband and the stop band.

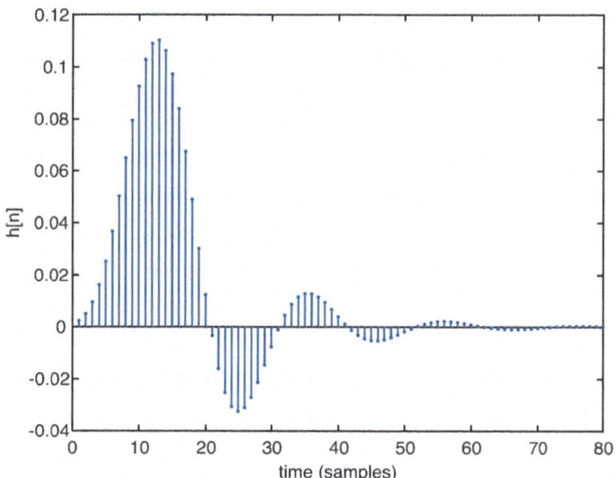

Fig. A.18 The discrete-time impulse response corresponding to the frequency response of Fig. A.17. This response was evaluated by sampling the response $H(e^{j2\pi f})$ of Fig. A.17 at N uniformly spaced points around the whole unit circle and then performing an inverse DFT to yield the time domain waveform

the filter at rejecting unwanted frequencies. However, with higher values of q, the more computations there are that must be performed within one sample period, thus placing more demands on the processor which executes this operation. Also, as q increases, the impulse response becomes longer with more "bumps" in the response.

An example of an output signal that corresponds to a filtered white noise input signal with unit variance, where the filter is a third-order IIR digital Butterworth filter with normalised cutoff frequency of 0.1 Hz, is shown in Fig. 2.7.

A.3.3 Finite Impulse Response (FIR) Digital Filters

Conceptually, FIR filters are simpler than their IIR counterparts. The basic idea is straightforward: we construct an ideal version $\tilde{H}(e^{j2\pi f})$ of the filter response we wish to realise (e.g. the ideal rectangular low-pass frequency response of Fig. A.7) and then evaluate its inverse DTFT to yield the corresponding impulse $\tilde{h}[n]$. In general, $\tilde{h}[n]$ is infinite in length, and so to create an impulse response $h[n]$ that is realisable (i.e. of finite length N), we truncate $\tilde{h}[n]$ by multiplying it point by point in time by a window function $w[n]$ of finite length N. That is, we form the finite-length $h[n]$ as $h[n] = w[n]. * \tilde{h}[n]$, where the $.*$ operator implies point-by-point multiplication. This operation effectively forces $h[n]$ to be zero outside the range of $w[n]$.

Once $h[n]$ is available, the FIR filter output $y[n]$ is generated simply by convolving the input sequence $x[n]$ with $h[n]$ according to the regular definition of the convolution sum:

$$y[n] = \sum_{k=0}^{N-1} h[k]x[n-k], \tag{A.32}$$

where N is the order of the FIR filter (i.e. the length of $h[n]$). During each sample period, the entire sum in (A.32) is evaluated for each respective value of n.

The only difficulty with this process is that the windowing procedure introduces distortion in the actual frequency response $H(e^{j2\pi f})$ relative to $\tilde{H}(e^{j2\pi f})$. The most straightforward window function is a simple rectangle function of length N samples. However, this window causes severe distortion, which in most cases is intolerable. An example is shown later in this section. This distortion can be alleviated using a window function $w[n]$ which goes to zero smoothly at the window end points instead of abruptly as is the case with a rectangular window. The price paid for this reduction in distortion is that the transition band (i.e. the frequency band between the pass- and stop bands in a practical filter) becomes wider. This effect can be controlled by increasing N, the length of the $h[n]$ after truncation. The reader is invited to consult the references for more detail on this matter.

There are many window functions that can be used for this purpose. The idea is to choose the window function so that distortion in the response $H(e^{j2\pi f})$ is reduced to the minimum possible level while simultaneously minimising the width of the transition band. A commonly used window function is the Hamming window, the formula for which is

$$w(n) = 0.54 - 0.46\cos(2\pi n/N), \quad 0 \leq n \leq N. \tag{A.33}$$

This function is similar in appearance to a Gaussian pulse and is symmetric about its midpoint in time. There are many other window functions available, each with their own trade-off between passband ripple, stopband attenuation, and transition bandwidth. See the references for further details.

It is interesting to note that the FIR filter configuration is also representable through a difference equation, which corresponds to the case where all the a-coefficients (except a_0 which is one) in (A.30) are zero. Thus, the FIR flowchart configuration consists only of the left-hand delay chain of Fig. A.16. It may be observed that both the difference equation and the flowchart configuration for the FIR case reduce to the convolution sum of (A.32).

Now that we have discussed both IIR and FIR filters, which one is better? The answer depends on the application. IIR filters require fewer coefficients for the same stop band characteristics and thus require less demanding computer resources, but FIR filters are unconditionally stable because they are non-recursive, meaning there is no feedback from the output that can induce positive feedback. This fact makes life easier from the designer's perspective. A more subtle advantage of FIR filters is that they can be designed to eliminate group delay distortion. The group delay $g(f)$ of a filter is a frequency-dependent quantity. It is proportional to the negative derivative of the phase response of the filter as a function of frequency. The quantity $g(f_o)$ measures the delay of the frequency component at f_o through the filter. Thus, if the phase response is nonlinear, each frequency component propagates through the filter with a different delay. Thus, with nonconstant group delay, different frequency components of, e.g. a pulse waveform, arrive at the output at different times, thus smearing and distorting the waveform at the filter output. On the other

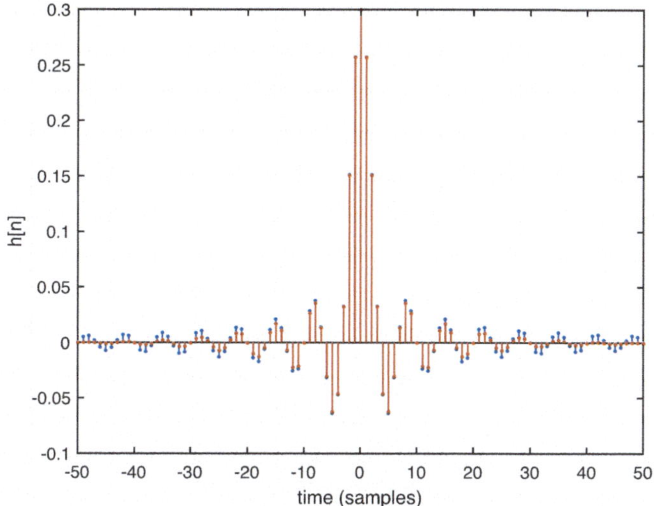

Fig. A.19 The FIR impulse response with the same specifications as the low-pass IIR filer from Example A.6. The filter cutoff is 0.15 Hz where the length of the filter is $N = 101$ samples. The impulse response follows a sinc function multiplied in time by a window function. As is evident from the figure, both functions are similar near $t = 0$, but the tail regions are noticeably different and decay more smoothly to zero in the Hamming window case. The rectangular window samples are shown in blue, whereas the Hammimg-weighted samples are in red

hand, FIR filters can be readily designed to have a linear phase response and thus have constant group delay with frequency, thus suppressing group delay distortion. From Properties 16 and 20, all that is required for this linear phase condition is that the impulse response be symmetric, as shown in Fig. A.19.

Example A.7 (The FIR Approximation Corresponding to Example A.6) Here, we implement an FIR version of the same low-pass IIR filter configuration that was shown in Example A.6, where the normalised cutoff frequency is again 0.15 Hz. The ideal frequency response in this case is rectangular of total width $2 * fc$, where $f_c = 0.15$ Hz. As we have seen, the corresponding impulse response $h[n]$ of infinite length is given by

$$h[n] = 2f_c\text{sinc}(2f_cn).$$

We have chosen the value $N = 101$. This sinc-function impulse response is then truncated to this finite length by multiplication using two different window functions for the sake of comparison. These are the rectangular and Hamming windows. The resulting impulse responses are shown in Fig. A.19. We see that the Hamming window case decays to zero more quickly than the rectangular case. The figure shows $h[n]$ symmetrically spaced about the time origin, which is the true representation of the sinc function. In practice, however, this situation is impossible to realise, since the presence of negative time values implies the filter is noncausal.

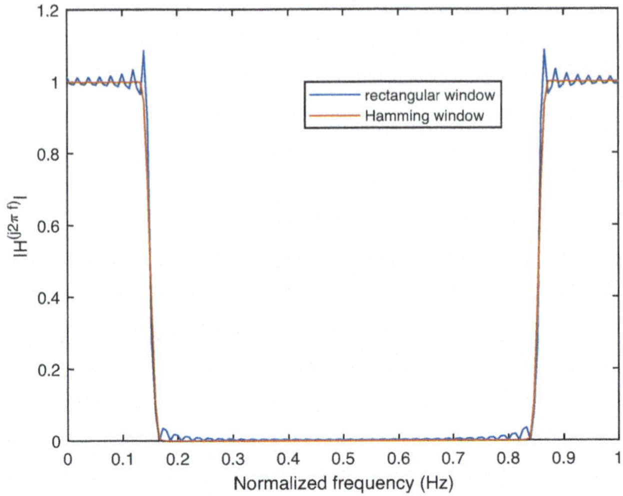

Fig. A.20 The FIR magnitude frequency responses corresponding to the impulse responses of Fig. A.19. Notice the distortion in the rectangular-windowed case at the passband edges and also at the edges of the stop band. This distortion virtually disappears in the Hamming window case. In fact, the stopband attenuation in the Hamming window case is at least 30 dB better than the rectangular case

That means the filter must respond before the input is applied. This situation is easily remedied by delaying the response so that the first sample occurs at time 0. The result is that, according to Property 16, the magnitude response is not changed, but a phase function, which is linear in frequency equal to $-2\pi f \tau$ radians, where τ is the delay in samples, is imposed on the resulting frequency response.

Figure A.20 shows the corresponding frequency responses for both window functions. It is evaluated by the inverse FFT function in Matlab®. We observe that the frequency response for the rectangular (blue) case has significant ripple at the edge of the passband, as well as ripple and less attenuation in the stopband near the cutoff frequency. On the other hand, we see that for the Hamming window (red) case, the passband ripple is greatly suppressed, and the stopband attenuation is greatly improved. Thus, we see that the favourable impact of applying a suitable window function is clearly evident.

References

1. A. Feuer, G. Goodwin, *Sampling in Digital Signal Processing and Control* (Springer Science & Business Media, New York, 2012)
2. S. Haykin, M. Moher, *Introduction to Analog & Digital Communications* (Wiley, New York, 2007)

3. W. Kester, *Data Conversion Handbook* (Newnes, Sydney, 2005)
4. H.Y.F. Lam, *Analog and Digital Filters: Design and Realization* (Prentice-Hall, Upper Saddle River, NJ, 1979)
5. B. Lathi, *Modern Digital and Analog Communication Systems* (Oxford University Press, Oxford, 2010)
6. P.D. Lax, M.S. Terrell, *Calculus with Applications*, vol. 4 (Springer, New York, 2014)
7. A.V. Oppenheim, R.W. Schafer, *Discrete-Time Signal Processing* (Pearson Higher Education Inc., Upper Saddle River, NJ, USA, 2010)
8. J.G. Proakis, D.G. Manolakis, *Digital Signal Processing: Pearson New International Edition* (Pearson Higher Ed., Upper Saddle River, NJ, 2013)

Index

J. Reilly, *Fundamentals of Linear Algebra for Signal Processing*, https://doi.org/10.1007/978-3-031-68915-4

The manufacturer's authorised representative in the EU is Springer Nature Customer Service Centre GmbH, Europaplatz 3, 69115 Heidelberg, Germany. If you have any concerns regarding our products, please contact ProductSafety@springernature.com

Printed and bound by CPI Group (UK) Ltd, Croydon, CR0 4YY
07/07/2026
02160911-0003